74

Fortschritte der Chemie organischer Naturstoffe

Progress in the Chemistry of Organic Natural Products

Founded by
L. Zechmeister

Edited by
W. Herz, G. W. Kirby,
R. E. Moore, W. Steglich,
and Ch. Tamm

Authors:
S. Garai, A. M. P. Koskinen,
S. B. Mahato, L. A. Otsomaa

Springer-Verlag Wien GmbH

Prof. W. Herz, Department of Chemistry,
The Florida State University, Tallahassee, Florida, U.S.A.

Prof. G. W. Kirby, Chemistry Department,
The University of Glasgow, Glasgow, Scotland

Prof. R. E. Moore, Department of Chemistry,
University of Hawaii at Manoa, Honolulu, Hawaii, U.S.A.

Prof. Dr. W. Steglich, Institut für Organische Chemie der Universität München,
München, Federal Republic of Germany

Prof. Dr. Ch. Tamm, Institut für Organische Chemie der Universität Basel,
Basel, Switzerland

Originally published by Springer-Verlag Wien New York in 1998

Library of Congress Catalog Card Number AC 39-1015

Typesetting: Thomson Press (India) Ltd., New Delhi

Graphic design: Ecke Bonk
Printed on acid-free and chlorine-free bleached paper
SPIN: 10643533

With 5 Figures

ISSN 0071-7886
ISBN 978-3-7091-7335-0 ISBN 978-3-7091-6496-9 (eBook)
DOI 10.1007/978-3-7091-6496-9

Contents

List of Contributors

GARAI, Dr. S., Indian Institute of Chemical Biology, 4, Raja S. C. Mullick Road, Jadavpur, Calcutta 700032, India

KOSKINEN, Prof. Dr. A. M. P., Department of Chemistry, University of Oulu, Linnanmaa, 90570 Oulu, Finland

MAHATO, Prof. Dr. S. B., Indian Institute of Chemical Biology, 4, Raja S. C. Mullick Road, Jadavpur, Calcutta 700032, India

OTSOMAA, Lic. Phil. L. A., Department of Chemistry, University of Oulu, Linnanmaa, 90570 Oulu, Finland

Triterpenoid Saponins

S. B. Mahato and S. Garai
Indian Institute of Chemical Biology, Jadavpur, Calcutta, India

Contents

1. Introduction

Saponins are complex molecules made up of sugars linked to a triterpenoid or a steroid or a steroidal alkaloid. These natural products are attracting much attention in recent years because of the host of biological activities they exhibit. The diversity of structural features, the challenges of isolation because of their occurrence as complex mixtures, the pharmacological and biological activities still to be discovered, and the prospect of commercialization – these all are driving the study of saponins. Triterpenoid saponins are dominating constituents of this class and occur widely throughout the plant kingdom including some human foods *e.g.* beans, spinach, tomatoes, and potatoes, and animal feed *e.g.* alfalfa and clover. Saponins were initially a rather neglected area of research primarily because of great difficulties in their isolation and characterization. With the advent of more sophisticated methods of isolation and structure elucidation through the last two decades, there has been increased interest in these natural products. Besides structure determination, research activities are now moving forward to clarify structure–activity relationships. Our previous reviews on triterpenoid saponins (*1*, *2*) covered literature from 1979 to mid-1989. The literature on triterpenoid saponins up to 1988 has also been covered by two reviews by HILLER *et al.* (*3*, *4*). This review incorporates newer trends in isolation and structure determination of triterpenoid saponins, new triterpenoid saponins isolated and biological properties of these products reported during the period late 1989–mid 1996.

2. Isolation

Saponins generally occur as complex mixtures and the usual methods of solvent extraction, column chromatography and preparative TLC are often found to be inadequate for isolation of the pure individual constituents. Special techniques are, therefore, employed to achieve the objective. As an example the saponins of the Chinese medicinal plant *Ardisia crenata* were successfully isolated as follows: The dried and powdered roots of the plant were first defatted with petrol and then extracted with $CHCl_3$ and MeOH under reflux. The MeOH extract was applied to a column of Diaion HP-20 and washed with water, 30, 50, 70 and 100% MeOH to give 50 fractions. The saponin-containing fractions were combined according to their TLC behavior. Each of the combined fractions was purified by an ODS column followed by further separation by HPLC (*5*). A similar procedure was adopted by XU *et al.* for the separation of the new saponins from *Mussaenda pubescens* (*6*). Ground air dried whole plants were extracted by cold percolation with EtOH. The extract was concentrated and partitioned between water – petroleum ether, water – EtOAc and water – *n*-BuOH (saturated with water). The *n*-BuOH extract was applied to a DA-201 column and eluted successively with H_2O, 40% and 60–70% EtOH. The crude saponin obtained from the last fraction was repeatedly separated by silica gel column chromatography. Massiot *et al.* (*7*) isolated the saponins from aerial parts of Alfalfa (*Medicago sativa*) by ether precipitation of the saponin mixture from MeOH solution of the *n*-BuOH extract followed by purification with flash chromatography and thick layer chromatography. Dried and powdered leaves were boiled with a mixture of MeOH and water (4:1) for 3 h. After filtration MeOH was removed and the aqueous layer was extracted three times with *n*-BuOH. The organic layers were combined and evaporated. The residue was dissolved in MeOH, the volume of MeOH concentrated and then diluted with ether. The precipitate was filtered, dried and further purified by flash chromatography on silica gel (particle size: 40–63 μm) under a pressure of 2 bar and thick layer chromatography.

Holothurinosides, new antitumor triterpenoid glycosides from the sea cucumber *Holothuria forskalii*, were isolated (*8*) as follows: Body walls and Cuverium tubules of 19 specimens were collected and extracted with MeOH. The dried MeOH extract was partitioned between water and hexane and the water layer further partitioned between water and *n*-butanol. The *n*-butanol extract was concentrated and passed through a column of a Amberlite XAD-2 which was washed with water and MeOH. The MeOH elute was chromatographed on Sephadex LH-20 eluting with methanol-water (2:1). The fractions thus obtained were further purified by droplet counter current chromatography (DCCC) (ascending mode) and HPLC on a $C_{18}\mu$ Bondapack column.

A somewhat different procedure was adopted for separation and isolation of the bioactive saponins from the fruit pericarps of *Acacia auriculiformis* (*9*). The air dried and powdered fruit of the plant was extracted with petroleum ether, $CHCl_3$ and MeOH. The MeOH extract was partitioned between water and *n*-BuOH. The organic layer was concentrated at reduced pressure, adsorbed on silica gel, dried, and extracted successively in a soxhlet on a water bath with $CHCl_3$, ethyl acetate and a $CHCl_3$-MeOH (80:20) mixture. The ethyl acetate extract on chromatographic purification yielded three relatively non-polar saponins. The $CHCl_3$-MeOH extract was chromatographed on silica gel and a Sephadex LH-20 column followed by preparative TLC and preparative HPLC (S-10-ODS column) to yield three polar acylated saponins.

3. Structure Elucidation

Structures of the isolated pure saponins are generally investigated by a combination of chemical and spectroscopic methods. However, the present favorable trend is to determine structures by spectroscopic methods alone which have the advantage of allowing one to examine a small amount of the intact saponin prior to any treatment which might produce artifacts. The saponins are composed of an aglycone to which are attached one or more sugar chains. In the usual method acid and alkaline hydrolysis experiments are performed to liberate the sugars, acyl constituents and aglycones which are separately investigated for characterization. The sugar and acyl constituents are identified by GC analysis of suitable derivatives and aglycones are characterized by spectroscopic methods. If a saponin contains an acid labile aglycone milder hydrolysis techniques are needed which are described in the previous review (*2*).

3.1. Mass Spectroscopy

The molecular masses of saponins are conveniently determined by soft-ion mass spectroscopic methods such as fast-atom bombardment mass spectrometry (FAB-MS) (*10*, *11*) in the positive and/or negative mode. Other desorption ionization techniques are field desorption (*12*), plasma desorption (*13*) and laser desorption (*14*). More recently, liquid chromatography/mass spectrometry and collision-induced dissociation of doubly charged molecular ions have been employed for structural elucidation (*15*). The molecular masses of the triterpenoid saponins gymnemic acids and their congeners were determined by Imoto *et al.* (*16*) by high performance liquid chromatography combined with atmospheric pressure ionization mass spectrometry (API-MS). The crude saponin isolated from the leaves was chromatographed on an ocatadecyl silica column and eluted with an aqueous methanol solution containing ammonium acetate. The fractions thus separated were directly introduced into an atmospheric pressure ionization mass spectrometer connected with a liquid chromatograph by an interface consisting of a nebulizer and a vaporizer through a PTFE tube (Hitachi, Japan). The vaporized sample and solvent molecules at 300 °C were introduced into the ion source of the API system. The drift voltage of the spectrometer was set at 160 V. Quasimolecular ions of gymnemic acids were detected as ammonium adduct ions and/or proton adduct ions. Molecular masses of 13 gymnemic acids and 5 compounds not containing glucuronic acid as part of the structure were determined. Three pairs of geometrical isomers of gymnemic acids were also detected. Moreover, acyl residues such as acetyl, tiglyl and benzoyl in the gymnemic acids were identified by interpretation of the fragmentation patterns.

Several workers have presented interesting results of their use of mass on spectrometric techniques in structure elucidation of saponins in a symposium "Saponins: Chemistry and Biological Activity" recently held in Chicago. For example, papers on the application of tandem mass spectral approaches to structure determination of saponins (*17*), structure determination of saponins from mungbean sprouts by tandem mass spectrometry (*18*), saponins from alfalfa, clover and mungbeans analyzed by electrospray ionization mass spectrometry (ESI MS) compared with positive and negative FAB mass spectrometry (*19*) and structure confirmation of alfalfa saponins by LSIMS and B/E LSIMS/MS (*20*) demonstrated the great potential of these ionization techniques.

The usefulness of tandem mass spectrometry in the structure elucidation of oleanene-type triterpene bisdesmosides was discussed by Arao *et al.* (*21*). In the MS/MS of $[M\text{-}H]^-$, $[M+H]^+$ ions of the bisdesmosides, the ions which originated from the cleavage of glycosidic bonds, were mostly observed. On the other hand the MS/MS of an $[M+Na]^+$ ion displayed various fragment ions together with those given by glycosidic bond cleavage. The ion derived *via* an retro Diels-Alder fission also appeared. The ESI MS of bellidiastroside C_2 (see Table), a oleanene-type triterpene bisdesmoside generated $[M+H]^+$ and $[M+Na]^+$ and m/z 1091 and 1113 respectively (*22*). MS/MS of $[M+H]^+$ afforded ions which indicated that the saponin had a terminal pentose, a terminal hexose and two inner deoxyhexoses. Appearance of an ion at m/z 425 $[\text{pent}+\text{dhex}+\text{dhex}-H_2O+H]^+$ suggested that three of the sugars were present as a trisaccharide unit.

3.2. NMR Spectroscopy

Of all the physical methods, the NMR technique has changed most during the last two decades, first with the introduction of the Fourier transform (FT) method and more recently as a result of the growth of multiple pulse and 2D NMR. The developments consequent on the pulse technique permit enormously greater control and manipulation of the sample's magnetization. Consequently, the structure information which is gleaned through pulse NMR is probably greater and more readily obtained than by any other single technique.

High-field NMR experiments, *viz.* COSY, HETCOR, TOCSY, NOESY, ROESY and 2D INADEQUATE techniques, coupled with computerized spectral analysis were used for the determination of the complete structure of a novel triterpenoid tetrasaccharide zizyphoiside A (**1**) (code name PT-2) isolated from *Alphitonia zizyphoides* (Rham-

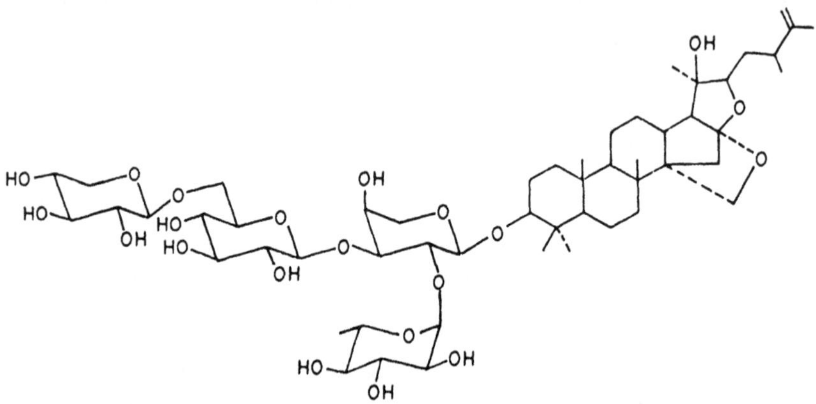

Zizyphoiside A (**1**)

naceae) (*23*). The 2D INADEQUATE technique (coupled with a computerized spectral analysis) was successfully employed to determine the structure of a fairly large saponin using a small amount of sample (60 mg). The 1D ^{13}C spectrum displayed 54 carbon signals. A DEPT experiment revealed 8-CH_3, 13-CH_2, 26-CH and 7 quaternary carbon atoms. Correlation of ^{13}C signals with those of directly bonded protons was achieved by means of 2D HETCOR experiment. The proton and carbon signals for the sugar units were assigned by means of HETCOR, COSY and TOCSY experiments. Starting from the anomeric carbon atom of each of the four sugar units all hydrogens within each sugar were identified using COSY and TOCSY data. Using HETCOR results, each assigned hydrogen was assigned to the corresponding carbon signal. The four sugars, rhamnose, xylose, glucose and galactose were identified by comparison with published chemical shift data for methyl glycosides. The FAB-MS fragmentation pattern indicated that both xylose and rhamnose are terminal sugars. However, the sugar-sugar and sugar-aglycone linkages were indicated by ROESY data which were used to obtain spatial correlations. The observed ROESY coupling between H-1 of the galactose and H-3 of the aglycone suggested that the galactose is linked to the aglycone at C-3. This was also confirmed by the downfield shift of C-3. A ROESY peak for H-1 of glucose and H-3 of galactose disclosed that the glucose and galactose were 1,3-linked. The other intersugar linkages were also determined by the observation of ROESY coupling between H-1 of rhamnose and H-2 of galactose, and between H-1 of xylose and H-6 of glucose.

The 2D INADEQUATE spectral data required analysis by the program CC Bond because the signals were too weak to be identified visually. The computer analysis permitted identification of 35 of the 53 C–C bonds in the saponin from ^{13}C–^{13}C connectivities. The structure of the aglycone moiety was also revealed by standard HETCOR and long-range correlation experiments, COSY and TOCSY data as well as comparison of the ^{13}C chemical shift assignments with those of a similar reference compound, jujubogenin. The stereochemistry at the C-1 position of the sugars was deduced from a comparison of the ^{13}C values with those of sugars of known structure and from the magnitude of the corresponding anomeric ^{1}H–^{1}H coupling constants.

The structures of three medicagenic acid bisdesmosides, one monodesmoside of the same acid and one soyasapogenol B monodesmoside were elucidated on the peracetylated derivatives of the saponins using the techniques such as COSY, relayed COSY, HOHAHA, HMQC, HMBC and ROESY (*7*). For example the assignments of the ^{13}C signals of medicagenic acid in saponin (**2**) were made using ^{1}H–^{13}C correlation experiments in the reverse mode such as HMQC for ^{1}J and HMBC for ^{2}J and ^{3}J. These experiments permitted assignments of most of the

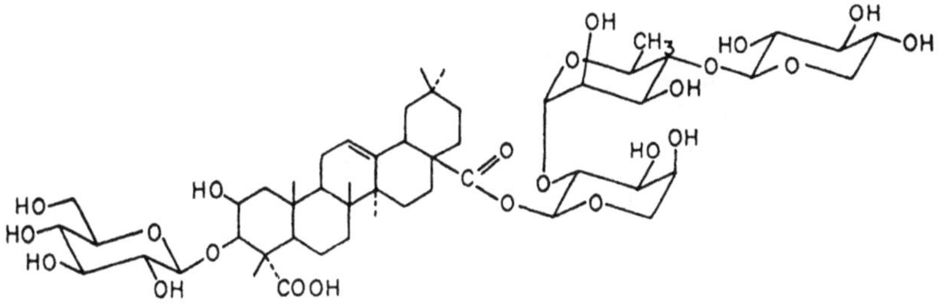

Saponin (**2**)

signals of the aglycone by observation of correlations with the angular methyl protons. The spin network of peracetylated (**2**) was identified by COSY, relayed COSY, HMQC and HMBC experiments. The HMBC experiment also allowed sequencing of all the elements of the molecule. The configuration and conformation of the arabinose unit were revealed from ^{3}J-H-1-H-2 which was found to be 6 Hz. The value indicated the α-L-configuration. The corresponding value for the β-L configuration in $^{4}C_{1}$ conformation is 2.3 Hz. The α-L configuration was also inferred from ROESY experiments. ROEs were found between arabinose H-1 and H-3 indicating α-L-arabinose in ${}^4C_1$ conformation but not ruling out the presence of some ${}^1C_4$ conformation isomer. The ROESY experiment also disclosed the β-D-glucose and the β-D-xylose configuration by H-1-H-5 ROEs. The α-L-rhamnose configuration was deduced from long range H-1-H-5 coupling in LR COSY. The ROEs, aglycone H-3 to glucose H-1, rhamnose H-1 to arabinose H-2 and xylose H-1 to rhamnose H-4 provided sequential information.

The structures of three new dammarane-type triterpenoid saponins, bacopasaponins A, B and C isolated from the reputed Indian medicinal plant *Bacopa monniera*, were elucidated by spectroscopic methods and some chemical transformations (*24*). The ${}^1H$ and ${}^{13}C$ signals of all the saponins were assigned and ring sizes of the sugars determined by DEPT, ${}^1H$–${}^1H$ COSY, HSQC and HMBC experiments. The configurations at C-20 and C-22 of the aglycone pseudojujubogenin in bacopasaponin C (**3**) were determined by phase-sensitive ROESY.

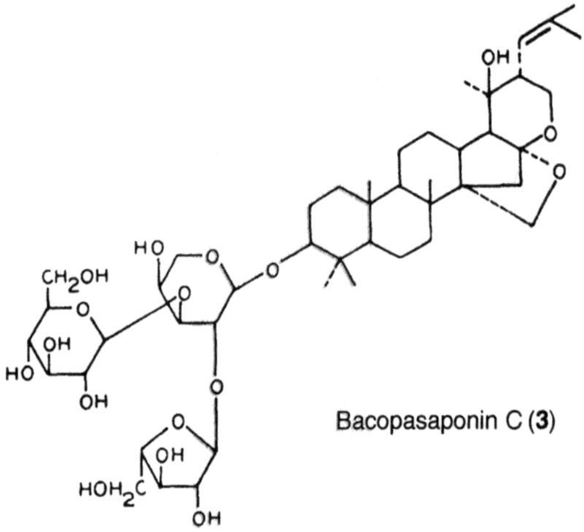

Bacopasaponin C (**3**)

The structures of two novel triterpenoid saponins, ardisicrenoside A and ardisicrenoside B, were determined by 2D NMR COSY, HOHAHA, HETCOR, HMBC and ROESY experiments (*5*). For example, ardisicrenoside A (**4**) showed in its ${}^{13}C$ NMR spectrum four anomeric carbon signals and its new aglycone displayed six sp^3 quaternary carbon atoms. The ${}^{13}C$ data of the aglycone part were similar to that of the known triterpene, cyclamiretin A (*25*). These data suggested that ardisicrenoside A is a triterpenoid tetrasaccharide. The assignments were confirmed by long-range coupling in HMBC and by spatial interaction in ROESY. The spatial proximities observed between H-3 and H-23, H-3 and H-5, H-16

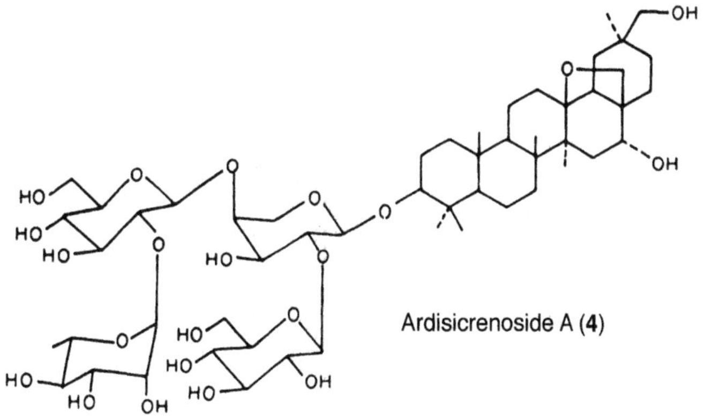

Ardisicrenoside A (**4**)

and H-28 suggested β and α configurations at C-3 and C-16 respectively. A correlation between H-18 and H-30 allowed assignment of the hydroxymethyl group to C-30.

The nature of the monosaccharides and their sequence were determined by means of H COSY, HOHAHA, HETCOR, HMBC and ROESY experiments. Starting from the anomeric protons of each sugar unit, all the hydrogens within each spin system were identified using COSY aided by the 2D HOHAHA spectrum. On the basis of the assigned hydrogens, the ^{13}C resonances of each sugar unit were assigned by HETCOR and further ascertained by an HMBC experiment.

A novel arjunolic acid tetrasaccharide (**5**) with an unusual carbohydrate chain was isolated from *Heteropappus biennis*. Its structure was established mainly by a combination of 1D selective and 2D NMR techniques such as COSY, TOCSY, ROESY, HMQC and HMBC. Molecular modelling calculations revealed that the oligosaccharide chain in the molecule is rather rigid (*26*). The structure of the complex carbohydrate chain was determined by NMR pulse experiments. The characteristic ^{13}C values of the anomeric carbons indicated four different monosaccharide units. The proton connectivities of the individual sugars were determined by H, H-COSY; 2D H,H-COSY and 1D TOCSY experiments were used to determine coupling constants. Using *Gaussian* pulses, the transitions of the anomeric protons of the individual monosaccharide units were selectively excited and then the magnetization was transferred within one monosaccharide residue to H-C(2), H-C(3), H-C(4), H-C(5) and in case of the glucose to $CH_2(6)$ depending on the mixing time used. The carbon atoms were identified by an HMQC spectrum. The HMBC technique was used to determine the sequence of the carbohydrate chain which was also confirmed by the ROESY spectra (1D, 2D).

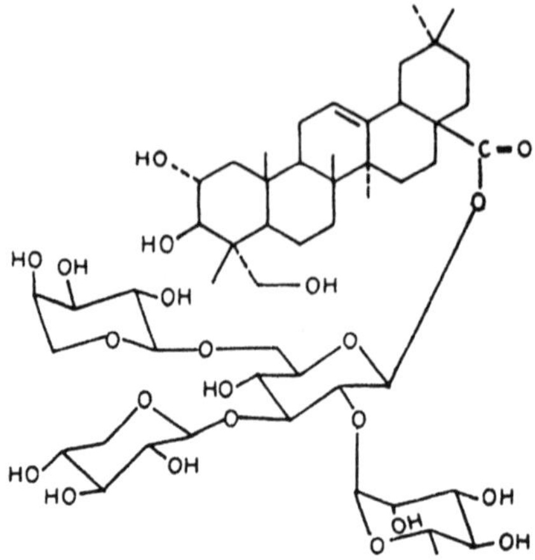

(5)

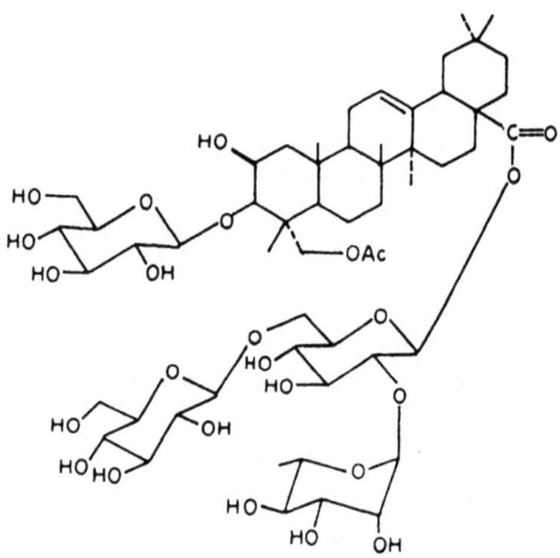

Asterbatanoside F (6)

NMR techniques including COSY, HETCOR, COLOC, HOHAHA, ROESY and selective INEPT were used for elucidation of the structure of four novel triterpenoid saponins, asterbatanosides F, G, H and I, from the roots of *Aster batagensis* (27). For example the COLOC spectrum of asterbatanoside F (6) displayed a correlation contour between the H-23 signal and the carbonyl carbon signal of the acetyl group suggesting presence of an acetyl group at the C-23 position of the aglycone. The 2D COSY and HOHAHA spectra helped to assign all of the proton signals in each monosaccharide and the HETCOR spectrum permitted assignment of all carbon signals of the sugar units. In a selective INEPT experiment, irradiation of the anomeric proton signal of the rhamnose at δ 6.47 enhanced the carbon resonance at δ 75.3 of C-2 of the inner glucose in the 28-O-sugar units suggesting a (1 → 2) linkage between the rhamnose and the 28-O-inner glucose unit. These conclusions were verified by a ROESY experiment which showed NOE correlations between H-1 of the rhamnosyl unit and H-2 of the inner glucosyl unit, and between H-1 of the outer glucosyl unit and H-6 of the inner glucosyl unit. Moreover, each glucose H-1 showed NOE with H-3 and H-5, and the rhamnose H-1 showed NOE with H-4 which confirmed the configuration of the sugar units.

4. Biological Activity

Triterpenoid saponins are widely distributed throughout the plant kingdom. Saponins in general have been in use as natural detergents, fish poisons, arrow poisons and foaming agents from the early stages of civilization. Earlier studies of the biological activities of saponins were limited to crude extracts containing saponins as well as other polar constituents. However, with the introduction of more and more sophisticated methods of isolation and structure determination, there has been increased interest in the study of structure-activity relationships among triterpenoid saponins. The results published so far provide a growing body of information about their diverse effects, particularly in health-related areas. Saponins are present in many animal feedstuffs and also in some human foods. Although many saponins are highly toxic when given intravenously to higher animals, the toxicity is very much lower when they are administered orally. This is because of their almost complete failure to cross the gut and enter the blood stream, and because the hemolytic effect is very much reduced in the presence of plasma.

4.1. Antifungal Activity

Many saponins exhibit antifungal activity under experimental conditions. The antifungal action of glycosides of polygalacic acid has been reported (*28*). The bisdesmosides virgaureasaponins 1 and 2, bellissaponin 1 and the corresponding mono-desmosides (prosapogenins) isolated from *Solidago virgaurea* and *Bellis perennis* inhibited the growth of *Candida* and *Cryptococcus* species *in vitro*. The bisdesmosides were more active than prosapogenins. Structure-activity relationships of α-hederin from *Hedera rhombea* was investigated by comparing its hemolytic and antifungal activities with analogues in which the terminal rhamnose was absent and in which the carboxyl group was methylated (*29*). The results demonstrated that the terminal rhamnose is more important for antifungal activity than for hemolytic activity, whereas the free carboxylic acid is more important for the latter than for the former. Antifungal activity was also detected in the saponin fraction obtained from the bottom cut of *Asparagus officinalis* (*30*). The activity was specific to certain fungi, *e.g. Candida, Cryptococcus, Trichophyton, Microsporum* and *Epidermophyton.* A new saponin (AS-1) was isolated from this fraction and its structure elucidated. The antifungal activity of the saponins isolated as a byproduct from the defatted cake of *Madhuca butyracea* oil seed was reported (*31*). Inhibitory concentrations against plant pathogenic fungi ranged from 500 to 2000 ppm. Maximum sensitivity to saponins was shown by *Penicillium expansum, Cephalosporium acremonium, Helminthosporium oryzae*, and *Trichoderma viride.* The saponins caused leakage of cell components and underwent degradation by the fungus, *T. viride.*

Zehavi *et al.* (*32*) extended the study of structure-antifungal activity relationships of medicagenic acid saponins to include synthetic glycosides of mannose, galactose, cellobiose, and lactose as well as a 23-hydroxymethyl analogue of medicagenic acid, namely, methyl 2β, 3β-dihydroxy-23-hydroxymethyl -Δ^{12}-oleanene-28-carboxylate, against *Sclerotium rolfsii, Rhizoctonia solani, Trichoderma viride, Aspergillus niger*, and *Fusarium oxysporum.* The native glucose-containing saponin was a more effective antifungal agent than the above-mentioned saponins except for the cellobiose-containing derivative and *F. oxysporum.* A 23-carboxyl substituent of the sapogenin displayed higher fungistatic activity than a methyl carboxylate. However, the latter was more effective than a hydroxymethyl group at the same position. The authors opined that in this series of compounds, the difference in antifungal activity could be interpreted *inter alia* by differences in penetration into the fungal cells, extent of interaction with membrane sterols and cell components, hydrolytic and detoxification activities of the fungi and the host tissues.

Bower *et al.* (*33*) observed that some fungal pathogens can enzymatically detoxify host plant saponins which suggests that saponin detoxification may determine the host range of these fungi. A gene encoding a saponin detoxifying enzyme was cloned from the cereal-infecting fungus *Gaeumannomyces graminis.* The fungal mutants generated by targeted gene disruption were no longer able to infect the saponin-containing host oats but retained full pathogenicity to wheat which does not contain saponins. It was evident that the ability of a phytopathogenic fungus to detoxify a plant saponin can determine its host range.

The antifungal activity of triterpenoid saponins, with hederagenin or oleanolic acid as aglycone, was investigated *in vitro* by the agar dilution method. Monodesmosidic hederagenin derivatives were shown to exhibit a broad spectrum of activity against yeast as well as dermatophyte species. α-Hederin was the most active compound, and *Candida glabrata* was the most susceptible strain. The structure-activity relationships were discussed (*34*).

4.2. Immunomodulatory Activity

Saponins from *Quillaja saponaria* (soapbark tree) have been identified as potent adjuvants. A plant extract called Quil A is used in veterinary vaccines and has been studied most thoroughly (*35*). The bark of *Q. saponaria* contains about 10% saponin and has been used as a source of commercial saponin as well as a foaming agent in beverages, confectionary, baked goods and dairy desserts. These saponins have drawn much attention in recent years for their use as adjuvants for human vaccines. The adjuvant activity of a single highly purified saponin from *Q. saponaria* was evaluated by using it as a component in an experimental vaccine containing rHIV-1 envelope protein (HIV-1 160D) adsorbed to alum (*36*). BALB/c mice immunized with experimental vaccine formulation containing the saponin adjuvant QS-21 produced significantly higher titers of antibodies than mice vaccinated with only the alum-adsorbed HIV-1 160D. Potent amnestic antibody responses to HIV-1 viral proteins were also induced. Antigen-specific (Ag-specific) proliferative responses to recombinant proteins and to three variants of HIV-1 were increased significantly using QS-21 as an adjuvant. Alum-adsorbed HIV-1 160D failed to induce measurable proliferative responses to inactivated HIV-1 viruses, but group-specific proliferative responses were raised when the QS-21 adjuvant was used in the vaccine formulation. MHC class 1 restricted CTL specific for immunodominant V-3 loop were induced but only when the QS-21 adjuvant was included in the vaccine formulation. QS-21 augmented cell-mediated immune-responses specific for epitopes outside of the V-3 loop. Moreover, QS-21 adjuvant appeared to induce recognition of weakly immunogenic epitopes that were not recognized using only alum-adsorbed HIV-1 160D. The ability of QS-21 to augment both antibody and cell-mediated immune responses suggested that this adjuvant could be a valuable component in subunit vaccines.

WHITE *et al.* reported that the purified saponin QS-21 from *Q. saponaria* acts as an adjuvant for a T-independent antigen (*37*). The ability of QS-21 to induce ovalbumin (OVA)-specific, class I MHC antigen-restricted cytotoxic lymphocites (CTL) was investigated by NEWMAN and co-workers (*38*) using different forms of soluble OVA and OVA adsorbed into alums as immunogens. The results demonstrated the ability of the QS-21 adjuvant to induce class I MHC Ag-restricted CTL after immunization with soluble proteins. QS-21 was found to be a more potent adjuvant than alum when the antibody response to either the peptide hapten, HGP-30, or the carrier, keyhole limpet hemocyanin was examined (*39*). QS-21 was well tolerated by the immunized mice. There were no differences in reactions at the injection site of QS-21, QS-21 plus alum and alum alone. The addition of alum to QS-21 modestly augmented the antipeptide titer, but it did not have a significant effect on the response generated by QS-21. The adjuvant activity and immunostimulating complex (ISCOM) formation by a series of saponins and glycoalkaloids were investigated (*40*). Saponins from *Gypsophila* and *Saponaria* besides those of *Quillaja* were adjuvant active. The common features of these saponins are that they contain branched sugar chains attached to positions 3 and 28 of the aglycone.

The purified saponin QS-21 was tested in juvenile rhesus macaques for adjuvant activity and toxicity (*41*). It was tested alone or as part of an experimental subunit HIV-1 vaccine containing a truncated recombinant HIV-1 envelope protein (gp 160D) adsorbed on alum. No toxic effects were observed. The results demonstrated that the QS-21 adjuvant augmented both antibody responses and cell-mediated immunity and established immunological memory. The potent adjuvant activity and lack of toxicity suggested that this adjuvant should be safe and effective for use in HIV-1 vaccines. It was observed (*42*) that induction of antigen-specific Killer T lymphocyte responses using subunit SIV mac 251 gag and env vaccines containing QS-21 saponin adjuvant is possible. Subunit vaccines based on recombinant proteins have proved useful for inducing antibody responses and they are safe for widespread use because they do not contain any live component. However, they do not typically induce the types of cell-mediated immune responses required to control viral pathogens. The authors used subunit vaccine formulations containing recombinant p55 gag or gp 120 env protein from the mac 251 strain of the simian immunodeficiency virus (SIV mac 251) and the QS-21 adjuvant to immunize rhesus macaques. These formulations induced SIV gag or env–specific cellular immunity that was detectable in vitro and included Killer cell activity. Despite the presence of these Killer cells, all of the animals became infected with the SIV mac 251 on experimental challenge. These findings demonstrated that antigen-specific Killer cell responses could be induced by a subunit vaccine formulated with the QS-21 saponin adjuvant. However, these types of cellular immune responses could not protect rhesus macaques from infections (SIV mac 251 challenge).

The saponins of *Panax quinquefolium* enhanced the stimulation effect of Con A for interleukin-2 and interferon formation by mouse T-cells for mouse splenocytes proliferation, and for natural killer cell activity. When injected *s.c.* into mice the saponin promoted the primary antibody response to sheep erythrocytes (*43*). Immune function stimulatory and regulatory action of ginseng saponins on chronic pulmonary heart diseases is also reported (*44*).

The purified quillaja components and the ISCOM matrix formulations were examined for their adjuvant activity in a model system consisting of purified influenza virus antigen and quillaja saponins (*45*). It was demonstrated that a quillaja component, designated QH-C, either as a 'free' component or in an ISCOM matrix, has strong adjuvant activity but little or no toxicity in the doses tested. In addition QH-C in the form of ISCOM matrix does not induce any local reaction at the site of injection. Thus ISCOMs containing the QH-C component devoid of toxicity, but with strong adjuvant activity, may be useful in adjuvant formulations for human use.

4.3. Molluscicidal Activity

Schistosomiasis is a disease linked with certain species of aquatic snails because they serve the parasite as intermediate hosts. This disease is endemic in several countries in Asia, Africa and South America and affects millions of people. It has been known for a long time that saponin-containing plants are toxic to schistosomiasis transmitting snails *Biomphalaria globrata*. Molluscicidal activities of plants are of much importance as they are less expensive than synthetic compounds. The previous review described the molluscicidal activity of a number of monodesmosidic and bisdesmosidic saponins isolated from various plant species (2). TANAKA *et al.* (*46*) screened thirty-four extracts of crude drugs and medicinal plants against *Oncomelania nosophora*, the intermediate host of the Japanese strain of *Schistosoma japonicum*. Strong molluscicidal activity was found in the MeOH extract of *Anemarrhena rhizoma*.

Timosaponin A-III, one of the main saponins of the plant, showed very strong killing activity. The monodesmosidic saponins from

Phytolacca dodecandra are the most promising molluscicide of plant origin (*47*, *48*). The acylated saponins from *Sapindus rorak* are reported to possess strong molluscicidal activity (*49*). *Catunarelgum nilotica*, a lowland shrub or tree is widespread in the Sudan and is also reported from lowland habitats in Central and East Africa as well as Cameroon and Nigeria. Initial molluscicidal screening of the crude water and ethanol extracts revealed 100% snail mortality at concentrations of 100 and 50 ppm respectively. The haemolytic activity of the molluscicidal saponins was determined as well and the HC_{50} values towards bovine erythrocytes were found to be 3 ppm for the new monodesmosidic saponin and 16 and 2 ppm respectively for the two known saponins (*50*).

4.4. Spermicidal Activity

A mixture of two partially characterized triterpenoid saponins containing acaciaside A and acaciaside B with the aglycone structure of acacic acid lactone isolated from the abundantly available plant *Acacia auriculiformis* showed sperm-immobilizing activity (*51*). The lowest concentration (ED) required for obvious immobilization of human sperm by using a modified Sander-Cramer test was 0.35 mg/ml. No permanent lesion was observed after application of 1.25 mg/ml saponin solution in physiological saline to the eye of rabbits for consecutive days. The spermicidal activity of purified neem seeds extracts, reetha saponins and quinine hydrochloride was studied individually and in combination (*52*). Minimum effective spermicidal concentrations for neem extract, reetha saponins and quinine hydrochloride were 25%, 0.05% and 0.346% respectively. At these concentrations, 100% of the sperms were immobilized within 20 seconds. The selected combination formulated into a suitable dosage form is likely to offer dual benefit of a potent contraceptive and an antimicrobial preparation. The antifungal saponin mollugogenol-A from *Mullugo pentaphylla* showed maximal spermicidal effect (4–5 fold decrease in motility and viability) with 300 μg/ml dose (*53*).

4.5. Hypoglycemic Activity

Gymnemic acid is a mixture of a number of triterpene saponins from the Indian medicinal plant *Gymnema sylvestre* (*54*). The effect of gymnemic acid on the elevation of blood glucose concentration induced with oral sucrose in streptozotocin-diabetic rats was studied by KANG *et al.* (*55*). Rats with streptozotocin induced diabetes mellitus and loaded orally with 4 g sucrose/kg were given one to four doses of 400 ng gymnemic acid/kg around the time of sucrose administration. It was observed that gymnemic acid has dose-dependent hypoglycemic activity. The saponin isolated from the leaves of *Acanthopanax senticosus* (100, 200 mg/kg, i.p.) is reported to decrease various cases of experimental hyperglycemias induced by the injection of adrenalin, glucose and alloxan, without affecting the levels of blood sugar in normal mice (*56*). Livers of streptozotocin-diabetic rats had decreased activities of glucose-6-phosphate, acetyl CoA carboxylase and 6-phosphogluconate dehydrogenase and these activities were increased by *in vivo* treatment with ginseng saponins, which also possess hypoglycemic action. Insulin biosynthesis by the liver also appeared to be stimulated by the saponins (*57*). The hypoglycemic effect of total saponins of *Aralia decaisneana* in rat and mice models was investigated (*58*). The saponins decreased normal euglycemic level to some extent and decreased adrenaline-induced hyperglycemia and alloxan-induced diabetic hyperglycemia but not glucose induced hyperglycemia in mice. The saponins also had no effect on glucose tolerance in alloxan diabetic rats.

4.6. Antitumor Activity

Tokuda *et al.* (*59*) reported inhibition of 12-O-tetradecanoylphorbol-13-acetate (TPA)-promoted mouse skin papilloma by saponins. Papillomas in the mouse skin were initiated with 7,12-dimethylbenz[a]anthracene. One week later, they were promoted with TPA. Five saponin related compounds used as potential antiinflammatory agents were then applied. The compounds effectively inhibited tumor formation even when given 1 h prior to TPA treatment. There was a general correlation between the antiinflammatory and antitumor-promoting activities of saponins. The tumoricidal activity of murine macrophage against K562 tumor cells was studied in the presence of lipopolysaccharide (LPS) and ginseng saponin. The tumoricidal activity was increased more by LPS plus ginseng total saponin than by LPS alone (*60*). The result suggested that ginseng saponins increase the tumoricidal activity against K562 tumor cells through the tumoricidal activity of the macrophage. Hasegawa *et al.* reported the inhibitory effect of triterpenoid saponins on glucose transport in tumor cells and its application to *in vitro* cytotoxic and antiviral activities (*61*). Saponins have been suggested as possible anticarcinogens. The proposed mechanisms of anticarcinogenic properties of saponins include direct cytotoxicity, immune modulatory effects, bile acid binding and normalization of carcinogen-induced cell proliferation (*62*). The effects of soybean saponins and gypsophila saponins on the growth and viability of colon tumor (HCT-15) cells in culture were studied. Cells were incubated in various concentrations of saponins for 1 h (short term) or 48 h (long term). Cell growth and viability were monitored at 24 h and 48 h. Soybean saponins and gypsophila saponins inhibited cell growth and reduced cell viability in a dose-dependent manner in long term treatment. The viability of cells was also reduced by short term treatment with gypsophila saponins (*63*).

4.7. Hypocholesterolemic Effect

Elevated plasma cholesterol levels are believed to be a significant risk factor in the etiology of cardiovascular disease. The hypocholesterolemic effect of some dietary saponins has therefore attracted considerable attention. A recent review deals with hypocholesterolemic effects of dietary saponins and mechanisms of these effects (*64*). *Panax notoginseng* saponins (100 or 200 mg/kg) administered intragrastically to hyperlipidemic rat and quail models for 7 days markedly lowered the serum total cholesterol and triglyceride contents (*65*). *P. quinquefolium* saponin (50–200 mg/kg/day orally for 12 days) decreased serum lipoproteincholesterol and liver and serum lipid peroxidase and increased serum HDL-C and HDL_2-C and liver and blood GSH-peroxidase in rats with hyperlipidemia. Thus the saponin was thought to be effective against formation and development of atherosclerosis (*66*). A patent has been registered on a preparation of poultry food containing Quillaja saponin and/or Yucca saponins (*67*). Chickens fed with this food produced low-cholesterol eggs.

4.8. Antiaging Effect

Ginseng (*Panax ginseng* C.A. Meyer) saponin which is a complex mixture of a large number of dammarane and oleanolic acid saponins is renowned for its antiaging property. The root extract of the plant has been used in oriental countries for centuries for increasing mental efficiency, recovering physical balance and stimulating metabolic function. Continuous use of the root extract in the form of tea leads to longevity with reduced weight. At present the wild plant is very rare and the drug is very expensive. Most of the commercial Ginseng roots are the products of cultivation in China, Korea and Japan. A number of *Panax* species are known besides *P. ginseng* and these are used as substitutes to a varying degree.

The possible antiaging effect of ginseng stem-leaf saponin was studied in terms of the free radical theory of aging. The saponin, at 50 and 100 mg/kg given intragastrically to mice for 15 days, inhibited the formation of lipid peroxide in the brain. However, at 100 mg/kg it only had inhibitory effect in liver. When 100 mg/kg were given orally for 30 days the lipofuscin content in rat cerebral cortex and liver was decreased. The saponin at 50 and 100 mg/kg increased the content of superoxide dismutase. At 100 mg/kg it also increased the catalase activity in mouse blood. The results suggest that the stem-leaf saponin has antiperoxidative action and may act as an antiaging factor (*68*).

The saponins from the stalk and leaf of *Panax notoginseng* given to *D. melanogaster* prolonged the life span and flying capability and lowered the lipofusion content in the head. The saponins inhibited lipid peroxide formation in tissues and elevated blood and brain superoxide dismutase activity. These results indicated that antiaging activity of the saponin is related to its free radical scavenging action (*69*). The antioxidative effect of *Panax quinquefolium* saponin on myocardium injury induced by doxorubicin in rats was examined (*70*). The authors concluded that the saponin processes antioxidation activity which may be related to the glutathione peroxidase (GSH-Pα) and superoxide dismutase (SOD) activities. Yang *et al.* (*71*) reported that *Panax quinquefolium* saponin could antagonize the action of xanthine and xanthine oxidase and protect against oxidative damage to myocardial cells. Wang and co-workers (*72*) observed that xanthine and xanthine oxidase caused free radical damage in intact rat heart and induced decreases in heart function parameters. Panaxadiol and panaxatriol saponins obtainable from ginseng could attenuate the free radical damaging action on myocardial contractibility and other functional parameters of heart isolated from drug-treated rats. Diploid fibroblasts from human embryonic lung and rat liver were used to study the antiaging effects of rhodosin and ginseng stem-leaf saponins (GSLS). General light microscope, fluorescence microscope, quantitative cytochemistry and phagocytosis of macrophages of the abdominal cavity in mice were used to observe the effects of rhodosin and GSLS on the morphology, growth proliferation and life span of 2BS cells (*73*). Both could prolong their life span, promote growth of the cells, regulate their metabolism, increase the vitality of the cells and decrease the cellular death rate. They not only reduced the activity of Ac pase but also promoted DNA formation, enhanced the activity of ATPase and phagocytosis of macrophages. Both rhodosin and GSLS have an obvious antiaging effect, with the former more pronounced than the latter.

4.9. Cardiovascular Activity

Using simultaneous recording of action potential and contractible force in right ventricle papillary muscle of guinea pig and measurement of ^{45}Ca uptake by cultured myocardial cells of neonatal rat, the effects of total saponin of *Panax notoginseng* on the Ca^{2+} influx into myocardial cells were studied (*74*). The results indicated that the saponin can inhibit Ca^{2+} influx into myocardial cells. A study on the cholesterol-fed atherosclerosis of quails suggested that *Smilax glabra* may have preventive effects on atherosclerosis (*75*). The effect of *P. notoginseng* saponins on myocardial ischemia and reperfusion injury in conscious rabbit was studied. The results suggested that the saponins have protective effects against myocardial ischemia and reperfusion injury (*76*). The study on the effects of *P. notoginseng* saponins on acute cerebral ischemia indicated that the anti-cerebral ischemic effect of a saponin Rb_1 may be related to its calcium antagonism (*77*). Panaxatriol saponins isolated from *P. notoginseng* demonstrated remarkable antiarrythmic activities in coronary artery ligation-induced ischemic and reperfused arrythmias in rats (*78*). Comparative effects of *P. notoginseng* saponins, verapamil and norepinephrine on cerebral circulation in anesthetized rats and rabbits were studied (*79*). The results indicated that the saponins and verapamil are vasodilators of brain blood vessels, which would be beneficial to cerebral circulation, whereas norepinephrine is a vasoconstrictor of the vessels. The Blocking effect of *P. notoginseng* saponins on calcium channels of culture rat myocardiocyte was reported (*80*). HAN *et al.* (*81*) reported protective effects of *P. notoginseng* and *P. japonicus* saponins and gypenosides on myocardial ischemia and reperfusion injury. The results suggested that the underlying protective mechanisms of *P. japonicus* and *P. notoginseng* saponins are related to the prevention of calcium overload and that gypenosides have an action of anti-lipid peroxidation. Astragalus saponins were able to improve the myocardial contractibility significantly, attenuate the coronary blood flow and thus play a protective role on the cardiac functions (*82*).

Panaxadiol and panaxatriol saponins decreased the action potential parameters in cultured rat venticular myocytes in a dose dependent manner via calcium channel blocking (*83*). Effects of saponins isolated from the leaves of *Acanthopanax senticasus* on myocardial infarct size were studied in acute ischemic dogs. The results showed that the saponins could significantly reduce the size of acute myocardial infarcts (*84*).

4.10. Antiviral Activity

The saponins from Chinese and American ginseng stem and leaf showed protective effect against herpex simplex 1 virus (HSV-1) in vitro and in vivo in oral HSV-1 infection in humans (*85*). The results also indicated that ginsenosides Rb, especially Rb_2, are the active principles. The *in vitro* antiviral activity of triterpenoid saponins from *Calendula arvensis* was investigated (*86*). An inhibitory effect against vesicular stomatitis virus (VSV) and rhinovirus (HRV) was observed for all the compounds tested while HRV replication was significantly affected only by a hydrolyzed product. As an *in vitro* model for human immunodeficiency virus (HIV), LINN *et al.* (*87*) observed the inhibitory effect of *Cimifuga dahurica* saponins (Cd-S) on simian immunodeficiency virus (SIV) in Hut-78-SIV culture *in vitro*, and compared it with AZT. The inhibition rate of Cd-S at 200 μg/ml was 24%. The free virus titre of SIV was reduced 2–3 units and the syncytia was ameliorated compared with AZT with an inhibition rate of 91.30%, Cd-S could inhibit SIV only slightly. However, the authors studied the effect of Cd-S on ^{3}H-TdR incorporation into PHA-stimulated lymphocytes of human *in vitro*. Cd-S at a dose of 175.00 μg/ml significantly inhibited ^{3}H-TdR incorporation by 93.85% which indicated that Cd-S inhibits SIV through inhibition of nucleotide transportation into SIV host cells. The synthesis rate of SIV DNA was lowered and the products of SIV were reduced.

4.11. Antisweet Activity

The acylated saponins isolated from the leaves of *Zizyphus jujuba* showed antisweet activity (*88*). Acyl groups are believed to play an important role in generation of the antisweet activity. However, the results of a study by YOSHIKAWA *et al.* (*89*) using nonacylated antisweet principles from *Gymnema sylvestre* suggested that the acyl groups only increase the antisweet activity rather than playing the essential role. Several new dammarane glycosides were isolated from the fresh leaves of *Hovenia dulcis*. All the compounds showed antisweet activity (*90*). Sweet taste sensation is believed to be induced by adsorption of sweet substances on the receptor protein in taste receptor membranes. In spite of extensive studies by various workers, the receptor mechanism of sweet substances is still not clear. The suppression of sweetness by gymnemic acids and the effects on glucose absorption in the small intestine and on glucan formation by bacterial glycosyl-transferase has been reviewed (*91*).

4.12. Analgesic Activity

The effects of orally administered ginseng leaf saponins (GLS) on the analgesic action of morphine, the development of morphine-induced tolerance and physical dependence and the hepatic glutathione contents in mice were investigated. GLS antagonized the analgesic action of morphine and inhibited the development of morphine-induced tolerance and physical dependence. It also inhibited the decrease in hepatic glutathione level induced by multiple injections of morphine (*92*). Kim *et al.* studied the blocking by ginseng total saponin (GTS) of the development of methamphetamine reverse tolerance and dopamine receptor supersensitivity in mice (*93*). Repeated administration of methamphetamine (2 mg/kg) caused the development of reverse tolerance to the ambulation-accelerating effect of the drug. I.P. administration of GTS (200 mg/kg body wt.) prior to and during chronic administration of methamphetamine inhibited the development of reverse tolerance. Dopamine receptor supersensitivity developed in reverse tolerant mice which was also prevented by GTS. These results indicated that GTS may be useful for prevention of the adverse actions of methamphetamine. Daily repeated administration of cocaine (15 mg/kg over a 7-day period) developed reverse tolerance to the ambulatory-enhancing effect of cocaine. I.P. administration of GTS (100 and 200 mg/kg body wt.) prior to and during chronic administration of cocaine inhibited the development of dopamine receptor supersensitivity induced by chronic administration of cocaine (*94*). These results suggested that GTS may be useful for the prevention and therapy of the adverse action of cocaine. The relationship between the brain monoamines and morphine tolerance was examined in ginseng total saponins treated mice (*95*). Daily treatment with ginseng total saponins (100 mg/kg) did not affect the brain levels of noradrenaline, dopamine and serotonin for 5 days but inhibited the development of morphine tolerance.

4.13. Antileishmanial Activity

Antileishmanial activity was reported for the first time for saponins of ivy, *Hedera helix*, *in vitro* on promastigote forms of *Leishmania infantum* and *L. tropica* (*96*). The compounds tested were an extract containing 60% of saponin complex, the bisdesmosides hederasaponin B, C, and D, the corresponding monodesmosides, and α-hederin and hederegenin. Monodesmosides were as effective on promastigote forms as the reference compound, pentamidine. Against amastigote forms, only hederagenin exhibited a significant activity which was equivalent to that of the reference compound N-methylglucamine antimonate.

4.14. Miscellaneous Effects

The active principle from the funicles of *Acacia auriculiformis*, consisting of two triterpenoid saponins, acaciaside A and acaciaside B, killed in vitro 97% microfilaria of *Setaria cervi* in 100 min at 4 mg/ml concentration and 100% of adults in 35 min. The drug, when administered orally at 100 mg/kg to adult rats with implanted intraperitoneally *S. cervi* increased blood microfilariae (mf) count by 1.5-fold after the first phase of treatment for 10 days. Following the third phase of treatment and thereafter, the mf density was reduced by more than 80%. No toxic effects of the saponins was observed in rats. As this saponin is water soluble, nontoxic and effective by oral administration it holds promise for future use against human filariasis (*97*). The allelopathic activity of root saponins from alfalfa (*Medicago sativa*) on weeds and wheat was studied (*98*). Bioassays were developed for increasing the allelopathic effects on dandelion (*Taraxacum vugare*), coffee weed (*Sesbania exaltata*), pig weed (*Amaranthus retroflexus*), barnyard grass (*Echinocloacrus galli*), and cheat (*Bromus secalinus*) using pure alfalfa root saponins containing primarily medicagenic acid type glycosides. The allelopathic effects of the saponins were most effective toward barnyard grass and cheat. Less so for pig weed and coffee weed with little effect on dendelions. The saponins were allelopathic toward wheat. A wheat seedling bioassay was used to indicate the relationship between the chemical structure of alfalfa saponins and their allelopathic activity (*99*). The most active were medicagenic acid, its glycosides substituted at C-3 position with glucose, and hederagenin monoglycoside. Gymnemic acid (a saponin mixture from *Gymnema sylvestre*) has been tested as a preventive of dental caries (*100*). The decomposition of sugar and production of glucan by *Streptococcus mutans* which causes dental caries are prevented by gymnemic acid as a cariostatic agent. The plant *G. sylvestre* can be used as a cariostatic food and a patent has been taken on using it to prepare saponin-containing beverages, ice cream, tablets etc. (*101*). EtOH-absorption inhibitors useful for prevention of hangover contain saponin of tea or quillaja. Oral administration of tea seed extract (containing $\geqslant$70% saponin) or quillaja extract (containing $\geqslant$80% saponin) at 0.1 g/kg or 0.5 g/kg respectively inhibited EtOH absorption in rats.

Sweet-tasting saponins are drawing attention for use as natural sweeteners. People have a love-hate relationship with sugar. Many people like sucrose's taste but abhor its calories as well as the damage it does to teeth. Synthetic sweeteners exist but apprehension of unwanted effects is common Glycyrrhizin, an oleanane glycoside has long been known as a sweet-tasting saponin. *Abrus precatorius* L is a weedy subtropical vine and its leaves are known to be sweet-tasting. Five cycloaratane glycosides, abrusosides A-E containing a common aglycone, abrusogenin have been isolated from the leaves (*102*). While abrusosides A-D have been rated as being 30–100 times more sweet than sucrose on a weight basis, abrusoside E has been found to be only marginally sweet. However, the monomethyl ester of abrusoside E proved to be more potent. These sweeteners are not acutely toxic for mice, or mutagenic for bacteria, and may be rendered water soluble by conversion to their ammonium salts. Abrusosides A-D were also isolated from the leaves of *Abrus fruticulosus* occurring in Thailand (*102*).

5. Production of Saponins by Tissue Culture

There has been considerable interest in recent years in plant cell cultures as a potential alternative to traditional agriculture for large scale production of secondary plant metabolites. Considerable effort in this direction is being made and encouraging results have been reported. *In vitro* cultures of four species of *Gypsophila* (*G. paniculata*, *G. petraea*, *G. muralis* and *G. repens*) obtained from seedling organs showed various patterns of triterpenoid saponins biosynthesis as measured by gypsogenin-3-O-glucuronide content. Such different biosynthetic behavior may be a model for comparative studies on the regulation of saponin biosynthesis (*103*). Saponins were extracted from callus and suspension cultures of alfalfa (*Medicago sativa* and *M. truncatula*). Acid hydrolysis of the saponins provided soyasapogenol B and medicagenic acid as the main genins (*104*). Callus tissues from *Gynostemma pentaphyllum* leaves were grown in 14 culture media. Highest level of saponin formation was observed with the medium containing 0.5 ppm NAA and 0.5 ppm 6-BA (*105*). Tissue culture of *Bupleurum falcarum* L. was carried out with several kinds of media and plant hormones to produce saikosaponins. Gamborg's B-5 medium containing 0.5 ppm kinetin and 1.0 ppm 3-indolebutyric acid was the most effective medium and hormone for production of saikosaponins (*106*). A liquid culture medium for saponin production in adventitious root of *Panax japonicus* was examined. The root from seedling calli, cultivated on Murashige-Skoog

(MS) solid medium containing 2,4-dichlorophenoxy-acetic acid and on MS medium containing 1-naphthaleneacetic acid (I), was cultivated in liq. media, MS or Gamborg B5, by addition of (I) or indole-3-butyric acid (II). The maximum saponin production, 0.37% was obtained by 28 days of 2 mg II/L addition in Gamborg B5 (*107*). Five saponins were separated from the cell cultures of *Panax notoginseng*. The saponin contents were different in the cell cultures and the cultivated plants (*108*). The production of bioactive triterpene saponins of *Astragalus homosus* was optimized in cells and hairy root cultures (*109*). Callus growth rate was the highest in the dark and in the presence of 3% sucrose in B5 and Murasige-Skoog media with 2,4-D 1.0+Kinetin 0.1 mg/L. Hairly root growth was more rapid and required no hormone. The effect of chemical composition of a culture medium and of plant growth regulators on the growth of ginseng (*Panax ginseng*) and the production of saponins was studied (*110*). The content of gingsenosides in the selected cell lines was significantly higher than in the parent line. Production of a number of triterpene oligo-glycosides in the hairy root cultures of *Astragalus membranaceous* and their characterization has been reported (*111*).

6. Future Possibilities

Considerable progress has already been made in the isolation, structure elucidation and evaluation of diverse biological activities of saponins. Further developments relating to their use in health related areas and agriculture are expected. The accumulated evidence showing that saponins from a number of dietary plant species can reduce plasma cholesterol levels in humans is likely to encourage the development of pharmaceutical preparations and saponin-containing hypocholesterolenic diets. Besides the safety aspect, the quality aspect of the saponin-rich diets will require considerable product evaluation because of the bitterness associated with many saponins. Such evaluations will require involvement of sensory scientists, analysts, physiologists and processors.

The study of saponins has by now provided enough material for scientists to extract structural information that can be used to make designed compounds. The synthesis of antifungal modified medicagenic acid saponins by ZEHAVI *et al.* (*32*) is an attempt in this direction. Although chemical synthesis of saponins has been seriously hampered by their complex structural features, increasing activities to synthesize simpler bioactive saponins are expected. Production of saponins by plant tissue culture is another aspect which may receive greater attention. Cost analyses indicate that production of a secondary metabolite in plant cell culture is economical for cultures producing more than 1 gram of compound per litre of cell culture for compounds with a market value of at least $1000 per kg. Extensive studies already have been made on the production of saponins by cell suspension culture of ginseng. The total saponin content in the cell suspension culture based on these studies was about 4 times higher than that in the parent plant (*112*). The production of ginseng and other costly saponins on an industrial scale by cell culture technology seems to be a distinct possibility.

7. Reports of New Triterpenoid Saponins

New triterpenoid saponins isolated during the period late 1989 – early 1996 along with their natural distribution, available physical data and various spectra for their characterization are listed in Table 1. Structures **7–335** are aglycones of the various saponins and structures **336–341** are those of some special saponins.

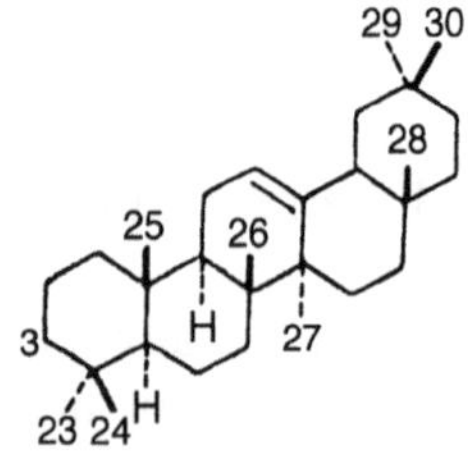

(**7**) OH-3β, CO_2H-28, oleanolic acid
(**8**) OH-3β, 24, 22-oxo, CO_2H-30
(**9**) OH-3β, 16β, 28 → 21β lactone, acacic acid lactone
(**10**) OH-3β, 16β, 21α, 23, 28, 11-OMe
(**11**) OH-3β, 23, CO_2H-28, hederagenin
(**12**) OH-3β, OMe-11α, CO_2H-28, 11α-methoxyoleanolic acid
(**13**) OH-3β, 23-oxo, CO_2H-28, gypsogenin
(**14**) OH-3β, 23, CO_2H-28, 20:29-ene, 30-nor
(**15**) OH-3β, 16α, CO_2H-28, echinocystic acid
(**16**) OH-2β,3β, CO_2H-28, 2β-hydroxyoleanolic acid
(**17**) OH-2β,3β, 23-oxo, CO_2H-28, 20:29-ene, 30-nor
(**18**) OH-2β,3β, 23, 11-oxo, CO_2H-28, CO_2Me-30
(**19**) OH-2β,3β, 23-oxo, CO_2H-28
(**20**) OH-2β,3β, CO_2H-23, 28, 20:29-ene, 30-nor
(**21**) OH-3β, 27, CO_2H-28, 27-hydroxyoleanolic acid
(**22**) OH-3β, CO_2Me-28
(**23**) OH-3β, 23, CO_2Me-28
(**24**) OH-2β,3β, 16α, CO_2H-28, astergenic acid
(**25**) OH-2β,3β, 23, CO_2H-28, bayogenin

(26) OH-3β, 16α, 22α, 21β-O-tigloyloxy, 28-O-isobutyryloxy
(27) OH-2β,3β, 16α, 23, CO_2H-28, polygalacic acid
(28) OH-3β, CO_2H-28, 20:29-ene, 30-nor
(29) OH-3β, 16α, 28, 22α-angeloyloxy, 23-oxo
(30) OH-3β, 16α, 28, 22α-tigloyloxy, 23-oxo
(31) OH-3β, 16α, 23, 28, 22α-tigloyloxy
(32) OH-2β, 3β, 28
(33) OH-2β, 3β, 23, CO_2Me-28
(34) OH-3β, 16α, 21β, 22α, 28, barringtogenol
(35) OH-3α, 16α, 21α, 22α, 28
(36) OH-3β, 23, CO_2H-28, CO_2Me-29, phytolaccagenic acid
(37) OH-2β, 3β, 6β, 23, CO_2H-28, protobassic acid
(38) OH-3β, CO_2H-28, 29
(39) OH-2α, 3β, CO_2H-28, maslinic acid
(40) OH-3β, 6β, 16α, OAc-28
(41) OH-3β, 16α, CO_2H-23, 28
(42) OH-2β, 3β, CO_2H-28, 20:29-ene, 30-nor
(43) OH-3β, 24, CO_2H-28
(44) OH-3β, 22α, 23, CO_2H-28, 22α-hydroxyhederagenin
(45) OH-3β, 21β, CO_2H-28, machaerinic acid
(46) OH-3β, 16α, 23-oxo, CO_2H-28, quillaic acid
(47) OH-3β, 16α, 23, CO_2H-28, caulophyllogenin
(48) OH-2β, 3β, CO_2H-23, 28, medicagenic acid
(49) OH-3β, 24, 22-oxo, soyasapogenol E
(50) OH-3β, CO_2H-28, CO_2Me-30
(51) OH-3β, 24, 28-oxo, 22β-O-[2,3-dihydro-2,5-dihydroxy-6-methyl-4H-pyran-4-one) (2′ →)]
(52) OH-2β, 3β, 27, CO_2H-23, 28
(53) OH-3β, 11-oxo, CO_2H-30, glycyrrhetic acid
(54) OH-3β, OAc-22β, CO_2H-30
(55) OH-3β, 11-oxo, 30 → 22β lactone, glabrolide
(56) OH-3β, 30 → 22β lactone, 11-deoxyglabrolide
(57) OH-3β, 24, 11-oxo, CO_2H-30, 24-hydroxy, glycyrrhetic acid
(58) OH-3β, 11-oxo, CO_2H-29, liquiritic acid
(59) OH-3β, 24, CO_2H-30
(60) OH-3β, 24, OAc-22β, CO_2H-30
(61) OH-3β, 16β, 22α, 23, 28, 21β-tigloyloxy
(62) OH-3β, 16β, 23, 28, 21β, 22α-ditigloyloxy
(63) OH-3β, 16β, 21β, 23, 28, gymnestrogenin
(64) OH-3β, 16β, 22α, 23, 28, 21β-2-methyl butyloyloxy
(65) OH-3β, 28, erythrodiol
(66) OH-2α, 3β, 23, CO_2H-28, arjunolic acid
(67) OH-3β, 19α, CO_2H-28, siaresinolic acid
(68) OH-3β, 19α, 23, CO_2H-28, ilexosapogenin A
(69) OH-3β, 22β, 24, soyasapogenol B
(70) OH-2β, 3β, 16α, CO_2H-23, 28
(71) OH-3β, 21β, 22β, 24, 29, kudzusapogenol A
(72) OH-2α, 3β, 23, 24, CO_2H-28, belleric acid
(73) OH-3β, 23, CO_2H-28, 30
(74) OH-2β, 3β, 23, CO_2H-28, 30

(75) OH-3β, 24, 22β-O-[3′-hydroxy-2′-methyl-5′,6′-dihydro-4′-pyrone (6′→)]
(76) OSO_3^--3β, CO_2H-28
(77) OH-3β, 16α, 22α, 21β-O-2-methylbutyroyloxy, 23-oxo, OAc-28
(78) OH-3β, 16α, 22α, 21β-angeloyloxy, OAc-28, 23-oxo
(79) OH-3β, 16α, 21β-tigloyloxy, OAc-28, 23-oxo
(80) OH-3α, CO_2H-23, 28
(81) OH-3β, 6β, 16α, 28
(82) OH-3β, 11α, 23, 28
(83) OH-3β, 23, 28, 11α-OMe
(84) OH-3β, 16β, 22β, 24
(85) OH-3β, 15α, 16α, 22α, 28, 21β-O-angeloyloxy
(86) OH-3β, 15α, 16α, 22α, 28, 21β-O-tigloyloxy
(87) OH-3β, 22, CO_2H-28, 17, 22-seco, 16-ene
(88) OH-3β, 21β, 24, kudzusapogenol C
(89) OH-2β, 3β, 6β, 16α, 23, CO_2H-28, 16α-hydroxyprotobassic acid
(90) OH-3β, 24, 22-oxo, CO_2Me-29
(91) OH-3α, 23, CO_2H-28, epihederagenin
(92) OH-3β, 24, OAc-22β, CO_2Me-30
(93) OH-3β, 16α, 28, 21β-O-benzoyl
(94) OH-3β, 21β, 22β, 24, soyasapogenol A
(95) OH-3β, 16β, 21β-O-[(6′*S*)-2′-trans,2′,6′-dimethyl-6′-hydroxy-2′,7′-octadienoyl], CO_2H-28
(96) OH-2α, 3β, 24, CO_2H-28
(97) OH-2β, 3β, OAc-23, CO_2H-28
(98) OH-3β, 23, CO_2H-28, CO_2Me-30
(99) OH-3β, 22β, 24, 29, oxytrogenol
(100) OH-3β, 16α, 28, CO_2H-30
(101) OH-3β, 22β, 24, CO_2H-29
(102) OH-2β, 3β, 16β, 23, 17β-oxo
(103) OH-3β, 24, 30, 22-oxo, wistariasapogenol A
(104) OH-3β, 16β, 23, 28, 11α-OMe
(105) OH-3β, 11α, 16β, 23, 28
(106) OH-3β, 16α, 23, 28, 22-angeloyloxy
(107) OH-3α, 21α, 22α, 28
(108) OH-3β, CO_2H-28, 23-O-(R)-1,2-propanediol-(1 → 23)-gypsogenic acid
(109) OH-3β, CO_2H-28, 23-O-(S)-1,2-propanediol-(1 → 23)-gypsogenic acid
(110) OH-3β, 16α, 23-oxo, CO_2Me-28
(111) OH-3β, 22β, 25-oxo
(112) OH-3β, 22β, 24, 30, wistariasapogenol B
(113) OH-3β, 16α, CO_2Me-28
(114) OH-3β, 22β, 24, CO_2H-29
(115) OH-3β, CO_2H-28, CO_2Me-29
(116) OH-2β, 3β, 23, CO_2H-28, CO_2Me-30, phytolaccagenin
(117) OH-2β, 3β, 6α, 23-oxo, CO_2H-28
(118) OH-2α, 3β, 19α, CO_2H-28
(119) OH-2α, 3β, 7α, 23, CO_2H-28 bellericagenin A
(120) OH-2α, 3β, 19α, 23, 24, CO_2H-28 bellericagenin B
(121) OH-3β, 16β, 21β, 22α, 23, 28-2S methyl-butyroyloxy
(122) OH-3β, 16β, 21β, 22α, 23, 28-tigloyloxy
(123) OH-3β, 16β, 21β, 22α, 23, OAc-28

(124) OH-3β, 16β, 22α, 23, 21β, 28-ditigloyloxy
(125) OH-3β, 16β, 22α, 23, 21β-tigloyloxy, OAc-28
(126) OH-3β, CO_2H-30
(127) OH-3α, CO_2H-28
(128) OH-3α, 23-oxo, CO_2H-28
(129) OH-3α, CO_2H-28, 29, 3-episerratagenic acid
(130) OH-3α, 23-oxo, CO_2H-28, 29
(131) OH-3α, 23, CO_2H-28, 29
(132) OH-3β, 11α, 16β, 28
(133) OH-3β, 22β
(134) OH-2α, 3β, 6β, 23, CO_2H-28, terminolic acid
(135) OH-3β, 16β, 28, longispinogenin
(136) OH-3β, 21α, 24
(137) OH-3β, 16β, 23, 28
(138) OH-3β, 16β, 23, 28, 21β-2-methylbutyroyloxy, 22α-methylcrotonoyloxy
(139) OH-3β, 21β, 23, 28, 16β, 22α-O-bis-2-methylcrotonoyloxy
(140) OH-3β, 16β, 22α, 23, 28, 21β-O-benzoyl
(141) OH-3β, 16β, 21β, 22α, 23, 28-O-benzoyl
(142) OH-2α, 3β, 23-oxo, CO_2H-28
(143) OH-3β, CO_2H-23, 28
(144) OH-3β, 16β, 23, 28, 22α-tigloyloxy
(145) OH-3β, 16β, 22α, 23, 28, gymnemanol
(146) OH-3β, 15α, 16α, 21β, 22α, 28
(147) OH-3β, 15α, 16α, 22α, 21β-tigloyloxy, OAc-28
(148) OH-3β, 15α, 16α, 28, 21β-tigloyloxy, OAc-22
(149) OH-3β, 15α, 16α, 22α, OAc-28, 21β-2-methyltigloyloxy
(150) OH-3β, 15α, 16α, 28, 21β-2-methylbutyroyloxy, OAc-22
(151) OH-3β, CO_2H-27, 28
(152) OH-3β, 23-oxo, CO_2Me-28
(153) OH-3β, CO_2H-24, 28
(154) OH-3β, 27, CO_2H-23, 28
(155) OH-3β, 21α, 23, yunganogenin
(156) OH-3β, 19α, 24, CO_2H-28
(157) OH-2β, 3β, 23, 30, CO_2H-28
(158) OH-3β, 16β, 28, 22α-O-N-methylanthranilyloxy
(159) OH-3β, 16β, 22α, 28-O-N-methylanthranilyloxy
(160) OH-3β, 16β, 28, 22α-tigloyloxy
(161) OH-3β, 16β, 22α, 28, 21-O-N-methylanthranilyloxy

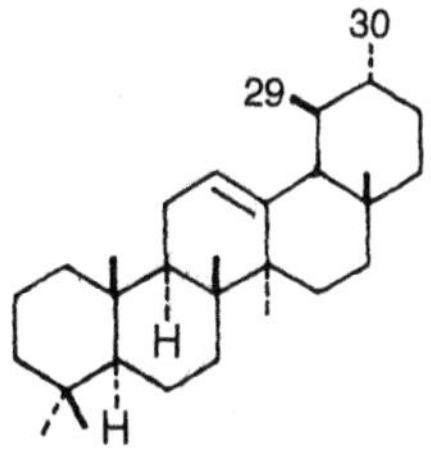

(162) OH-3β, 24, CO_2H-28, 24-hydroxyursolic acid
(163) OH-3β, 19α, CO_2H-28, pomolic acid
(164) OH-3β, 19α, CO_2H-23, 28, rotundioic acid

(**165**) OH-1α, 3β, 19α, 23, CO_2H-28
(**166**) OH-1β, 2α, 3β, 19α, 23, CO_2H-28
(**167**) OH-2α, 3β, 19α, 23, CO_2H-28
(**168**) OH-2α 3β, 23-oxo, CO_2H-28
(**169**) OH-3β, 23, CO_2H-28, 23-hydroxyursolic acid
(**170**) OH-3β, 19α, 23, CO_2H-28, rotundic acid
(**171**) OH-3β, CO_2H-27, 28, quinovic acid
(**172**) OH-2α, 3β, 6β, 23, CO_2H-28
(**173**) OH-2α, 3β, CO_2H-28
(**174**) OH-19α, SO_3Na-3β, CO_2H-28
(**175**) OH-3β, CO_2H-28, ursolic acid
(**176**) OH-2α, 3β, 6β, 19α, 23, CO_2H-28
(**177**) OH-3α, 19α, CO_2H-28
(**178**) OH-2α, 3β, 19α, CO_2H-28, tormentic acid
(**179**) OH-2α 3α, CO_2H-28, 19:29-ene
(**180**) OH-3β, 19α, 2-oxo, CO_2H-28
(**181**) OH-3α, CO_2H-23, 28
(**182**) OH-3β, 19α, 28 → 20 lactone, 11:12, 13:18-ene
(**183**) OH-3β, 11β, 19α, 28 → 20 lactone, 13:18-ene
(**184**) OH-3β, 27, CO_2H-28, 27-hydroxyursolic acid
(**185**) OH-3β, 19α, 24, 23-oxo, CO_2H-28, 23-oxorotungenic acid
(**186**) OH-3β, 19α, 23, 30, CO_2H-28, 30-hydroxyrotundic acid
(**187**) OH-3β, 23, 30, CO_2H-28
(**188**) OH-3β, 19α, 24, CO_2H-23, 28, 24-hydroxyrotundioic acid

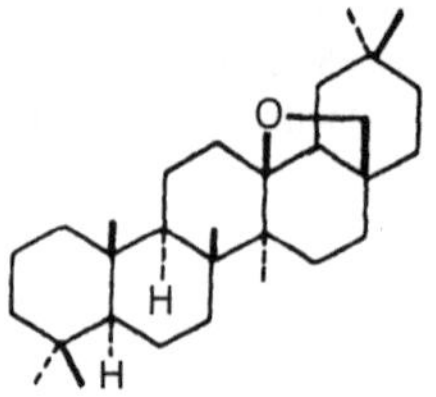

(**189**) OH-3β, 16α, 28β
(**190**) OH-3β, 16β, 23, 11-ene, saikogenin F
(**191**) OH-3β, 16β, 11:12-ene
(**192**) OH-3β, 16α, 29-oxo
(**193**) OH-3β, 16α, protoprimulagenin A
(**194**) OH-3β, 16α, OAc-22α, priverogenin B-22 acetate
(**195**) OH-3β, 16α, 22α, 28α
(**196**) OH-3β, 16α, 30-oxo, cyclamiretin A
(**197**) OH-3β, 16α, 28-oxo
(**198**) OH-3β, 16α, 30
(**199**) OH-3β, 16α, CO_2H-29
(**200**) OH-3β, 23, 16β-propanoyloxy, 11:12, 21:22-ene
(**201**) OH-3β, 16α, 30-$CH(OMe)_2$
(**202**) OH-3β, 16α, 22α, 28, anagallogenin A
(**203**) OH-3β, 16α, 23, 11:12-ene, epi-saikogenin F
(**204**) OH-3β, 16α, 22α
(**205**) OH-3β, 23, 11:12-ene
(**206**) OH-3β, 16α, 23, 28, OAc-22, anagallogenin A-22-acetate

(**207**) OH-3β, 16α, 23, anagallogenin B
(**208**) OH-3β, 23, 16-oxo
(**209**) OH-3β, 16β, 22
(**210**) OH-3β, 16β, 23, OAc-22
(**211**) OH-3β, 21β, 28α, 16α-propanoyloxy, 22α-angeloyloxy
(**212**) OH-3β, 21β, 28α, 16α-butanoyloxy, 22α-angeloyloxy
(**213**) OH-3β, 21β, 28α, 16α, 22α-diangeloyloxy
(**214**) OH-3β, 21β, 28α, 16α-2-methyl-butanoyloxy, 22α-angeloyloxy
(**215**) OH-3β, 28, 16α-propanoyloxy, OAc-21β, 22α-angeloyloxy
(**216**) OH-3β, 28, 16α-butanoyloxy, OAc-21β, 22α-angeloyloxy
(**217**) OH-3β, 16β, 23, 12-oxo, 9:11-ene

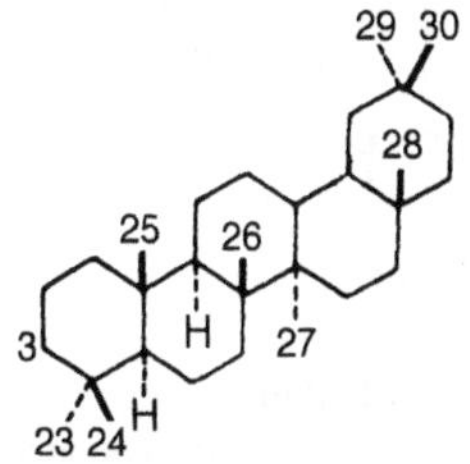

(**218**) OH-3β, 16β, 21β, 23, 28, 11:12, 13:18-ene
(**219**) OH-3β, 23, 28, 11:12, 13:18-ene
(**220**) OH-3β, 16β, 23, 28, 11:12, 13:18-ene, saikogenin A
(**221**) OH-3β, 16β, 23, 28, 30, 11:12, 13:18-ene
(**222**) OH-3β, 16α, 23, 28, 11:12, 13:18-ene, saikogenin D
(**223**) OH-2β, 3β, 23, CO_2H-28, CO_2Me-30, 9:11, 12:13-ene
(**224**) OH-3β, 24, CO_2Me-29
(**225**) OH-3β, 24, $CONH_2$-29
(**226**) OH-3β, 13, 23, 28, 11-ene
(**227**) OH-3β, 30, 22β-syringloyl, 25-oxo, 18:19-ene
(**228**) OH-3β, 16α, 23, 28, 30, 11-ene
(**229**) OH-3β, 24, CO_2H-30, 11:12, 13:18-ene
(**230**) OH-3β, 21α, CO_2H-29, 11:12, 13:18-ene
(**231**) OH-3β, 22β, 24
(**232**) OH-3β, 22β, 25-oxo, 18:19-ene
(**233**) OH-3β, 22β, 24, 11:12, 13:18-ene
(**234**) OH-3β, 22-oxo, CO_2H-30, 11:12, 13:18-ene
(**235**) OH-3β, 22β, CO_2H-30, 11:12, 13:18-ene
(**236**) OH-3β, CO_2H-28, 13:14-ene, pyrocincholic acid
(**237**) OH-3β, CO_2H-30, 11:12, 13:18-ene
(**238**) OH-3β, CO_2H-29, 11:12, 13:18-ene
(**239**) OH-3β, 23, CO_2H-28, 12:13, 21:22-ene

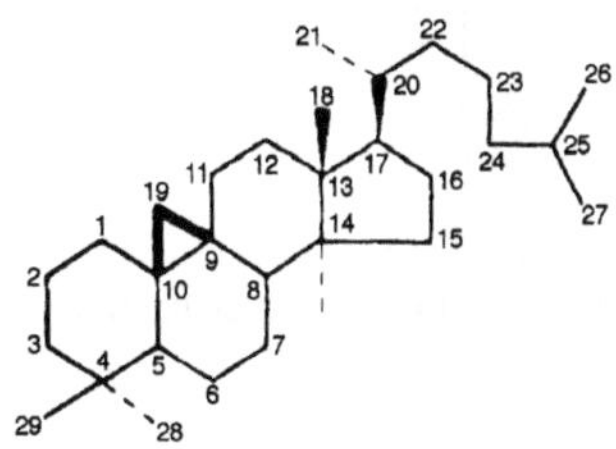

(**240**) OH-3β, 16β, 20(S), 24(S), 25
(**241**) OH-3β, 6α, 16β, 25, 20S:24R-epoxy, cycloastragenol
(**242**) OH-1α, 3β, 26, 24:25-ene
(**243**) OH-1α, 12β, 26, OAc-3β, 24:25-ene
(**244**) OH-3β, 6α, 16β, 24β, 25
(**245**) OH-1α, 3β, 15α, OAc-23, 16-oxo, 24:25-epoxy
(**246**) OH-3β, 22ξ, 24, CO_2H-21, 24:25-ene
(**247**) OH-3β, 22(S), 27, 24:25-ene

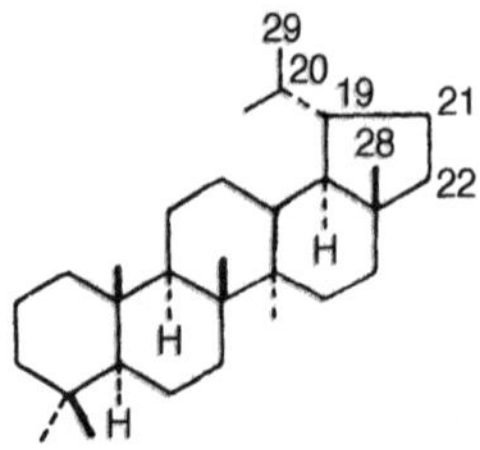

(**248**) OH-3β, CO_2H-28, 20:29-ene, betulinic acid
(**249**) OH-3β, 27, CO_2H-28, 20:29-ene, cylicodiscic acid
(**250**) OH-2α, 3β, 20:29-ene
(**251**) OH-3β, 28, 12:13, 20:29-ene
(**252**) OH-3α, CO_2H-28, 20:29-ene, 3-epibetulinic acid
(**253**) OSO_3-3α, CO_2H-28, 20:29-ene
(**254**) OH-3α, 11α, CO_2H-23, 28, 20:29-ene

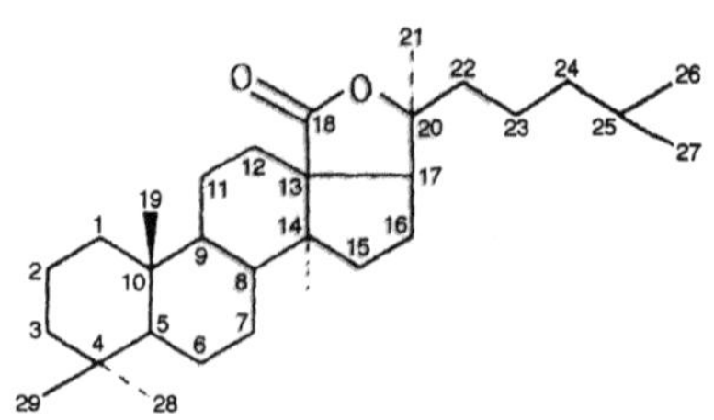

(**255**) OH-3β, 12α, 17α, 22:25-epoxy, 9:11-ene, holothurigenin
(**256**) OH-3β, 12α, 17α, OAc-25, 9:11, 22:23-ene
(**257**) OH-3β, 12α, 22:25-epoxy, 9:11-ene

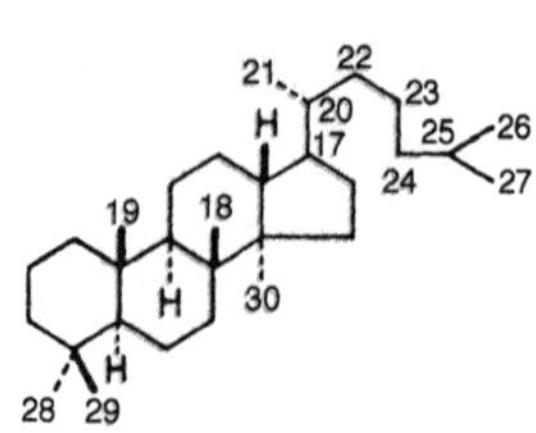

(**258**) OH-1β, 3β, 12β, 20(S), 26, 24-ene
(**259**) OH-3β, 12β, 20(S), 24-ene, 20(S)-protopanaxadiol
(**260**) OH-3β, 20, 19-oxo, 24-ene
(**261**) OH-3β, 12β, 20(S), 24(S), 25:26-ene

(262) OH-2α, 3β, 12β, 20, 25, 23-ene
(263) OH-3β, 20(S), 24-ene
(264) OH-3β, 20(R), 16α:30, 16β:22R-diepoxy, 24-ene
(265) OH-3β, 6α, 12β, 20(S), 25, 22-ene
(266) OH-3β, 20(R), OAc-15α, 16α:30, 16β:22(R)-diepoxy, 24(24′)-methylene
(267) OH-3β, 12β, 25, 30, 20:24-epoxy, capsugenin
(268) OH-3β, 6α, 20(S), 24-ene
(269) OH-3β, 15α, 16β, 20(S), 16:22-epoxy, 24-ene
(270) OH-3β, 24α, OSO_3^--20(S), 16β:30, 16β:23-diepoxy, 25:26-ene
(271) OH-3β, 12β, 23(S), 25, 20(S): 24(S) epoxy
(272) OH-3β, 6α, 12β, 25, 20:24-epoxy
(273) OH-3β, 23, 19-oxo, 16β:22-epoxy
(274) OH-3β, 6α, 12β, 24α, 20:25-epoxy
(275) OH-3β, 6α, 12β, 20(S), 24ξ, 25
(276) OH-3β, 6α, 12β, 25ξ, 26, 20(S):24(S)-epoxy
(277) OH-3β, 20, 16α:30, 16β:22-epoxy, 24-Me, 25:26-ene
(278) OH-3β, 6α, 12β, 20:21, 23:24-ene
(279) OH-3β, 12β, 20R:25-epoxy, panaxadiol
(280) OH-3β, 12β, 20, 25, 23-ene
(281) OH-3β, 12β, 20(S), 24ξ, 25
(282) OH-2β, 16α, 20(S), 25, 3, 11, 22-trioxo, 5-ene
(283) OH-3β, 24, 25(R), 11-oxo, 5-ene
(284) OH-3β, 20, 21, 24, 25:27-ene
(285) OH-3β, 20, 25, 27, 23-ene
(286) OH-3β, 20, 25, 30, 16-oxo, 23-ene
(287) OH-3β, 12β, 23(S), 25, 28, 20(S):24(S)-epoxy
(288) OH-3β, 12β, 25, 20(S):24(S)-epoxy
(289) OH-3β, 25, 12-oxo, 20(S):24(R)-epoxy
(290) OH-3β, 25, 26, 20(S):24(R)-epoxy
(291) OH-3β, 25, 20(S):24(R)-epoxy
(292) OH-3β, 12β, 20(S), 24(R), 25
(293) OH-3β, 12β, 20(S), 24(R), 25, 28
(294) OH-3β, 20(S), 24(R), 25
(295) OH-3β, 20(S), 24(R), 25, 12-oxo
(296) OH-3β, 12β, 23(S), 24(R), 20(S):25-epoxy
(297) OH-3β, 12β, 25, 20(S):24(R)-epoxy
(298) OH-3β, 12β, 20(S), 25, 22-ene
(299) OH-3β, 6α, 12β, 20(R), 25, 22-ene
(300) OH-3β, 11α, 25, 20(S):24(R)-epoxy
(301) OH-3β, 25, OAc-11α, 20(S):24(R)-epoxy
(302) OH-3β, 11α, 20(S), 24-ene

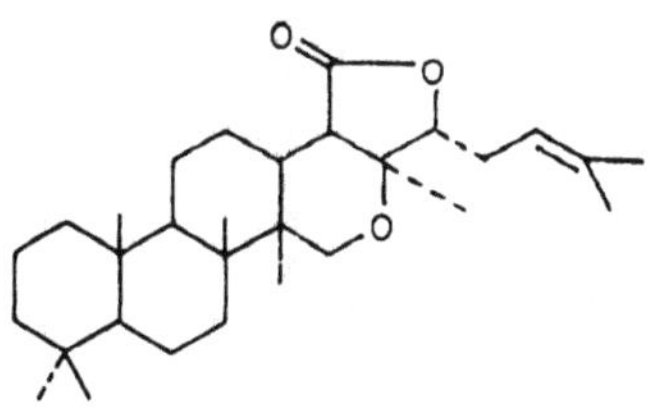

(**303**) OH-3β, 15α, mabiogenin
(**304**) OH-3β, 21β, 15-oxo

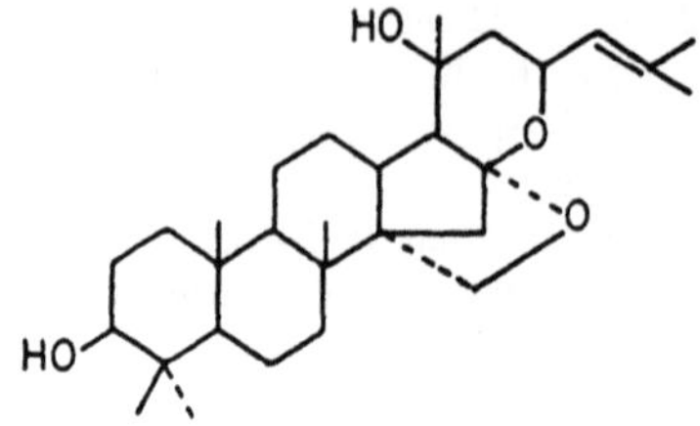

(**305**) Jujubogenin

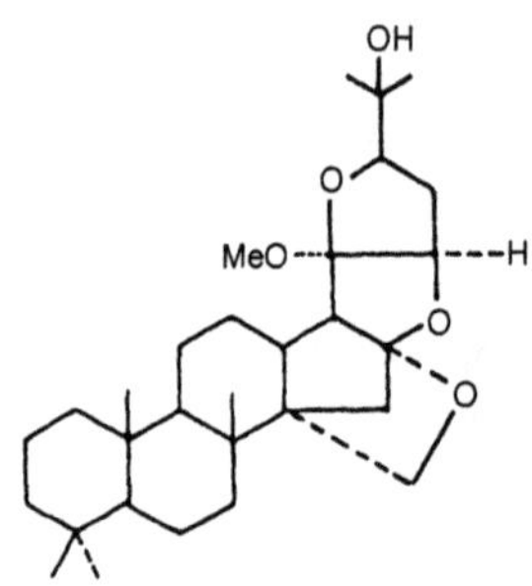

(**306**) Trevoagenin D

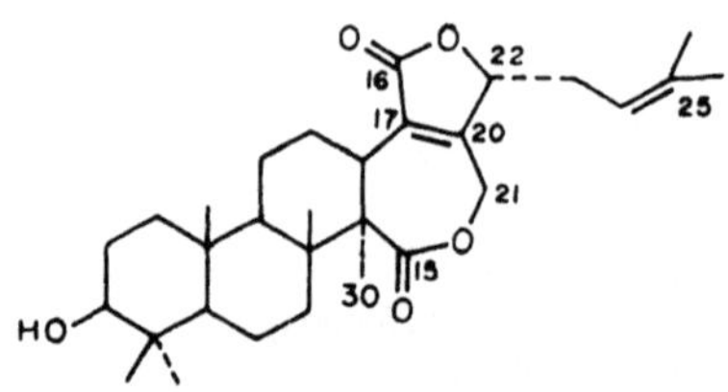

(**307**)

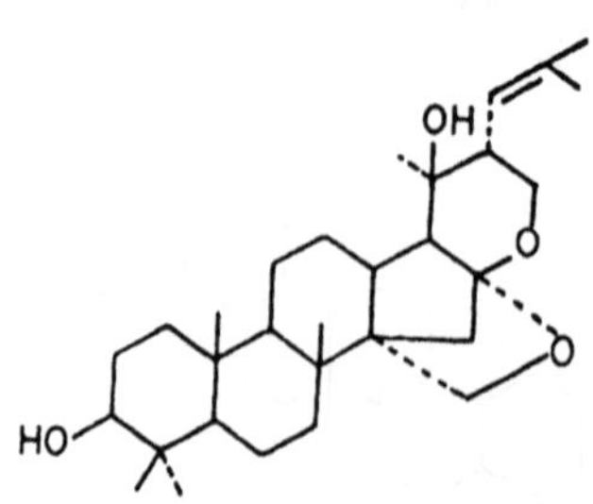

(**308**) Pseudojujubogenin

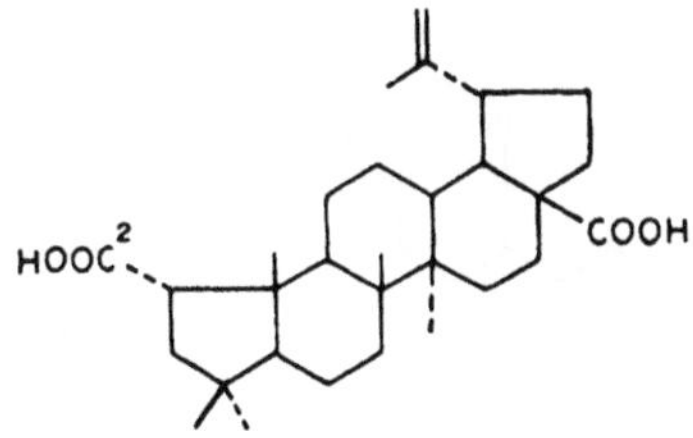

(**309**) OH-3β, ceanothic acid
(**310**) OH-3α, isoceanothic acid

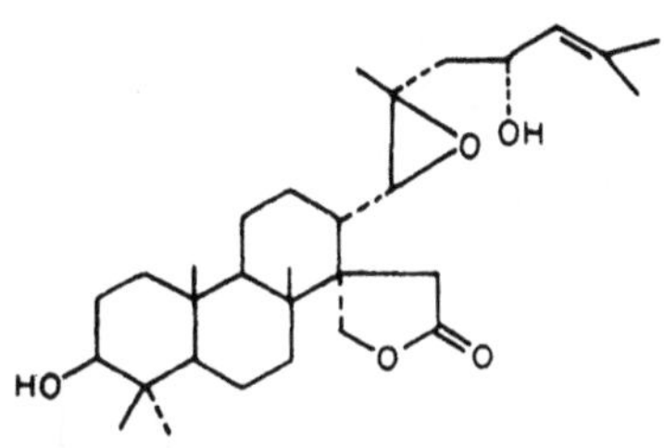

(**311**) Hovenolactone

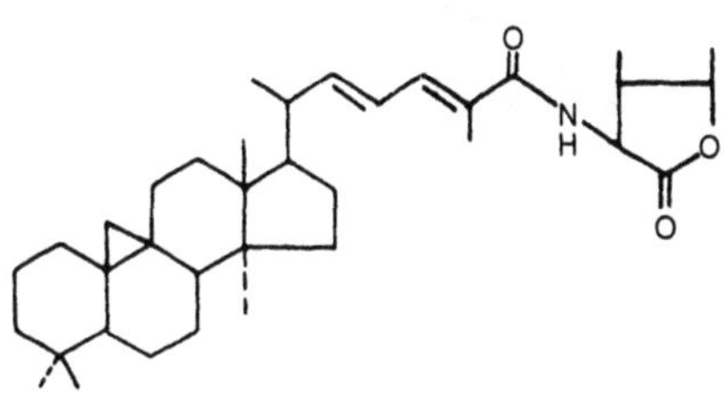

(**312**) OH-3β, heinsiagenin A
(**313**) OH-2α, 3β

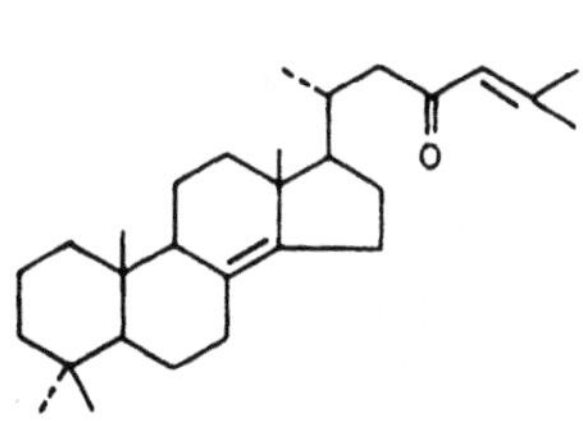

(**314**) OH-3β, 5α, 12
(**315**) OH-3β, 5α, 12β-OMe

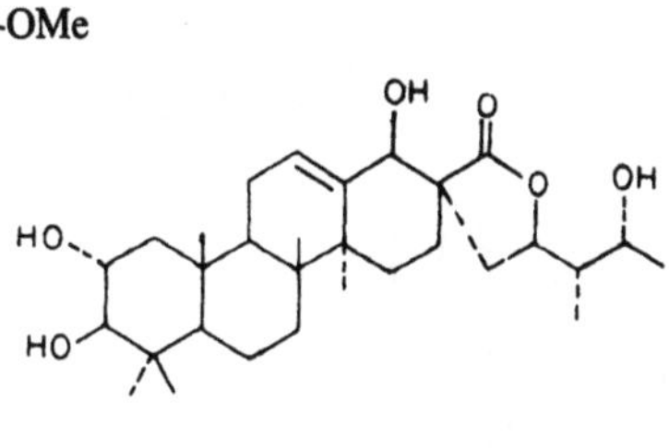

(**316**)

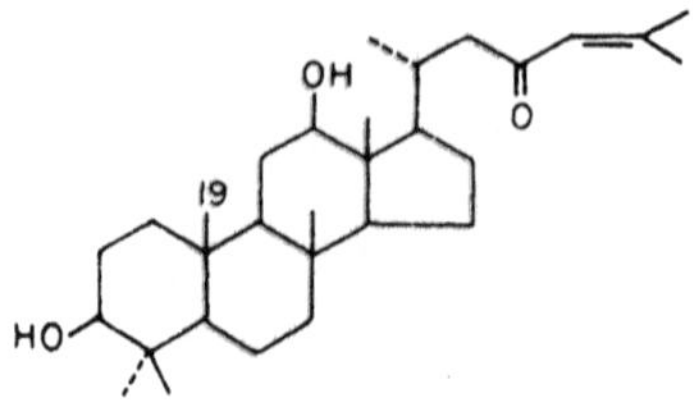

(**317**)

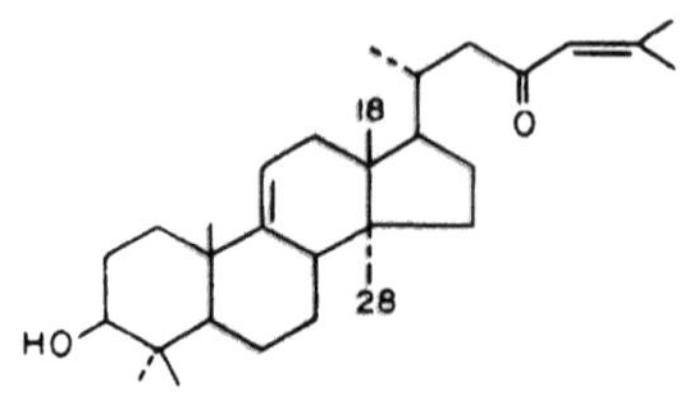

(**318**)

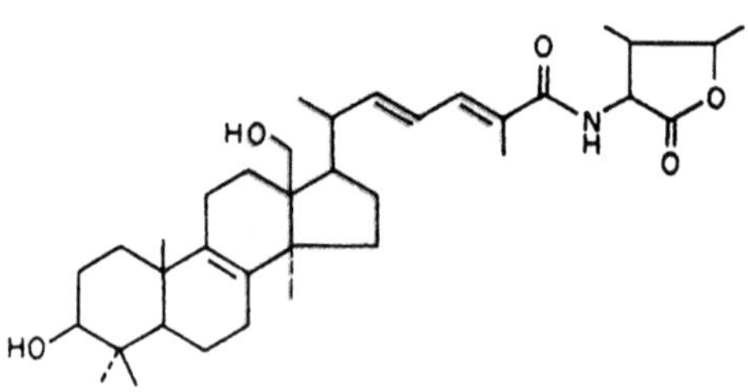

(**319**)

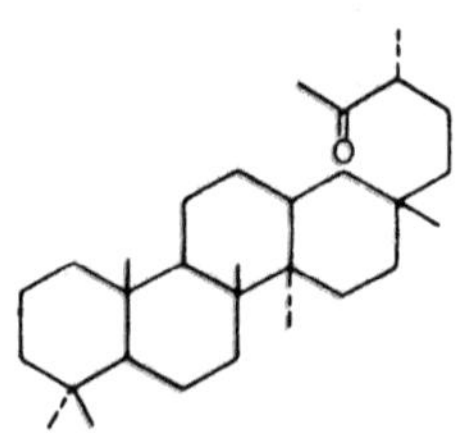

(**320**) OH-3β, CO_2H-17, 11:12, 13:18-ene
(**321**) OH-3β, 12β, CO_2H-17, 13:18-ene

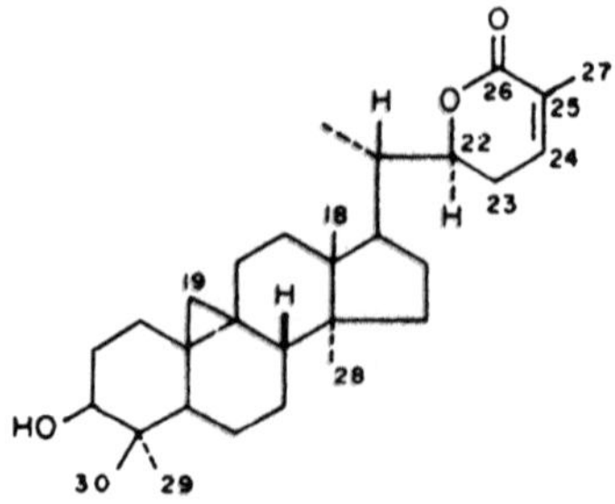

(**322**) OH-3β, CO_2H-29
(**323**) OH-3β

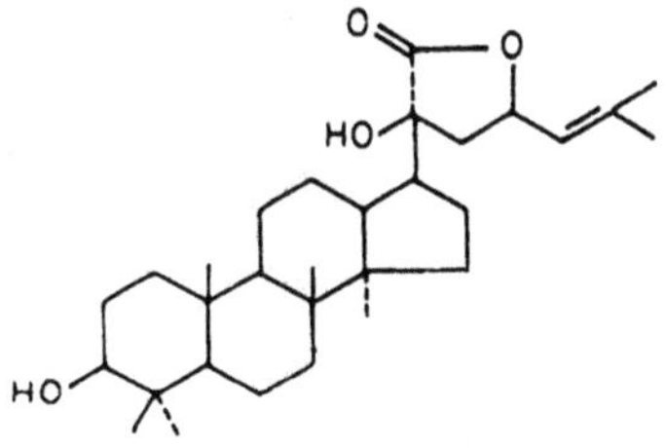

(**324**)

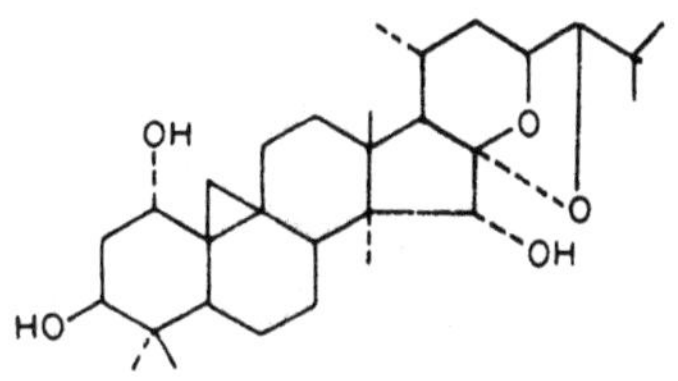

(**325**) OAc-25
(**326**) OH-25

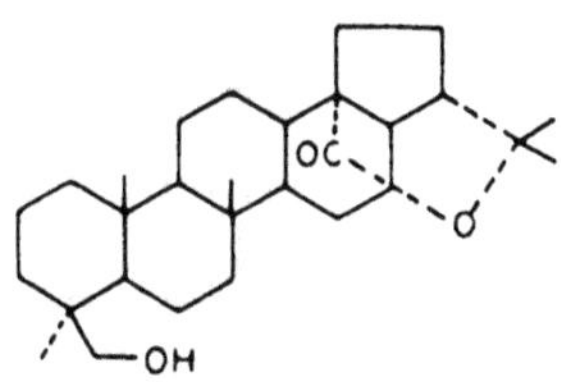

(**327**) OH-3β
(**328**) OH-3β, CO_2H-29

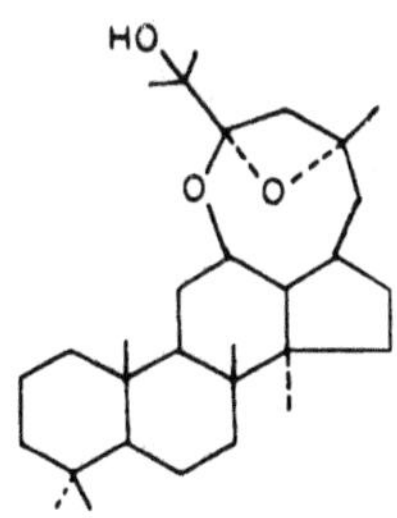

(**329**) OH-3β
(**330**) OH-3β, 23(S)

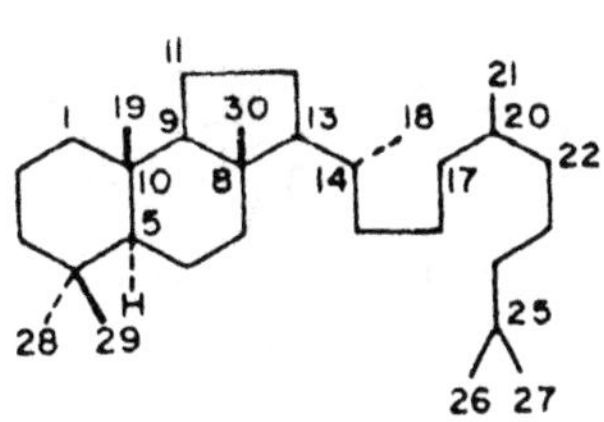

(**331**) OH-3β, 12α, 25, 30, 14R:17R, 20R:24S-diepoxy
(**332**) OH-3β, 12α, 14, 17, 25, 20R:24S-epoxy
(**333**) OH-3β, 12α, 20R, 24S, 25, 14R:17R-epoxy

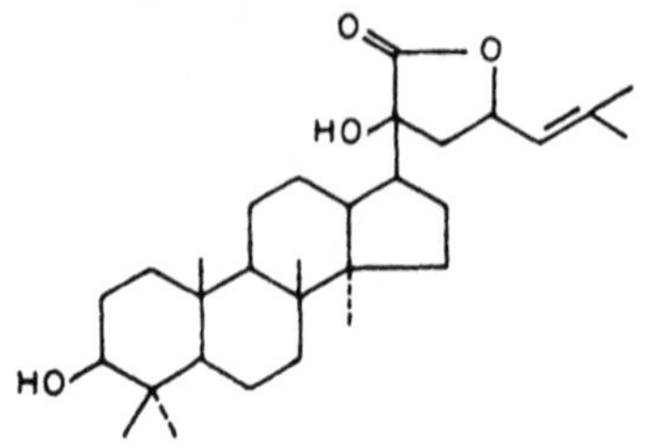

(**334**)

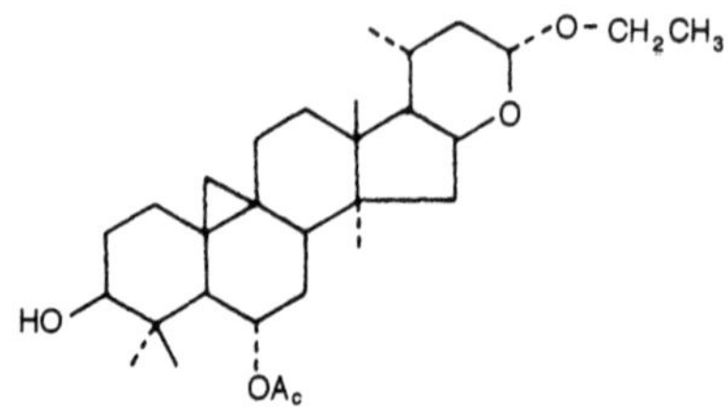

(**335**)

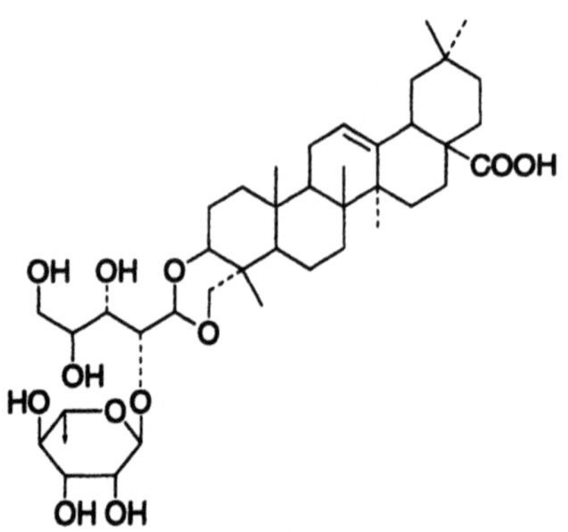

(**336**)

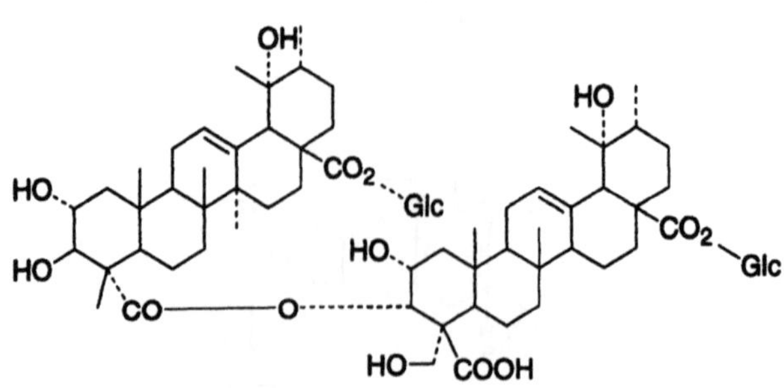

(**337**)

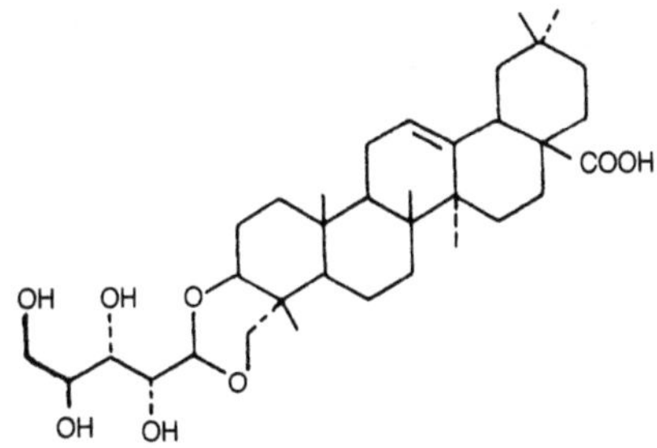

(**338**)

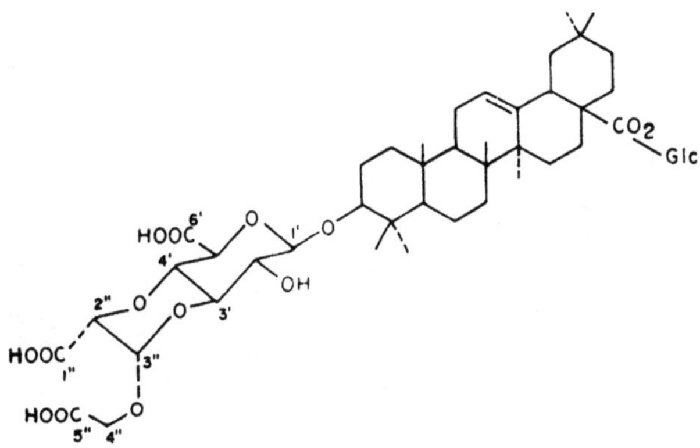

(**339**) OMe-2″
(**340**) OH-2″

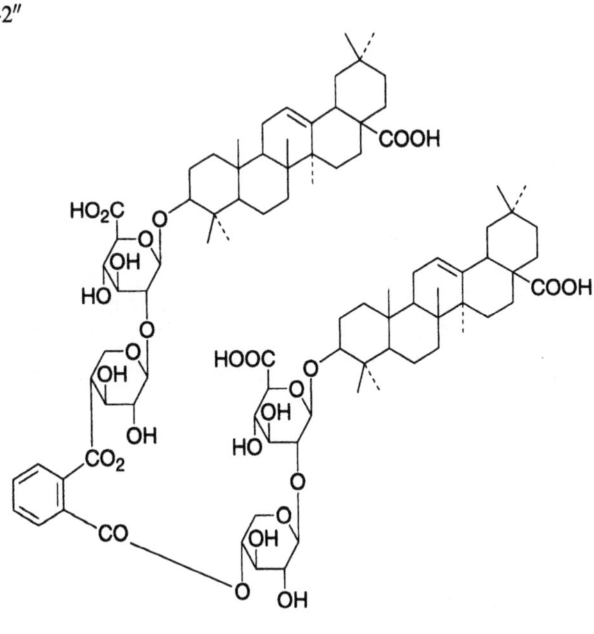

(**341**)

Table 1. *Triterpenoid Saponins Isolated from Mid-1989 to Mid-1996*

Source (1)	Saponin mp, $[\alpha]_D$, spectra recorded (2)	Structure (3)	Ref. (4)
Abrus cantoniensis (Fabaceae)	Abrisaponin I ^1H, ^{13}C, EIMS	Aglycone (**8**) Rha-^{2}Gal-^{2}GlcA (OH-3β)	*113*
A. precatorius	Abrusoside A 278–280°, +11.2° UV, IR, ^1H, ^{13}C, FABMS	Aglycone (**322**) Glc (OH-3β)	*114*
	Abrusoside B 243–245°, +5.8° UV, IR, ^1H, ^{13}C, FABMS	Aglycone (**322**) Glc-^{2}(Me-ester-6′) GlcA (OH-3β)	*114*
	Abrusoside C 260–262°, +31.4° UV, IR, ^1H, ^{13}C, FABMS	Aglycone (**322**) Glc-^{2}Glc (OH-3β)	*114*
	Abrusoside D 237–239°, +9.9° UV, IR, ^1H, ^{13}C, FABMS	Aglycone (**322**) Glc-^{2}GlcA (OH-3β)	*114*
	Abrusoside E 265°, +2° UV, IR, FABMS	Aglycone (**322**) GlcA^2-Glc (OH-3β)	*102*

Acacia auriculiformis (Leguminosae)	Acaciaside A 240–242°, −19.5° UV, IR, ^{1}H, ^{13}C, FABMS	Aglycone (**95**) Glc-^{6}[Ara-^{2}]Glc (OH-3β) [(6′S)-2′-trans-2′, 6′-dimethyl-6′-hydroxy-2′, 7′-octadienoyl] Glc (OH-6′ β) Rha-^{6}[Xyl-^{2}]Glc (CO_2 H-28)	*115*
	Acaciaside B 257°, −26.2° UV, IR, ^{1}H, ^{13}C, FABMS	Aglycone (**95**) Glc-^{6}[Ara-^{2}]Glc (OH-3 β) {(6′S)-2′-trans-2′,6′-dimethyl-6′-hydroxy-2′,7′-octadienoyl} Glc (OH-6′β) (Glc substituted at 2 by Xyl) Rha-^{6}[Xyl-^{2}]Glc (CO_2 H-28)	*115*
Acanthopanax hypoleucus (Araliaceae)	Hypoleucoside A +2.8°, ^{1}H, ^{13}C, FABMS	11α-Methoxyoleanolic acid (**12**) Glc (OH-3β) Glc (CO_2H-28)	*116*
	Hypoleucoside B +2.1° ^{1}H, ^{13}C. FABMS	Oleanolic acid (**7**) Glc-^{2}Ara-^{4}Glc (OH-3β) Glc-^{6}Glc (CO_2H-28)	*116*
A. siebodianus	Sieboldianuside A −24.2°, IR, ^{1}H, ^{13}C, FABMS	Oleanolic acid (**7**) Xyl-^{3}Rha-^{2}Ara (OH-3β)	*117*

Table 1. *(continued)*

Source (1)	Saponin mp, $[\alpha]_D$, spectra recorded (2)	Structure (3)	Ref. (4)
	Sieboldianuside B −29.4°, IR, ^{1}H, ^{13}C, FABMS	Hederagenin (**11**) Rha-^{4}Glc-^{6}Glc (OH-3β)	*117*
A. spinosus	Spinoside C_1 −15.5°, IR, ^{1}H, ^{13}C, FABMS	Aglycone (**127**) Rha-^{4}Glc-^{6}Glc (CO_2H-28)	*118*
	Spinoside C_4 −22.1°, IR, ^{1}H, ^{13}C, FABMS	Aglycone (**128**) Rha-^{4}Glc-^{6}Glc (CO_2H-28)	*118*
	Spinoside C_5 −17.3°, IR, ^{1}H, ^{13}C, FABMS	Epi-hederagenin (**91**) Rha-^{4}Glc-^{6}Glc (CO_2H-28)	*118*
	Spinoside D_1 −19.2°, IR, ^{1}H, ^{13}C, FABMS	3-Episeratogenic acid (**129**) Rha-^{4}Glc-^{6}Glc (CO_2H-28)	*119*
	Spinoside D_2 −25.5°, IR, ^{1}H, ^{13}C, FABMS	Aglycone (**130**) Rha-^{4}Glc-^{6}Glc (CO_2H-28)	*119*
	Spinoside D_3 −17.0°, IR, ^{1}H, ^{13}C, FABMS	Aglycone (**131**) Rha-^{4}Glc-^{6}Glc (CO_2H-28)	*119*
Acanthophyllum squarrosum (Caryophyllaceae)	Squarroside A 2D	Gyposogenin (**13**) Xyl $\searrow^3$ GlcA (OH-3 β) Gal $\nearrow_2$	*120*

Achyranthes fauriei (Anearanthaceae)	Achyranthoside A +64.3°, ^{1}H, ^{13}C, 2D, FABMS	Ara —$^{3}_{2}$ (OAc-4′) Fuc (CO_2H-28) Xyl-^{4}Rha / (**339**)	*121*
	Achyranthoside B +51.7°, ^{1}H, ^{13}C, 2D, FABMS	(**340**)	*121*
	Achyranthoside C ^{13}C, 2D, FABMS	Oleanolic acid (**7**) ($CO_2HCHOHCHOCH_2CO_2H$-3′)GlcA(OH-3β) Glc (CO_2H-28)	*122*
	Achyranthoside D ^{13}C, 2D, FABMS	Oleanolic acid (**7**) Glc-^{2}($CO_2HCHOHCHOCH_2CO_2H$-3′) GlcA (OH-3β) Glc(CO_2H-28)	*122*
Adesmia aconaguensis (Leguminosae)	210–212°, −36.1° IR, ^{1}H, ^{13}C, FABMS	Aglycone (**331**) Glc (OH-3β)	*123*
	243–245°, +1.4° IR, ^{1}H, ^{13}C, FABMS	Aglycone (**332**) Glc (OH-3β)	*123*
	255-257°, +12.6° IR, ^{1}H, ^{13}C, FABMS	Aglycone (**333**) Glc (OH-3β)	*123*
Adina rubella (Rubiaceae)	Rubelloside A 246–247°, +16.2° IR, ^{1}H, ^{13}C, FABMS	Quinovic acid (**171**) Glc-^{4}Rha (OH-3β)	*124*
	Rubelloside B 252–253°, +36.2° IR, ^{1}H, ^{13}C, FABMS	Quinovic acid (**171**) Glc-^{4}Fuc (OH-3β)	*124*

Table 1. *(continued)*

Source (1)	Saponin mp, $[\alpha]_D$, spectra recorded (2)	Structure (3)	Ref. (4)
Akebia quinata (Lardizabalaceae)	Quinotoside A 256–260°, +78.3° ^{1}H, ^{13}C, EIMS	Aglycone (**14**) Ara (OH-3β)	*125*
	Quinotoside B 268–270°, +106.3° ^{1}H, ^{13}C, SIMS	Aglycone (**14**) Glc-^{3}Ara (OH-3β)	*125*
	Quinotoside C 248–252°, +86.2° ^{1}H, ^{13}C, SIMS	Aglycone (**14**) Xyl-^{2}Ara (OH-3β)	*125*
	Quinotoside D 290°, +43.7° ^{1}H, ^{13}C, SIMS	Aglycone (**28**) Xyl-^{2}Ara (OH-3β)	*125*
A. trifoliata	Trifoside A −0.62°, ^{1}H, ^{13}C, FABMS	Aglycone (**38**) Ara —^{3} Glc (OH-3β) Xyl —^{2}	*126*
	Trifoside B +72.8°, ^{1}H, ^{13}C, FABMS	Aglycone (**28**) Glc —^{3} Ara (OH-3β) Xyl —^{2}	*126*
	Trifoside C +65.1°, ^{1}H, ^{13}C, FABMS	Aglycone (**14**) Glc —^{3} Ara (OH-3β) Xyl —^{2}	*126*

Albizia *lebbeck* (Leguminosae)	Albiziasaponin A 200–202°, −22.0° ^{1}H, ^{13}C, FABMS	Acacic acid lactone (**9**) Xyl-^{2}Ara-^{6}Glc (OH-3β)	*127*
	Albiziasaponin B 260–262°, −40.0° ^{1}H, ^{13}C, FABMS	Acacic acid lactone (**9**) Ara ╲ $^{6}_{2}$ Glc (OH-3β) Glc ╱	*127*
	Albiziasaponin C 198–200°, −23.5° ^{1}H, ^{13}C, FABMS	Acacic acid lactone (**9**) Xyl-^{2}Ara ╲ $^{6}_{2}$ Glc (OH-3β) Glc ╱	*127*
A. lucida	Saponin 1 246°, −3.27° ^{1}H, ^{13}C, FABMS	Echinocystic acid (**15**) Ara ╲ $^{6}_{2}$ Glc (OH-3β) Glc ╱	*128*
	Saponin 2 275°, −14.75° ^{1}H, ^{13}C, FABMS	Echinocystic acid (**15**) Xyl-^{2}Ara ╲ $^{6}_{2}$ Glc (OH-3β) Glc ╱	*128*
	Saponin 3 254°, 0° ^{1}H, ^{13}C, FABMS	Aglycone (**113**) (2′-acetylamino-2′-deoxy) Glc (OH-3β) 6| Fuc^{2}-Xyl	*128*
Alphitonia *Zizyphoides* (Rhamnaceae)	Zizyphoiside A 2D	Aglycone (**279**) Xyl-^{6}Glc-^{3}Gal (OH-3β) |2 Rha	*23*
	Zizyphoiside C ^{1}H, ^{13}C	Jujubogenin (**305**) (AcO-3′) Rha-^{3}Glc-^{3}Ara (OH-3β)	*129*

Table 1. *(continued)*

Source (1)	Saponin mp, $[\alpha]_D$, spectra recorded (2)	Structure (3)	Ref. (4)
	Zizyphoiside D ^{1}H, ^{13}C	Jujubogenin (**305**) (OAc-2′) Rha-^{3}Glc-^{3}Ara (OH-3β)	*129*
	Zyzyphoiside E ^{1}H, ^{13}C	Jujubogenin (**305**) (OAc-4′) Rha-^{3}Glc-^{3}Ara (OH-3β)	*129*
Amaranthus caudatus (Amarantheceae)	Saponin 1	2β-Hydroxyoleanolic acid (**16**) Rha-^{2}Glc (OH-3β) Glc (CO_2H-28)	*130*
	Saponin 2	2β-Hydroxyoleanolic acid (**16**) Rha-^{2}(OMe-6′) Glc (OH-3β) Glc (CO_2H-28)	*130*
	Saponin 3	Aglycone (**19**) Rha-^{2}Glc (OH-3β) Glc (CO_2H-28)	*130*
	Saponin 4	Aglycone (**117**) Rha-^{2}Glc (OH-3β) Glc (CO_2H-28)	*130*
	Saponin 5	Aglycone (**20**) Glc (CO_2H-28)	*130*
	Saponin 6	Aglycone (**20**) Glc (CO_2H-23) Glc (CO_2H-28)	*130*
	Saponin 7	Aglycone (**20**) Glc (OH-3β) Glc (OH-23) Glc (CO_2H-28)	*130*

A. hypochon-driacus	Amaranthus-Saponin I +23.3° ^{1}H, ^{13}C, FABMS	2β-Hydroxyoleanolic acid (**16**) Rha-^{3}GlcA (OH-3β) Glc (CO_2H-28)	*131*
	Amaranthus-Saponin II +9.2°, ^{1}H, ^{13}C, FABMS	Aglycone (**19**) Rha-^{3}GlcA (OH-3β) Glc (CO_2H-28)	*131*
	Amaranthus-Saponin III +22.0°, ^{1}H, ^{13}C, FABMS	Aglycone (**42**) Rha-^{3}GlcA (OH-3β) Glc (CO_2H-28)	*131*
	Amaranthus-Saponin IV +71.9°, ^{1}H, ^{13}C, FABMS	Aglycone (**17**) Rha-^{3}GlcA (OH-3β) Glc (CO_2H-28)	*131*
Ampelozizphus amazonicus (Rhamnaceae)	saponin 235°, IR, ^{1}H, ^{13}C, FABMS	Jujubogenin (**305**) Glc-^{2}Ara (OH-3β) Rha (OH-20β)	*132*
	Saponin 3 204–208°, −35.0°, ^{1}H, ^{13}C, FABMS	Aglycone (**266**) Rha-^{2}Glc (OH-3β)	*133*
Anagallis arvensis (Primulaceae)	Anagallosaponin I >300°, −11.1°, ^{1}H, ^{13}C, FABMS	Anagallogenin A (**202**) Glc ╲ $^{4}_{2}$ Glc—$^{4}_{2}$ Ara (OH-3β) Xyl ╱ ╱ Glc	*134*
	Anagallosaponin II 255–257°, −4.5°, ^{1}H, ^{13}C, FABMS	Anagallogenin A 22-acetate (**206**) Xyl-^{2}Glc-$^{4}_{2}$ Ara (OH-3β) Glc ╱	*134*
	Anagallosaponin III 246–247°, −17.3°, ^{1}H, ^{13}C, FABMS	Aglycone (**194**) Xyl-^{2}Glc-^{4}Ara (OH-3β)	*134*

Table 1. *(continued)*

Source (1)	Saponin mp, $[\alpha]_D$, spectra recorded (2)	Structure (3)	Ref. (4)
	Anagallosaponin IV 237–239°, −19.4°, ^1H, ^{13}C, FABMS	Aglycone (**194**) Xyl-^{2}Glc-$^4_2$ Ara (OH-3β) / Glc	*134*
	Anagallosaponin V 253–255°, −26.2°, ^1H, ^{13}C, FABMS	Aglycone (**194**) Glc ╲ $^4_2$ Glc-^{4}Ara (OH-3β) Xyl ╱	*134*
	Anagallosaponin VI 235–237°, −8.7°, IR, ^1H, ^{13}C, FABMS	Aglycone (**209**) Xyl-^{2}Glc-^{4}Ara (OH-3β)	*135*
	Anagallosaponin VII 258–259°, −13.6°, IR, ^1H, ^{13}C, FABMS	Aglycone (**209**) Xyl-$^2_4$ Glc ╲ ^{4}Ara (OH-3β) / Glc	*135*
	Anagallosaponin VIII 245–247°, −14.3°, IR, ^1H, ^{13}C, FABMS	Aglycone (**210**) Xyl-^{2}Glc-$^4_2$ Ara (OH-3β) | Glc	*135*
	Anagallosaponin IX 248–249°, −15.4°, IR, ^1H, ^{13}C, FABMS	Aglycone (**210**) Glc ╲ $^4_2$ Glc ╲ Xyl ╱ $^2_4$ Ara (OH-3β) Glc ╱	*135*

	Anagallisin A 244–246°, −5.81°, ^{1}H, ^{13}C, FABMS	Anagalligenin B (**207**) Xyl-^{2}Glc $\searrow^{4}$ Ara (OH-3β) Glc-^{4}Glc $\nearrow_{2}$	*136*
	Anagallisin B 236–238°, −3.2°, ^{1}H, ^{13}C, FABMS	Aglycone (**208**) Xyl-^{2}Glc $\searrow^{4}$ Ara (OH-3 β) Glc-^{4}Glc $\nearrow_{2}$	*136*
	Anagallisin D 256–260°, −6.9°, ^{1}H, ^{13}C, FABMS	Aglycone (**208**) Xyl-^{2}Glc $\searrow^{4}$ Ara (OH-3β) Glc $\nearrow_{2}$	*136*
	Anagallisin E 224–226°, −6.8°, ^{1}H, ^{13}C, FABMS	Anagalligenin B (**207**) Glc $\searrow^{4}$ Ara (OH-3β) Glc $\nearrow_{2}$	*136*
Anemocleama glaucifolium (Ranunculaceae)	Anemoclemoside A 194–198°, +55°, ^{13}C, FABMS	(**338**)	*137*
	Anemoclemoside B 220–230°, +22° ^{13}C, FABMS	(**336**)	*137*
Anemone hupehensis (Ranunculaceae)	Hupehensis saponin D 215–216°, −26.4° ^{13}C. EIMS, FABMS	Oleanolic acid (**7**) Glc-^{3}Ribo-^{3}Rha-^{2}Ara (OH-3β) Rha-^{4}Glc-^{6}Glc (CO_2H-28)	*138*
	Hupehensis saponin E 212–213°, −21.4° ^{13}C, EIMS	Hederagenin (**11**) Ribo-^{3}Rha-^{2}Ara (OH-3β) |3 Glc Rha-^{4}Glc-^{6}Glc (CO_2H-28)	*138*

Table 1. *(continued)*

Source (1)	Saponin mp, $[\alpha]_D$, spectra recorded (2)	Structure (3)	Ref. (4)
Aphloia madagascariensis (Flacourtiaceae)	Saponin 1	Aglycone (**176**) Glc (CO_2H-28)	*139*
Aralia armata (Araliaceae)	Saponin 3 ^1H, FABMS	Oleanolic acid (**7**) (Butyl-ester-6′) GlcA (OH-3β)	*140*
	Saponin 5 −17.9°, IR, ^1H, FABMS	Oleanolic acid (**7**) Ara(f)-^{4}(Me-ester-6′) GlcA (OH-3β)	*140*
	Saponin 8 +22.2°, IR, ^1H, FABMS	Oleanolic acid (**7**) Gal-^{3}(Me-ester-6′) Glc A (OH-3β)	*140*
	Saponin 9 +26.4°, ^1H, FABMS	Oleanolic acid (**7**) Gal-^{3}(Butyl-ester-6′) Glc A (OH-3β)	*140*
	Saponin 13 +6.0°, ^1H, FABMS	Oleanolic acid (**7**) (Butyl-ester-6′) Glc A (OH-3β) Glc (CO_2H-28)	*140*
	Saponin 15 +12.4°, ^1H, FABMS	Oleanolic acid (**7**) Gal-^{3}(Me-ester-6′) Glc A (OH-3β) Glc (CO_2H-28)	*140*
A. chinensis	Araliasaponin XII +15.3°, ^1H, ^{13}C, FABMS	Oleanolic acid (**7**) Glc —^{3} Ara (OH-3β) Glc —^{2} Glc (CO_2H-28)	*141*

Araliasaponin XIII $-7.1°$, ^1H, ^{13}C, FABMS	**Oleanolic acid (7)** Glc \\ 3 Ara (OH-3β)) Xyl / 2 Rha-^{4}Glc-^{6}Glc (CO_2H-28)	*141*
Araliasaponin XIV $+16.8°$, ^1H, ^{13}C, FABMS	Oleanolic acid (**7**) Glc \\ 3 Glc (OH-3β) Gal / 2 Glc (CO_2H-28)	*141*
Araliasaponin XV $+2.8°$, ^1H, ^{13}C, FABMS	Oleanolic acid (**7**) Glc \\ 3 Glc (OH-3β) Xyl / 2 Glc-^{6}Glc (CO_2H-28)	*141*
Araliasaponin XVI $+15.9°$, ^1H, ^{13}C, FABMS	Oleanolic acid (**7**) Glc \\ 3 Gal (OH-3β) Gal / 2 Glc (CO_2H-28)	*141*
Araliasaponin XVII $-19.8°$, ^1H, ^{13}C, FABMS	Aglycone (**22**) Ara(f) \\ 4 (Me-ester-6′) GlcA (OH-3β) Glc / 2	*141*
Araliasaponin XVIII $-33.3°$, ^1H, ^{13}C, FABMS	Oleanolic acid (**7**) Ara(f) \\ 4 (Me-ester-6′) GlcA (OH-3β) Glc / 2 Glc-^{6}Glc (CO_2H-28)	*141*

Table 1. *(continued)*

Source (1)	Saponin mp, $[\alpha]_D$, spectra recorded (2)	Structure (3)	Ref. (4)
A. cordata	Udosaponin A 135–137°, −11.2° ^{13}C, FABMS	Aglycone (**22**) Xyl-^{4}(Me-ester-6′) GlcA (OH-3β)	*142*
	Udosaponin B −2.4°, ^{13}C, FABMS	Oleanolic acid (**7**) Gal-^{2} (Me-ester-6′) Glc A (OH-3β) Glc (CO_2H-28)	*142*
	Udosaponin C −8.2°, ^{13}C, FABMS	Oleanolic acid (**7**) Xyl-$^4_2$ (Me-ester-6′) Glc A (OH-3β) / Gal Glc (CO_2H-28)	*142*
	Udosaponin D ^{13}C, FABMS	Aglycone (**23**) Xyl-^{4}(Me-ester-6′) Glc A (OH-3β)	*142*
	Udosaponin E −12.3°, ^{13}C, FABMS	Hederagenin (**11**)) Xyl-^{4}(Me-ester-6′) Glc A (OH-3β) Glc (CO_2H-28)	*142* *142*
	Udosaponin F +1.5°, ^{13}C, FABMS	Hederagenin (**11**) Gal-^{2}(Me-ester-6′) Glc A (OH-3β) Glc (CO_2H-28)	*142*
A. decaisneana	Araliasaponin I +17.6°, ^1H, ^{13}C, FABMS	Oleanolic acid (**7**) Xyl-^{3}Glc ╲^{3} Ara (OH-3β) Xyl ╱$_2$ Glc (CO_2H-28)	*143*
	Araliasaponin II +5.5°, ^1H, ^{13}C, FABMS	Oleanolic acid (**7**) Glc-^{3}Ara (OH-3β) Glc-^{6}Glc (CO_2H-28)	*143*

Araliasaponin III +6.4°, ^{1}H, ^{13}C, FABMS	Oleanolic acid (**7**) Glc-(1→3)[Xyl-(1→2)]-Ara (OH-3β) Glc-^{6}Glc (CO_2H-28)	*143*
Araliasaponin IV +11.9°, ^{1}H, ^{13}C, FABMS	Oleanolic acid (**7**) Glc-(1→3)[Xyl-(1→2)]-Glc (OH-3β) Glc (CO_2H-28)	*143*
Araliasaponin V +32.5°, ^{1}H, ^{13}C, FABMS	Oleanolic acid (**7**) Glc-(1→3)[Xyl-(1→2)]-Gal (OH-3β)	*143*
Araliasaponin VI +21.5°, ^{1}H, ^{13}C, FABMS	Oleanolic acid (**7**) Glc-(1→3)[Xyl-(1→2)]-Gal (OH-3β) Glc (CO_2H-28)	*143*
Araliasaponin VII +5.5°, ^{1}H, ^{13}C, FABMS	Oleanolic acid (**7**) Glc-(1→3)[Xyl-(1→2)]-Gal (OH-3β) Glc-^{6}Glc (CO_2H-28)	*143*
Araliasaponin VIII +22.6°, ^{1}H, ^{13}C, FABMS	Ursolic acid (**175**) Glc-(1→3)[Xyl-(1→2)]-Ara (OH-3β) Glc (CO_2H-28)	*143*

Table 1. *(continued)*

Source (1)	Saponin mp, $[\alpha]_D$, spectra recorded (2)	Structure (3)	Ref. (4)
	Araliasaponin IX +16.7°, ^{1}H, ^{13}C, FABMS	Ursolic acid (**175**) Glc $\searrow^{3}$ $_{2}$ Ara (OH-3β) Xyl $\nearrow$	*143*
A. elata	Saponin I 209–213°, −5.7° IR, ^{1}H, ^{13}C, FABMS	Hederagenin (**11**) Rha-^{2}Ara (OH-3β) Xyl-^{6}Glc (CO_2H-28)	*144*
	Saponin II 208–212°, −5.7° IR, ^{1}H, ^{13}C, FABMS	Oleanolic acid (**7**) Rha-^{2}Ara (OH-3β) Xyl-^{6}Glc (CO_2H-28)	*144*
	Saponin III 219–221°, −6.9° IR, ^{1}H, ^{13}C, FABMS	Hederagenin (**11**) Glc-^{3}Rha-^{2}Ara (OH-3β) Glc-^{6}Glc (CO_2H-28)	*144*
	Saponin IV 240–245°, +7.6° IR, ^{1}H, ^{13}C, FABMS	Hederagenin (**11**) Glc-^{3}Rha-^{2}Ara (OH-3β)	*144*
	Araloside G	Oleanolic acid (**7**) Glc $\searrow^{4}$ $_{3}$ Glc (OH-3β) Glc $\nearrow$ Glc (CO_2H-28)	*145*
	Terasoponin I 196–202°, −10.5° IR, ^{1}H, ^{13}C, FABMS	Aglycone (**22**) Ara(f) $\searrow^{4}$ $_{3}$ (Me-ester-6′) GlcA (OH-3β) Glc $\nearrow$	*146*

Terasaponin II +14.5°, IR, ^{1}H, ^{13}C, FABMS	Aglycone (**22**) Gal $^{3}_{2}$ (Me-ester-6′) GlcA (OH-3β) Xyl	
Terasaponin III +7.1°, IR, ^{1}H, ^{13}C, FABMS	Oleanolic acid (**7**) Gal-^{3}(Me-ester-6′) GlcA (OH-3β) Glc (CO_2H-28)	*146*
Terasaponin III 223–232°, +28.7° IR, ^{1}H, ^{13}C, FABMS	Aglycone (**22**) Glc $^{3}_{2}$ Ara (OH-3β) Xyl	*147*
Terasaponin IV 196–206°, −22.6° IR, ^{1}H, ^{13}C, FABMS	Oleanolic acid (**7**) Ara(f) $^{4}_{2}$ (Me-ester-6′) GlcA (OH-3β) Glc Glc (CO_2H-28)	*147*
Terasaponin V 235–245°, +5.1° IR, ^{1}H, ^{13}C, FABMS	Oleanolic acid (**7**) Glc $^{3}_{2}$ (Me-ester-6′) GlcA (OH-3β) Xyl Glc (CO_2H-28)	*147*
Terasaponin VI 218–230°, +48° IR, ^{1}H, ^{13}C, FABMS	Oleanolic acid (**7**) Gal $^{3}_{2}$ (Me-ester-6′) GlcA (OH-3β) Xyl Glc (CO_2H-28)	*147*
Terasaponin VII 249–258°, +18.1° IR, ^{1}H, ^{13}C, FABMS	Oleanolic acid (**7**) Glc $^{3}_{2}$ Ara (OH-3β) Xyl Glc (CO_2H-28)	*147*

Table 1. *(continued)*

Source (1)	Saponin mp, $[\alpha]_D$, spectra recorded (2)	Structure (3)	Ref. (4)
A. spinifolia	Araloside H 238–241°, −2.7° ^1H, ^{13}C, 2D, FABMS	Oleanolic acid (**7**) GlcA–^{3}Glc (OH-3β) Xyl–^{2}Glc	*148*
	Araloside J 208–210°, −31.1° ^1H, ^{13}C, 2D, FABMS	Oleanolic acid (**7**) Ara(f)-^{4}GlcA (OH-3β) Gal (CO_2H-28)	*148*
Ardisia crenata (Myrsinaceae)	Ardisicrenoside A 268–270°, −22.4° IR, ^1H, ^{13}C, 2D, FABMS	Aglycone (**198**) Rha-^{2}Glc–^{4}Ara (OH-3β) Glc–^{2}Ara	*5*
	Ardisicrenoside B 264–265°, −4.4° IR, ^1H, ^{13}C, 2D, FABMS	Aglycone (**198**) Xyl-^{2}Glc–^{4}Ara (HO-3β) Glc–^{2}Ara	*5*
	Ardisicrenoside C 234–236°, +4.80° IR, ^1H, ^{13}C, 2D, FABMS	Aglycone (**100**) Rha-^{2}Glc–^{4}Ara (OH-3β)) Glc–^{2}Ara Glc (CO_2H-30)	*149*
	Ardisicrenoside D 213–216°, +23.4° IR, ^1H, ^{13}C, 2D, FABMS	Aglycone (**100**) Xyl-^{2}Glc–^{4}Ara (OH-3β) Glc–^{2}Ara	*149*

		Glc (CO_2H-30)	
	Ardisicrenoside E 227–230°, +30.4° IR, ^1H, ^{13}C, 2D, FABMS	Aglycone (**100**) Rha-^{2}Glc ╲ $^4_2$ Ara (OH-3β) Glc ╱ (Me-ester-6′) GlcA-3′ Glycerol (1′ →) (CO_2H-30)	*150*
	Ardisicrenoside F 225–228°, +41.6° IR, ^1H, ^{13}C, 2D, FABMS	Aglycone (**100**) Xyl-^{2}Glc ╲ $^4_2$ Ara (OH-3β) Glc ╱ (Me-ester-6′) GlcA-3′ Glycerol (1′ →) (CO_2H-30)	*150*
	Saponin	Cyclamiretin A (**196**) Rha-^{4}Glc ╲ $^4_2$ Ara (OH-3β) Glc ╱	*151*
A. japonica	Saponin I −8.7°, ^1H, ^{13}C, 2D, FABMS	Cyclamiretin A (**196**) Glc ╲ $^4_2$ Ara (OH-3β) Rha-^{4}Glc ╱	*152*
	Saponin II −10.5°, ^1H, ^{13}C, 2D, FABMS	Aglycone (**201**) Glc ╲ $^4_2$ Ara (OH-3β) Rha-^{4}Glc ╱	*152*
	Saponin III −2.4°, ^1H, ^{13}C, 2D, FABMS	Aglycone (**199**) Glc ╲ $^4_2$ Ara (OH-3β) Rha-^{4}Glc ╱	*152*

Table 1. *(continued)*

Source (1)	Saponin mp, $[\alpha]_D$, spectra recorded (2)	Structure (3)	Ref. (4)
Argania spinosa (Sapotaceae)	Arganine A ^{1}H, ^{13}C, 2D, FABMS	16α-Hydroxyprotobassic acid (**89**) Glc-^{6}Glc (OH-3β) Rha-^{3}Xyl-^{4}Rha-^{2}Ara (CO_2H-28)	*153*
	Arganine B ^{1}H, ^{13}C, 2D, FABMS	16α-Hydroxyprotobassic acid (**89**) Glc-^{6}Glc (OH-3β) Apio(f)-^{3}Xyl-^{4}Rha-^{2}-Ara (CO_2H-28)	*153*
	Arganine D ^{1}H, ^{13}C, 2D, FABMS	Protobassic acid (**37**) Glc-^{6}Glc (OH-3β) Rha-^{3}Xyl-^{4}Rha-^{2}Ara (CO_2H-28)	*153*
	Arganine E ^{1}H, ^{13}C, 2D, FABMS	Protobassic acid (**37**) Glc-^{6}Glc (OH-3β) Apio(f)-^{3}Xyl-^{4}Rha-^{2}Ara (CO_2H-28)	*153*
	Arganine F ^{1}H, ^{13}C, 2D, FABMS	Protobassic acid (**37**) Glc (OH-3β) Apio(f)-^{3}Xyl-^{4}Rha-^{2}Ara (CO_2H-28)	*153*
Aster batangensis (Compositae)	Asterbatanoside D 1D, 2D, MS	Bayogenin (**25**) Glc (OH-3β) Glc-^{6}Glc (CO_2H-28)	*154*
	Asterbatanoside E 1D, 2D, MS	Bayogenin (**25**) (OAc-6′) Glc (OH-3β) Glc-^{6}Glc (CO_2H-28)	*154*
	Asterbatanoside F 218–220°, −4.11° ^{1}H, ^{13}C, 2D, FABMS	Aglycone (**97**) Glc (OH-3β) Glc —^{6} / Rha —^{2} Glc (CO_2H-28)	*27*

	Asterbatanoside G 232–234°, −7.7° ^{1}H, ^{13}C, 2D, FABMS	Bayogenin (**25**) Glc (OH-3β) Glc $\searrow^{6}$ Glc (CO_2H-28) Rha $\nearrow_{2}$	*27*
	Asterbatanoside H 229–231°, +8.31° ^{1}H, ^{13}C, 2D, FABMS	Bayogenin (**25**) Glc-^{3}Glc (OH-3β) Glc-^{6}Glc (CO_2H-28)	*27*
	Asterbatanoside I 225–227°, +23.66° ^{1}H, ^{13}C, 2D, FABMS	Aglycone (**97**) Glc-^{3}Glc (OH-3β) Glc-^{6}Glc (CO_2H-28)	*27*
	Asterbatanoside J 237–239°, −3.47° IR, ^{1}H, ^{13}C, 2D, FABMS	Medicagenic acid (**48**) Glc-^{6}Glc (OH-3β) Ara-^{3}Rha-^{2}Fuc (CO_2H-28)	*155*
	Asterbatanoside K 240–242°, +8.1° IR, ^{1}H, ^{13}C, 2D, FABMS	Medicagenic acid (**48**) Glc-^{6}Glc (OH-3β) Xyl $\searrow^{4}$ Rha-^{2}Fuc (CO_2H-28) Ara $\nearrow_{3}$	*155*
A. bellidia-strum	Bellidiastroside C_2 214–217°, −15.3°, ^{1}H, ^{13}C, FABMS	Polygalacic acid (**27**) Glc (OH-3β) Xyl-^{4}Rha-^{2}Fuc (CO_2H-28)	*22*
A. scaber	Scaberoside Ha −68.9°, ^{1}H, ^{13}C, FABMS	Echinocystic acid (**15**) (Me-ester-6′) GlcA (OH-3β) Rha $\searrow^{3}$ Xyl (CO_2H-28) Rha $\nearrow_{2}$	*156*

Table 1. *(continued)*

Source (1)	Saponin mp, $[\alpha]_D$, spectra recorded (2)	Structure (3)	Ref. (4)
	Scaberoside Hb_1 −71.0°, ^{1}H, ^{13}C, FABMS	Echinocystic acid (**15**) (Me-ester-6′) GlcA (OH-3β) Rha ⟍ $^{3}_{2}$ Xyl (CO_2H-28) Xyl-^{4}Rha ⟋	*156*
	Scaberoside Hb_2 271°, −64.0°, ^{1}H, ^{13}C, FABMS	Echinocystic acid (**15**) (Me-ester-6′) GlcA (OH-3β) Rha ⟍ $^{3}_{2}$ Xyl (CO_2H-28) Xyl-^{3}Xyl-^{4}Rha ⟋	*156*
	Scaberoside Hc_1 238–240°, −74.7° ^{1}H, ^{13}C, FABMS	Echinocystic acid (**15**) (Me-ester-6′) GlcA (OH-3β) Rha-$^{3}_{2}$Xyl (CO_2H-28) Xyl ⟍ $^{4}_{3}$ Rha (linked to Xyl above) Xyl-^{3}Xyl ⟋	*156*
	Scaberoside Hc_2 −67.4°, IR, ^{1}H, ^{13}C, FABMS	Echinocystic acid (**15**) Xyl-^{3}(Me-ester-6′) GlcA (OH-3β) Rha ⟍ $^{3}_{2}$ Xyl (CO_2H-28) Xyl-^{3}Xyl-^{4}Rha ⟋	*157*
	Scaberoside Hd −73.6°, IR, ^{1}H, ^{13}C, FABMS	Echinocystic acid (**15**) Xyl-^{3}(Me-ester-6′) GlcA (OH-3β) Rha-$^{3}_{2}$ Xyl (CO_2H-28) Xyl ⟍ $^{4}_{3}$ Rha (linked to Xyl above) Xyl-^{3}Xyl ⟋	*157*

	Scaberoside Hf −67.7°, IR, ^{1}H, ^{13}C, FABMS	Echinocystic acid (**15**) Gal-^{2}(Me-ester-6′) GlcA (OH-3β) Rha ╲^{3} Xyl (CO$_2$H-28) Xyl-^{4}Rha ╱$_{2}$	*157*
	Scaberoside Hg 260–261°, −67.3° IR, ^{1}H, ^{13}C, FABMS	Echinocystic acid (**15**) Gal-^{2}(Me-ester-6′) GlcA (OH-3β) Rha ╲^{3} Xyl (CO$_2$H-28) Xyl-^{3}Xyl-^{4}Rha ╱$_{2}$	*157*
	Scaberoside Hh −56.9°, IR, ^{1}H, ^{13}C, FABMS	Echinocystic acid (**15**) Xyl ╲^{3} (Me-ester-6′) GlcA (OH-3β) Gal ╱$_{2}$ Rha ╲^{3} Xyl (CO$_2$H-28) Xyl-^{3}Xyl-^{4}Rha ╱$_{2}$	*157*
	Scaberoside Hi −64.4°, IR, ^{1}H, ^{13}C, FABMS	Echinocystic acid (**15**) Rha ╲^{3} Xyl (CO$_2$H-28) Xyl-$^{4}_{3}$ Rha ╱$_{2}$ Xyl-^{3}Xyl	*157*
A. tataricus	Astersaponin Ha −19.3°, ^{1}H, ^{13}C, FABMS	Echinocystic acid (**15**) GlcA (OH-3β) Ara (CO$_2$H-28)	*158*
	Astersaponin Hb −54.3°, ^{1}H, ^{13}C, FABMS	Echinocystic acid (**15**) GlcA (OH-3β) Rha-^{2}Ara (CO$_2$H-28)	*158*

Table 1. *(continued)*

Source (1)	Saponin mp, $[\alpha]_D$, spectra recorded (2)	Structure (3)	Ref. (4)
	Astersaponin Hc 227–228°, −47.3° ^{1}H, ^{13}C, FABMS	Echinocystic acid (**15**) GlcA (OH-3β) Xyl-^{3}Xyl-^{4}Rha-^{2}Ara (CO_2H-28)	*158*
	Astersaponin Hd 235–237°, −62.8° ^{1}H, ^{13}C, FABMS	Echinocystic acid (**15**) GlcA (OH-3β) Xyl-^{3}Xyl ╲ $_4$ $_3$ Rha-^{2}Ara (CO_2H-28) Apio (f) ╱	*158*
	Foetidissimoside A 220–222°, −47.7° ^{1}H, ^{13}C, FABMS	Echinocystic acid (**15**) (Me-ester-6′) GlcA (OH-3β) Ara (CO_2H-28)	*158*
	Astersaponin G 235–237°, −27.8° ^{1}H, ^{13}C, FABMS	Asterogenic acid (**24**) Ara-^{6}Glc (OH-3β) Xyl-^{4}Rha-^{2}Xyl (CO_2H-28)	*159*
	Astersaponin E −33.7°, ^{1}H, ^{13}C, FABMS	Asterogenic acid (**24**) Ara-^{6}Glc (OH-3β) Xyl-^{3}Ara-^{4}Rha-^{2}Xyl (CO_2H-28)	*160*
	Astersaponin F −45.6°, ^{1}H, ^{13}C, FABMS	Echinocystic acid (**15**) Ara-^{6}Glc (OH-3β) Xyl-^{3}Ara-^{4}Rha-^{2}Xyl (CO_2H-28)	*160*
A. yunnanensis	Asteryunnanoside H 254–255°, −41.66° IR, ^{1}H, ^{13}C, FABMS	Echinocystic acid (**15**) Glc (OH-3β) Xyl-^{4}Xyl ╲ $_4$ $_3$ Rha-^{2}Ara (CO_2H-28) Ara ╱	*161*

	Asteryunnanoside A 238–239°, −3.16° IR, ^{1}H, ^{13}C, FABMS	Arjunolic acid (**66**) Rha-^{2}Glc (CO_2H-28)	*162*
	Asteryunnanoside B 221–223°, +17.12° IR, ^{1}H, ^{13}C, FABMS	Arjunolic acid (**66**) Glc-^{2}Glc (CO_2H-28)	*162*
	Asteryunnanoside C 216–217°, −19.69° IR, ^{1}H, ^{13}C, FABMS	Maslinic acid (**39**) Rha-^{2}Glc (CO_2H-28)	*162*
	Asteryunnanoside D 217–219°, +6.93° IR, ^{1}H, ^{13}C, FABMS	Maslinic acid (**39**) Glc-^{2}Glc (CO_2H-28)	*162*
	Asteryunnanoside E 240–242°, IR, ^{1}H, ^{13}C, 2D, FABMS	Bayogenin (**25**) Glc (OH-3β) Glc-^{2}Glc (CO_2H-28)	*163*
Asteropus sarasinosum (Sterculiaceae)	Sarasinoside D 207–211°, −12.7° ^{1}H, ^{13}C, 2D, FABMS	Aglycone (**317**) (2′-acetylamino-2′-deoxy) Gal —4 Xyl (OH-3β) Glc-^{2}Xyl-^{6}(2′-acetylamino-2′-deoxy)Glc —2	*164*
	Sarasinoside E 193–197°, −8.4° ^{1}H, ^{13}C, 2D, FABMS	Aglycone (**314**) (2′-acetylamino-2′-deoxy) Gal —4 Xyl (OH-3β) Xyl-^{6}(2′-acetylamino-2′-deoxy) Glc —2 Glc-^{2} (to the acetylaminoglucose) 6| Glc	*164*

Table 1. *(continued)*

Source (1)	Saponin mp, $[\alpha]_D$, spectra recorded (2)	Structure (3)	Ref. (4)
	Sarasinoside F 192–195°, −8.4° ^1H, ^{13}C, 2D, FABMS	Aglycone (**315**) (2′-acetylamino-2′-deoxy) Gal—$^4_2$ Xyl (OH-3β) (2′-acetylamino-2′-deoxy) Glc— 6| Xyl^2-Glc	*164*
	Sarasinoside G 203–206°, −29.9° ^1H, ^{13}C, 2D, FABMS	Aglycone (**318**) (2′-acetylamino-2-deoxy) Gal-$^4_2$ Xyl (OH-3β) (2′-acetylamino-2′-deoxy) Glc/ 6| Xyl^2-Glc	*164*
Astragalus alexandrinus (Leguminosae)	Alexandroside I 288–290°, +43.2° ^1H, ^{13}C, 2D, FABMS	Aglycone (**244**) Glc (OH-3β)	*165*
A. ernestii	Asternestioside C 204–207°, −13.2° ^1H, ^{13}C, 2D, FABMS	Cycloastragenol (**241**) Rha-^{2}(OAc-4′) Xyl (OH-3β) Glc (OH-25)	*166*
A. membranaceus	Agroastragaloside III 191–193°, +5.9° IR, ^1H, ^{13}C, FABMS	Cycloastragenol (**241**) (OAc-2′,3′) Xyl(OH-3β) Glc (OH-6α) Glc (OH-25)	*111*
	Agrostragaloside IV 187–189°, +13.9°	Cycloastragenol (**241**) (OAc-2′) Xyl (OH-3β)	*111*

	IR, ^{1}H, ^{13}C, FABMS	Glc (OH-6α) Glc (OH-25)	
A. mongholicus	Mongholicoside I 143–145°, +47.9° ^{1}H, ^{13}C, FABMS	Aglycone (**242**) Glc (OH-26)	*167*
	Mongholicoside II 128–130°, +42.1° ^{1}H, ^{13}C, FABMS	Aglycone (**243**) Glc (OH-26)	*167*
A. tomentosus	Tomentoside I 247–250°, −18.7° IR, ^{1}H, ^{13}C, FABMS	Aglycone (**335**) Xyl (OH-3β)	*168*
A. trigonus	Trigonoside I 226°, +25°, ^{1}H, ^{13}C, FABMS	Cycloastragenol (**241**) Xyl (OH-6α)	*169*
	Trigonoside II 243°, −30°, ^{1}H, ^{13}C, FABMS	Cycloastragenol (**241**) Ara-^{2}Xyl (OH-3β) Xyl (OH-6α)	*169*
	Trigonoside III 264°, −45°, ^{1}H, ^{13}C, FABMS	Cycloastragenol (**241**) Ara-^{2} (OAc-3′) Xyl (OH-3β) Xyl (OH-6α)	*169*
Astrantia major (Umbelliferae)	Saponin Ia ^{13}C, FABMS	Hederagenin (**11**) Rha-^{2}Ara (OH-3β) (OAc-3′) Rha-^{4}Glc-^{6}Glc (CO_2H-28)	*170*
	Saponin Ib ^{13}C, FABMS	Hederagenin (**11**) Rha-^{2}Ara (OH-3β) (OAc-2′) Rha-^{4}Glc-^{6}Glc (CO_2H-28)	*170*

Table 1. *(continued)*

Source (1)	Saponin mp, $[\alpha]_D$, spectra recorded (2)	Structure (3)	Ref. (4)
	Saponin V ^{13}C, FABMS	Hederagenin (**11**) Rha-^{2}Ara (OH-3β) Glc —^{6} Glc (CO_2H-28) Rha —^{2}	*170*
Bacopa monniera (Scrophulariaceae)	Bacoside A_1 240°, +168° IR, ^{1}H, ^{13}C, FABMS	Jujubogenin (**305**) Ara(f)-^{3}Ara (OH-3β)	*171*
	Bacoside A_3 IR, ^{1}H, ^{13}C, FABMS	Jujubogenin (**305**) Glc —^{3} Glc (OH-3β) Ara (f) —^{2}	*172*
	Bacopasaponin A 256°, −90°, IR, ^{1}H, ^{13}C, 2D, FABMS	Jujubogenin (**305**) Ara (OH-3β) Ara (OH-20β)	*24*
	Bacopasaponin B 283°, −65.4°, IR, ^{1}H, ^{13}C, 2D, FABMS	Pseudojujubogenin (**308**) Ara(f)-^{2}Ara (OH-3β)	*24*
	Bacopasaponin C 222°, −47.5°, IR, ^{1}H, ^{13}C, 2D, FABMS	Pseudojujubogenin (**308**) Glc —^{3} Ara (OH-3β) Ara (f) —^{2}	*24*
	Bacopasaponin D 250°, −42°, IR, ^{1}H, ^{13}C, 2D, FABMS	Pseudojujubogenin (**308**) Ara (f)-^{2}Glc (OH-3β)	*173*

Barringtonia acutaangula (Pleosporaceae)	Barringtoside A 258–260°, −1.0° ^{1}H, ^{13}C, 2D, FABMS	Barringtogenol (**34**) Xyl ╲ 3 (Me-ester-6′) GlcA (OH-3β) Gal ╱ 2	*174*
	Barringtoside B +12.6°, ^{1}H, ^{13}C, 2D, FABMS	Aglycone (**26**) Xyl ╲ 3 (Me-ester-6′) GlcA (OH-3β) Gal ╱ 2	*174*
	Barringtoside C 240–242°, +15.1° ^{1}H, ^{13}C, 2D, FABMS	Barringtogenol (**34**) Ara ╲ 3 (Me-ester-6′) GlcA (OH-3β) Gal ╱ 2	*174*
	Esculentoside M 219–221°, IR, ^{1}H, ^{13}C, FABMS	Aglycone (**18**) Glc-^{4}Xyl (OH-3β) Glc (CO_2H-28)	*175*
Bellis perennis (Asteraceae)	Bellissaponin BA_1 ^{1}H, ^{13}C, 2D, FABMS	Polygalacic acid (**27**) Rha (OH-3β)	*176*
	Bellissaponin BA_2 ^{1}H, ^{13}C, 2D FABMS	Polygalacic acid (**27**) Glc (OH-3β) (E-CH_3CH=CH-CO-) ╲ 4 Fuc (CO_2H-28) Rha-^{3}Xyl-^{4}Rha ╱ 2	*176*
	Bellissaponin BS_4 ^{1}H, ^{13}C, 2D	Asterogenic acid (**24**) Glc (OH-3β) Glc ╲ 6 Glc (CO_2H-28) Rha ╱ 2	*177*

Table 1. *(continued)*

Source (1)	Saponin mp,$[\alpha]_D$, spectra recorded (2)	Structure (3)	Ref. (4)
	Bellissaponin BS_5 ^1H, ^{13}C, 2D	Bayogenin (**25**) Glc (OH-3β) Glc —$^6_2$ Glc (CO_2H-28) Xyl —	*177*
	Bellissaponin BS_6 ^1H, ^{13}C, 2D,	Bayogenin (**25**) Glc —$^6_2$ Glc (CO_2H-28) Rha —	*177*
	Bellissaponin BS_7 ^1H, ^{13}C, 2D	Bayogenin (**25**) Rha (OH-3β) Glc —$^6_2$ Glc (CO_2H-28) Glc —	*177*
B. sylvestris	Besysaponin C_{12} ^1H, ^{13}C, 2D, FABMS	Polygalacic acid (**27**) Rha (OH-3β) Xyl-^{4}Rha-^{2}Fuc (CO_2H-28)	*178*
Bellium bellidioides (Asteraceae)	Bellidioside A 210–213°, +20.6° ^1H, ^{13}C, 2D	Aglycone (**97**) Glc —$^3_2$ (OAc-6′) Glc (CO_2H-28) Rha —	*179*
	Desacyl bellidioside B_4 −41.6°, ^1H, ^{13}C, 2D, ESIMS	Polygalacic acid (**27**) Rha (OH-3β) Ara (f) —$^3_2$ Fuc (CO_2H-28) Rha-^{3}Xyl-^{4}Rha —	*180*

Beta vulgaris (Chenopodiaceae)	Saponin	Oleanolic acid (**7**) Glc-^{2}Xyl-^{3}GlcA (OH-3β) Glc (CO_2H-28)	*181*
Betula ermanii (Betulaceae)	+22°, ^{1}H, ^{13}C, HRFABMS	Aglycone (**300**) Glc (OH-3β)	*182*
	+14°, ^{1}H, ^{13}C, HRFABMS	Aglycone (**300**) (OAc-2′) Glc (OH-3β)	*182*
	+25°, ^{1}H, ^{13}C, FABMS	Aglycone (**301**) (OAc-2′) Glc (OH-3β)	*182*
	+20°, ^{1}H, ^{13}C, HRFABMS	Aglycone (**302**) (OAc-2′) Glc (OH-3β)	*182*
Bhesa peniculata (Celastraceae)	Gongganoside A +18.3°, IR, ^{1}H, ^{13}C, FABMS	Quinovic acid (**171**) Xyl-^{3}Rha (OH-3β)	*183*
	Gongganoside B +31.6°, IR, ^{1}H, ^{13}C, FABMS	Quinovic acid (**171**) Rha (OH-3β) Glc(CO_2H-28)	*183*
	Gongganoside C +11.8°, IR, ^{1}H, ^{13}C, FABMS	Quinovic acid (**171**) Xyl-^{3}Rha (OH-3β) Glc (CO_2H-28)	*183*
Boussingaultia baselloids (Basellaceae)	Boussingoside D$_{1}$ ^{1}H, ^{13}C, FABMS	Aglycone (**28**) Xyl-^{3}Glc (OH-3β)	*184*
	Saponin I	Aglycone (**28**) GlcA (OH-3β)	*185*

Table 1. *(continued)*

Source (1)	Saponin mp, $[\alpha]_D$, spectra recorded (2)	Structure (3)	Ref. (4)
	Saponin II	Aglycone (**28**) GlcA (OH-3β) Glc (CO_2H-28)	*185*
	Saponin III	Aglycone (**14**) Glc (OH-3β) Glc (CO_2H-28)	*185*
	Saponin IV	Aglycone (**14**) GlcA (OH-3β)	*185*
Bryonia dioica (Cucurbitaceae)	Brydioside A 180–181.5°, −28.6° ^{1}H, ^{13}C, FABMS	Aglycone (**282**) Glc (OH-2β) Glc (OH-25)	*186*
	Brydioside B 164–165.5°, +60.1° ^{1}H, ^{13}C, FABMS	Aglycone (**283**) Glc (OH-3β) Glc (OH-25)	*186*
	Brydioside C 267–268°, +50.6°, IR, ^{1}H, ^{13}C, FABMS	Aglycone (**283**) Glc (OH-3β) Glc-^{6}Glc (OH-25)	*186*
Bupleurum falcatum (Umbelliferae)	Malonyl saikosaponin a +42.8°, IR, ^{1}H, ^{13}C, FABMS	Saikogenin F (**190**) (Malonate-6′)) Glc-^{3}Rha (OH-3β)	*187*
	Malonyl saikosaponin d +29.6°, IR, ^{1}H, ^{13}C, HRFABMS	Epi-saikogenin F (**203**) (Malonate-6′) Glc-^{3}Rha (OH-3α)	*187*
	Saponin	Aglycone (**228**) Glc-^{3}Fuc (OH-3β)	*188*

References, pp. 173–196

	Hydroxysaikosaponin a +4.4°, IR, ^{1}H, ^{13}C, FABMS	Aglycone (**105**) Glc-^{3}Rha (OH-3β)	*189*
	Hydroxysaikosaponin c −30.8°, IR, ^{1}H, ^{13}C, FABMS	Aglycone (**132**) Glc ^{6} / Rha ^{4} → Glc (OH-3β)	*189*
	Acetyl saikosaponin d +42.6°, IR, ^{1}H, ^{13}C, FABMS	Anagalligenin B (**207**) (OAC-4′) Glc-^{3}Fuc (OH-3β)	*189*
B. fruticosum	Malonylbuddlejasaponin IV	Saikogenin F (**190**) Glc-$^{3}_{2}$ Fuc (OH-3β) (Malonate-6′) Glc	*190*
	Saponin 1 255–258°, +28.84° ^{1}H, ^{13}C, 2D, FABMS	Saikogenin F (**190**) Glc ^{3} / Glc ^{2} → Fuc (OH-3β)	*191*
	Saponin 2 245–253°, +16.04° ^{1}H, ^{13}C, 2D, FABMS	Aglyone (**104**) Glc ^{3} / Glc ^{2} → Fuc (OH-3β)	*191*
	Saponin 3 230°, +16.34°, ^{1}H, ^{13}C, 2D, FABMS	Aglycone (**105**) Glc ^{3} / Glc ^{2} → Fuc (OH-3β)	*191*
B. smithii	Saikosaponin M UV, IR, ^{1}H, ^{13}C, FABMS	Aglycone (**219**) Glc-^{3}Fuc (OH-3β)	*192*

Table 1. *(continued)*

Source (1)	Saponin mp, $[\alpha]_D$, spectra recorded (2)	Structure (3)	Ref. (4)
	Saikosaponin N UV, IR, ^{1}H, ^{13}C, FABMS	Saikogenin A (**220**) Glc ╲ $^6_4$ Glc (OH-3β) Rha ╱	*192*
	Saikosaponin Q	Aglycone (**221**) Glc ╲ $^6_4$ Glc (OH-3β) Rha ╱	*193*
Calendula arvensis (Compositae)		Olenolic acid (**7**) Glc-^{3}GlcA (OH-3β) Glc (CO_2H-28)	*86*
		Oleanolic acid (**7**) Glc-^{3}Glc (OH-3β)	*86*
Calliandra anomala (Leguminosae)	Calliandra saponin A 204–210°, −22.1° ^{1}H, ^{13}C, FABMS	Echinocystic acid (**15**) ^{2}Ara-^{6}(2′-acetylamino-2′-deoxy) Glc (OH-3β) | Ara [(6′S)-2′-trans-2′ ,6′-dimethyl-6′-O-βD-xylopyranosyl-2,7-octadienoyl] ╲ ^{6}Glc (CO_2H-28) |2 Xyl ╲ $^4_3$ Rha Glc ╱	*194*

Calliandra saponin B 220–226°, −9.85° ^{1}H, ^{13}C, 2D, FABMS	Echinocystic acid (**15**) Ara-^{6}(2′-acetylamino-2′-deoxy) Glc (OH-3β) 2| Ara [(6S)-2-trans-2,6-dimethyl-6-O-β-D-quinovo- pyranosyl-2,7-octadienoyl]-$^{6}_{2}$ Glc (CO_2H-28) Xyl-^{3}Xyl-$^{4}_{3}$ Rha Glc	*195*
Calliandra saponin C 192–195°, −10.2°,	Echinocystic acid (**15**) Ara-^{6}(2′-acetylamino-2′-deoxy) Glc (OH-3β) 2| Ara [(6S)-2-trans-2,6-dimethyl-6-*O*-(6′S)-2′- trans-2′-6′-dimethyl-6′-*O*-β-D-xylopyran- osyl-2′,7′-octadienyl (1 → 2)-β-D-xylopy- ranosyl-2,7-octadienoyl]-$^{6}_{2}$ Glc (CO_2H-28) Xyl-^{3}Xyl-$^{4}_{3}$ Rha Glc	*195*
Calliandra saponin D 194–196°, −14.6°, ^{1}H, ^{13}C, 2D, FABMS	Echinocystic acid (**15**) Ara-^{6}(2′-acetylamino-2′-deoxy) Glc(OH-3β) 2| Ara [(6S)-2-trans-2,6-dimethyl-6-O-(6′S)-2′- trans-2′, 6′-dimethyl-6′-O-β-D-quinovo- pyranosyl-2′,7′-octadienoyl (1 → 2)-β-D- xylopyranosyl-2,7-octadienoyl]-^{6} Glc (CO_2H-28) |2 Xyl-^{3}Xyl $^{4}_{3}$ Rha Glc	*194*

Table 1. *(continued)*

Source (1)	Saponin mp, $[\alpha]_D$, spectra recorded (2)	Structure (3)	Ref. (4)
	Calliandra saponin E 193–197°, +4.4°, ^1H, ^{13}C, FABMS	Echinocystic acid (**15**) Ara-^{6}(2′-acetylamino-2′-deoxy) Glc (OH-3β) 2| Ara [(6′S)-2′-trans-2′,-6′-dimethyl-6′-O-(2-O-(6S)-2,7-octadienoyl)-β-D-xylopyranosyl-2′,7′-octadienoyl] | ^{6}Glc (CO_2H-28) |2 Xyl-^{3}Xyl —4 Rha Glc —3	*194*
	Calliandra saponin F 186–189°, −3.6°, ^1H, ^{13}C, 2D, FABMS	Echinocystic acid (**15**) Ara-^{6}(2′-acetylamino-2′-deoxy) Glc (OH-3β) 2| Ara [(6S)-2-trans-2,6-dimethyl-6-O-(6′S)-2′-trans-2′,6′-dimethyl-6′-O-β-D-Xylopyranosyl-2′, 7′-octadienoyl]-^{6}Glc (CO_2H-28) |2 Xyl-^{3}Xyl —4 Rha (OAc-6′) Glc —3	*194*

Camellia japonica (Theaceae)	Camelliasaponin B_1 209.6–211.1°, +23.7°, IR, ^1H, ^{13}C, 2D, FABMS	Aglycone (**29**) Glc-^{2}Ara $\searrow^3$ GlcA (OH-3β) Glc $\nearrow_2$	*196*
	Camelliasaponin B_2 233.5–235.6°, +20.7°, IR, ^1H, ^{13}C, 2D, FABMS	Aglycone (**30**) Glc-^{2}Ara $\searrow^3$ GlcA (OH-3β) Glc $\nearrow_2$	*196*
	Camelliasaponin C_1 165.8–167.2°, +4.3°, IR, ^1H, ^{13}C, 2D, FABMS	Aglycone (**106**) Glc-^{2}Ara $\searrow^3$ GlcA (OH-3β) Glc $\nearrow_2$	*196*
	Camelliasaponin C_2 +8.8°, IR, ^1H, ^{13}C, 2D, FABMS	Aglycone (**31**) Glc-^{2}Ara $\searrow^3$ GlcA (OH-3β) Glc $\nearrow_2$	*196*
Castanospermum australe (Fabaceae)	Castaraleside F +32.2°, ^1H, ^{13}C, FABMS	Bayogenin (**25**) Gal-^{4}GlcA (OH-3β)	*197*
	Castaraleside G +114.9°, ^1H, ^{13}C, FABMS	Aglycone (**32**) Rha-^{4}Gal-^{2}GlcA (OH-3β)	*197*
	Castaraleside H +43.3°, ^1H, ^{13}C, FABMS	Aglycone (**32**) Rha-^{4}Xyl-^{2}GlcA (OH-3β)	*197*
	Saponin 1 149–151°, +34°, ^1H, ^{13}C, FABMS	Aglycone (**33**) (Me-ester-6′) (4-deoxy-β-L-threo-hex-4-ene-pyranosiduronic acid (OH-3β)	*198*

Table 1. *(continued)*

Source (1)	Saponin mp, $[\alpha]_D$, spectra recorded (2)	Structure (3)	Ref. (4)
Centella asiatica (Umbelliferae)	Asiaticoside A ^1H, ^{13}C, FABMS	Aglycone (**172**) Rha-^{4}Glc-^{6}Glc (CO_2H-28)	*199*
	Asiaticoside B ^1H, ^{13}C, FABMS	Terminolic acid (**134**) Rha-^{4}Glc-^{6}Glc (CO_2H-28)	*199*
Centipeda minima (Compositae)	Compound 1 139°, IR, ^1H, ^{13}C, EIMS	Aglycone (**165**) Xyl (CO_2H-28)	*200*
	Compound 2 210°, IR, ^1H, ^{13}C, EIMS	Aglycone (**166**) Xyl (CO_2H-28)	*200*
	Compound 3 126°, IR, ^1H, ^{13}C, EIMS	Aglycone (**107**) Xyl (OH-28)	*200*
	Compound 4 145°, IR, ^1H, ^{13}C, EIMS	Aglycone (**35**) Xyl (OH-28)	*200*
Catunaregam nilotica (Rubiaceae)	Saponin ^1H, ^{13}C, FABMS	Oleanolic acid (**7**) Rha-^{3}Glc-^{3}Glc (OH-3β) Glc (CO_2H-28)	*50*
	Saponin +52.5°, ^1H, ^{13}C, FABMS	Oleanolic acid (**7**) Rha-^{3}Glc-^{3}Glc (OH-3β)	*50*
Cephalaria transylvanica (Dipsacaceae)	Cephalaria saponin A +9.1°, IR, ^1H, ^{13}C, FABMS	Hederagenin (**11**) Glc-^{4}Rha-^{4}Xyl (HO-3β) Glc (CO_2H-28)	*201*

	Cephalaria saponin B $-65.47°$, IR, ^1H, ^{13}C, FABMS	Hederagenin (**11**) Xyl-^{4}Rha-^{2}Xyl (OH-3β) Glc (CO_2H-28)	*202*
	Transsylvanoside B $-8.73°$, IR, ^1H, ^{13}C, FABMS	Hederagenin (**11**) Glc-^{2}Xyl-^{4}Rha-^{4}Xyl (OH-3β)	*203*
	Transsylvanoside E $-4.95°$, ^1H, ^{13}C, FABMS	Hederagenin (**11**) Xyl-^{3}Rha-^{4}Glc-^{4}Glc-^{2}Xyl (OH-3β)	*204*
	Transsylvanoside F $-3.69°$, ^1H, ^{13}C, FABMS	Hederagenin (**11**) Glc-^{3}Rha-^{4}Xyl (OH-3β) Glc-^{4}Glc (CO_2H-28)	*204*
Chenopodium quinoa (Chenopodiaceae)	Quinoa saponin 7 $+56.5°$, ^1H, ^{13}C, EIMS	Aglycone (**50**) Glc-^{2}Glc-^{3}Ara (OH-3β) Glc (CO_2H-28)	*205*
	Quinoa saponin 8 241–243°, $+26.9°$, ^1H, ^{13}C, EIMS	Oleanolic acid (**7**) Glc-^{2}Glc-^{3}Ara (HO-3β) Glc (CO_2H-28)	*205*
	Quinoa saponin 9 $+52.5°$, ^1H, ^{13}C, EIMS	Aglycone (**36**) Glc-^{2}Glc-^{3}Ara (OH-3β) Glc (CO_2H-28)	*205*
	Quinoa saponin 11 $+25.8°$, ^1H, ^{13}C, EIMS	Hederagenin (**11**) GlcA (OH-3β) Glc (CO_2H-28)	*205*
	Quinoa saponin 13 $+8.7°$, ^1H, ^{13}C, EIMS	Hederagenin (**11**) Xyl-^{3}GlcA (OH-3β) Glc (CO_2H-28)	*205*
Cimicifuga simplex (Ranunculaceae)	Glycoside I 245–247°, $-20.0°$, ^1H, ^{13}C, FABMS	Aglycone (**245**) Xyl (OH-3β)	*206*

Table 1. *(continued)*

Source (1)	Saponin mp, $[\alpha]_D$, spectra recorded (2)	Structure (3)	Ref. (4)
	Glycoside II 175–176°, +26.2°, ^{1}H, ^{13}C, FABMS	Aglycone (**325**) Xyl (OH-3β)	*206*
	Glycoside III 187–188°, +24.6°, ^{1}H, ^{13}C, FABMS	Aglycone (**326**) Xyl (OH-3β)	*206*
Clematis chinensis (Ranunculaceae)	Clematichinenoside A 198–200°, −35.5°, IR, ^{1}H, ^{13}C, FABMS	Oleanolic acid (**7**) Ribo-^{3}Rha-^{2}Ara (OH-3β) Glc (CO_2H-28)	*207*
	Clematichinenoside B 227–230°, −26.8°, IR, ^{1}H, ^{13}C, FABMS	Hederagenin (**11**) Glc-^{4}Ribo-^{3}Rha-^{2}Ara (OH-3β) Rha-^{4}Glc-^{6}Glc (CO_2H-28)	*207*
C. montana	Clemontanoside A 230–232°, −100°, IR, ^{1}H, ^{13}C, FABMS	Oleanolic acid (**7**) Glc (OH-3β) Glc-^{6}Glc(CO_2H-28) Rha╱^{2}	*208*
	Clemontanoside E	Oleanolic acid (**7**) Glc (OH-3β) Glc-^{6}Glc (CO_2H-28)	*209*
C. koreana		Oleanolic acid (**7**) Xyl-^{3}Ara (OH-3β) Rha^{4}Glc-^{6}Glc (CO_2H-28)	*210*

Clinopodium chinense (Zabiatae)	Clinopodiside	Saikogenin A (**220**) Glc ╲ ^{6} Glc (OH-3β) Glc ╱ ^{4}	*211*
	Clinopodiside B ^{1}H, ^{13}C, FABMS	Saikogenin F (**190**) Glc-^{4}Glc ╲ ^{3} Fuc (OH-3β) Glc ╱ ^{2}	*212*
	Clinopodiside C ^{1}H, ^{13}C, FABMS	Aglycone (**239**) Glc-^{3}Fuc (OH-3β) Glc (CO_2H-28)	*212*
	Clinopodiside D +44.6°, ^{1}H, ^{13}C, FABMS	Aglycone (**217**) Glc ╲ ^{3} Fuc (OH-3β) Glc ╱ ^{2}	*213*
	Clinopodiside E +68.8°, ^{1}H, ^{13}C, FABMS	Aglycone (**200**) Glc ╲ ^{3} Fuc (OH-3β) Glc ╱ ^{2}	*213*
	Clinopodiside F +17.5°, ^{1}H, ^{13}C, FABMS	Aglycone (**10**) Glc ╲ ^{3} Fuc (OH-3β) Glc ╱ ^{2}	*213*
	Clinopodiside G +7.37°, ^{1}H, ^{13}C, FABMS	Aglycone (**218**) Glc ╲ ^{3} Fuc (OH-3β) Glc ╱ ^{2}	*213*
	Clinoposaponin IX +64.0°, ^{1}H, ^{13}C, FABMS	Saikogenin F (**190**) Glc (OH-3β)	*214*

Table 1. *(continued)*

Source (1)	Saponin mp, $[\alpha]_D$, spectra recorded (2)	Structure (3)	Ref. (4)
	Clinoposaponin X +12.5°, ^{1}H, ^{13}C, FABMS	Saikogenin F (**190**) Glc-^{6}Glc (OH-3β)	*214*
	Clinoposaponin XI +41.3°, ^{1}H, ^{13}C, FABMS	Aglycone (**191**) Glc ＼$^3_2$ Fuc (OH-3β) Glc ／	*214*
C. gracile	Clinopodium saponin I +52.4°, ^{1}H, ^{13}C, FABMS	Saikogenin F (**190**) Glc-^{6}Glc-^{3}Fuc (OH-3β)	*215*
	Clinopodium saponin II +23.7°, ^{1}H, ^{13}C, FABMS	Saikogenin F (**190**) Glc-^{6}Glc-^{3}Fuc (OH-3β) 4| Glc	*215*
	Clinopodium saponin III +36.4°, ^{1}H, ^{13}C, FABMS	Saikogenin F (**190**) Glc-^{6}Glc ＼$^3_2$ Fuc (OH-3β) Glc ／	*215*
	Clinopodium saponin IV +55.7°, ^{1}H, ^{13}C, FABMS	Saikogenin F (**190**) Glc-^{2}Glc ＼$^3_2$ Fuc (OH-3β) Glc ／	*216*
	Clinopodium saponin V +27.0°, ^{1}H, ^{13}C, FABMS	Saikogenin F (**190**) Glc-^{4}Glc-^{6}Glc ＼$^3_2$ Fuc (OH-3β) Glc ／	*216*

C. micranthum	Clinoposaponin VI	Saikogenin F (**190**) Fuc (OH-3β) Glc (OH-16β)	*216*
	Clinoposaponin VII	Saikogenin F (**190**) (OAc-6′) Glc-^{3}Fuc (OH-3β)	*216*
C. polycephalum	Clinoposide A 249–251°, +10.7° ^{1}H, ^{13}C, FABMS	Saikogenin A (**220**) Glc \\ $^{6}_{4}$ Glc (OH-3β) Glc /	*217*
Collinsonia canadensis (Labiatae)	Collinsonin 266-267°, +26.5°, IR, ^{1}H, ^{13}C, FABMS	Hederagenin (**11**) Ara (OH-3β)	*218*
	Collinsonidin 250–252°, +55.6°, IR, ^{1}H, ^{13}C, FABMS	Hederagenin (**11**) Glc-^{3}Ara (OH-3β)	*218*
Colubrina elliptica (Rhamnaceae)	Mabioside A 230–234°, −23.7°, IR, ^{1}H, ^{13}C, 2D, FABMS	Mabiogenin (**303**) Rha-^{6}Glc (OH-3β) Glc (HO-15α)	*219*
	Mabioside B 2D	Aglycone (**304**) Rha \\ $^{6}_{2}$ Glc (OH-3β) Glc /	*220*
	Mabioside C −15.6°, IR, ^{1}H, ^{13}C, FABMS	Aglycone (**307**) Rha-^{6}Glc (OH-3β)	*221*
	Mabioside D −17°, IR, ^{1}H, ^{13}C, FABMS	Aglycone (**269**) Rha-^{6}Glc (OH-3β)	*221*

Table 1. *(continued)*

Source (1)	Saponin mp, $[\alpha]_D$, spectra recorded (2)	Structure (3)	Ref. (4)
	Mabioside E −19.8°, IR, ^1H, ^{13}C, FABMS	Aglycone (**270**) Rha $\searrow$ $^4_2$ Glc (OH-3β) Glc $\nearrow$	*221*
Corchorus capsularis (Tiliaceae)	190–191°, −13°, IR, ^1H, ^{13}C, 2D, FABMS	Capsugenin (**267**) Glc (OH-25β) Glc (OH-30)	*222*
Crocosmia crocosmiiflora (Iridaceae)	Crocosmioside A −33.6°, IR, ^1H, ^{13}C, FABMS	Polygalacic acid (**27**) Ara-^{6}Glc (OH-3β) (9-hydroxy-16α-L-rhamnopyranoxyloxy-2-β-D-xylopyranosyloxyhexadecanoate) $\vert^4$ Glc $\searrow$ $^3_2$ Fuc (CO_2H-28) Apio (f)-^{4}Xyl-^{4}Rha $\nearrow$	*223*
	Crocosmioside B −31.2°, IR, ^1H, ^{13}C, FABMS	Polygalacic acid (**27**) Ara-^{6}Glc (OH-3β) (9,16-dihydroxy-2-β-D-xylopyranosyloxyhexadecanoate) $\vert^4$ Glc $\searrow$ $^3_2$ Fuc (CO_2H-28) Apio (f)-^{4}Xyl-^{4}Rha $\nearrow$	*223*

Crocosmioside C −16.8°, IR, ^{1}H, ^{13}C, FABMS	Polygalacic acid (**27**) Ara-^{6}Glc (OH-3β) (9-oxo-16-hydroxy-2-β-D-xylopyrano- syloxyhexadecanoate)\ Glc-3 $^{4}_{2}$ Fuc (CO_2H-28) Apio (f)-^{4}Xyl-^{4}Rha /	*224*
Crocosmioside D −16.9°, IR, ^{1}H, ^{13}C, FABMS	Polygalacic acid (**27**)) Ara-^{6}Glc (OH-3β) (9,16-dihydroxy-2-β-D-xylopyrano- syloxyhexadecanoate)\ Glc-3 $^{4}_{2}$ Fuc (CO_2H-28) Apio (f)-^{4}Xyl-^{4}Rha /	*224*
Crocosmioside E −15.4°, IR, ^{1}H, ^{13}C, FABMS	Polygalacic acid (**27**) Ara-^{6}Glc (OH-3β) (2,9,16-trihydroxyhexadecanoate) | 4 Glc \ $^{3}_{2}$ Fuc (CO_2H-28) Apio (f)-^{4}Xyl-^{4}Rha /	*224*
Crocosmioside F −20.4°, IR, ^{1}H, ^{13}C, FABMS	Polygalacic acid (**27**) Ara-^{6}Glc (OH-3β) (2,9-dihydroxy-16-α-L-rhamno- pyranoxyloxyhexadecanoate)\ Glc-3 $^{4}_{2}$ Fuc (CO_2H-28) Apio (f)-^{4}Xyl-^{4}Rha /	*224*
Crocosmioside G −20.8°, IR, ^{1}H, ^{13}C, FABMS	Polygalacic acid (**27**) Ara-^{6}Glc (OH-3β) (9-oxo-16-α-L-rhamnopyranosyloxy- 2-β-D-xylopyranosyloxyhexadecanoate) | 4 Glc \ $^{3}_{2}$ Fuc (CO_2H-28) Apio (f)-^{4}Xyl-^{4}Rha /	*224*

Table 1. *(continued)*

Source (1)	Saponin mp, $[\alpha]_D$, spectra recorded (2)	Structure (3)	Ref. (4)
	Crocosmioside H −20°, IR, ^1H, ^{13}C, FABMS	Polygalacic acid (**27**) Ara-^{6}Glc (OH-3β) (9-hydroxyl-16-O-α-L-rhamnopyranosyl-2-O-β-D-xylopyranosylhexadecanoate) ⟶ ^{4} Fuc (CO_2H-28) Xyl-^{4}Rha ⟶ ^{2} Fuc	*223*
	Crocosmioside I −10.5°, IR, ^1H, ^{13}C, FABMS	Polygalacic acid (**27**) Ara-^{6}Glc (OH-3β) (2,9-dihydroxy-16-α-L-rhamnopyranosyloxyhexadecanoate) ⟶ Glc-3 ⟶ ^{4} Fuc (CO_2H-28) Xyl-^{4}Rha ⟶ ^{2} Fuc	*224*
C. masoniorum	Masonoside A −22°, IR, ^1H, ^{13}C, FABMS	Polygalacic acid (**27**) Ara-^{6}Glc (OH-3β) (2-hydroxy-9-oxo-16-α-L-rhamnopyranosyloxyhexadecanoate) ⟶ ^{4} Fuc (CO_2H-28) Glc ⟶ ^{3} Fuc Apio (f)-^{4}Xyl-^{4}Rha ⟶ ^{2} Fuc	*225*

Masonoside B −16.3°, IR, ^{1}H, ^{13}C, FABMS	Polygalacic acid (**27**) Ara-^{6}Glc (OH-3β) (2,16-dihydroxy-9-oxohexadocanoate) |4 Glc ╲ $^{3}_{2}$ Fuc (CO_2H-28) Apio (f)-^{4}Xyl-^{4}Rha ╱	*225*
Masonoside C −1.1°, IR, ^{1}H, ^{13}C, FABMS	Polygalacic acid (**27**) Ara-^{6}Glc (OH-3β) (2,16-dihydroxy-9-oxohexadecanoate) |4 Glc ╲ $^{3}_{2}$ Fuc (CO_2H-28) Xyl-^{4}Rha ╱	*225*
Desacylmasonoside 1 −20.8°, IR, ^{1}H, ^{13}C, FABMS	Polygalacic acid (**27**) Ara-^{6}Glc (OH-3β) Glc ╲ $^{3}_{2}$ Fuc (CO_2H-28) Apio (f)-^{4}Xyl ╱	*226*
Desacylmasonoside 2 −25.4°, IR, ^{13}C, FABMS	Polygalacic acid (**27**) Glc (OH-3β) Glc ╲ $^{3}_{2}$ Fuc (CO_2H-28) Apio (f)-^{4}Xyl-^{4}Rha ╱	*226*
Desacylmasonoside 3 −4.7°, IR, ^{1}H, ^{13}C, FABMS	Polygalacic acid (**27**) Ara-^{6}Glc (OH-3β) Glc ╲ $^{3}_{2}$ Fuc (CO_2H-28) Xyl-^{4}Rha ╱	*226*

Table 1. *(continued)*

Source (1)	Saponin mp, $[\alpha]_D$, spectra recorded (2)	Structure (3)	Ref. (4)
	Desacylmasonoside 4 −28.3°, IR, ^1H, ^{13}C, FABMS	Polygalacic acid (**27**) Fuc-^{6}Glc (OH-3β) Glc $^3_2$ Fuc (CO_2H-28) Apio (f)-^{4}Xyl-^{4}Rha	*227*
	Desacylmasonoside 5 +0.6°, IR, ^1H, ^{13}C, FABMS	Polygalacic acid (**27**) Ara-^{6}Glc (OH-3β) Glc $^3_2$ Fuc (CO_2H-28) Rha	*227*
Crossopteryx febrifuga (Rubiaceae)	Saponin 1 213–214°, −48.0°, ^1H, ^{13}C, 2D, FABMS	16α-Hydroxyprotobassic acid (**89**) Glc (OH-3β) Rha-^{3}Xyl-^{4}Rha-^{2}Ara (CO_2H-28)	*228*
	Saponin 2 221–222°, −56.3°, ^1H, ^{13}C, 2D, FABMS	16α-Hydroxyprotobassic acid (**89**) Apio (f)-^{3}Glc (OH-3β) Rha-^{3}Xyl-^{4}Rha-^{2}Ara (CO_2H-28)	*228*
Crotalaria albida (Leguminosae)	+161.8°, ^1H, ^{13}C, FABMS	Aglycone (**133**) Xyl-^{2}Gal-^{2} (Me-ester-6′) GlcA (OH-3β)	*229*
	−11.1°, ^1H, ^{13}C, FABMS	Soyasapogenol B (**69**) Glc $^6_2$-Gal-^{2} (Me-ester-6′) GlcA (OH-3β) Rha	*229*

Cyclamen graecum (Primulaceae)	Isocyclamin C ^{1}H, ^{13}C, FABMS	Aglycone (**192**) Glc ╲ $^6_2$ Glc Xyl ╱ ╲ $^4_2$ Ara (OH-3β) Glc ╱	*230*
Cylicodiscus gaebunensis (Leguminosae)	Cylicodiscoside IR, ^{1}H, ^{13}C, FABMS	Cylicodiscic acid (**249**) Ara-^{2}Ara-^{3}Glc (OH-3β)	*231*
	Saponin S_1 +32°, IR, ^{1}H, ^{13}C, FABMS	Maslinic acid (**39**) Ara-^{2}Ara-^{3}Glc (OH-3β) Glc-^{6}Glc-^{2}Rha (CO_2H-28)	*232*
	Saponin S_2 +33°, IR, ^{1}H, ^{13}C, FABMS	Maslinic acid (**39**) Ara-^{2}-Ara-^{3}Glc (OH-3β) Glc-^{2}Rha (CO_2H-28)	*232*
	Saponin S_3 +55°, ^{1}H, ^{13}C	Maslinic acid (**39**) Ara-^{2}Ara-^{3}Glc (OH-3β)	*232*
	Saponin S_4 +17°, IR, ^{13}C, FABMS	Cylicodiscic acid (**249**) Ara-^{3}Glc (OH-3β)	*232*
Decaisnea fargessii (Lardizabalaceae)	Decaisoside A 250°, −26°, ^{1}H, ^{13}C, FABMS	Oleanolic acid (**7**) Gal ╲ $^3_2$ Ara (OH-3β) Rha ╱	*233*
	Decaisoside B 228–231°, −12°, ^{1}H, ^{13}C, FABMS	Oleanolic acid (**7**) Gal ╲ $^3_2$ Ara (OH-3β) Rha ╱ Glc-^{6}Glc (CO_2H-28)	*233*

Table 1. *(continued)*

Source (1)	Saponin mp, $[\alpha]_D$, spectra recorded (2)	Structure (3)	Ref. (4)
	Decaisoside C 234–236°, −48°, ^{1}H, ^{13}C, FABMS	Oleanolic acid (**7**) Gal ╲ ^{3} Ara (OH-3β) Rha ╱ ^{2} Rha-^{4}Glc-^{6}Glc (CO_2H-28)	*233*
	Decaisoside D 226–228°, −10°, ^{1}H, ^{13}C, FABMS	Hederagenin (**11**) Xyl-^{3}Rha-^{2}Ara (OH-3β) Glc (CO_2H-28)	*233*
	Decaisoside E 222–225°, −23°, ^{1}H, ^{13}C, FABMS	Hederagenin (**11**) Xyl-^{3}Rha-^{2}Ara (OH-3β) Glc-^{6}Glc (CO_2H-28)	*233*
Desmodium styracifolium (Leguminosae)	Soyasaponin II 272–280°, −43.2°, IR, ^{1}H, ^{13}C,	Soyasapogenol E (**49**) Rha-^{2}Gal-^{2}GlcA (OH-3β)	*234*
Deutzia corymbosa (Saxifragaceae)	Deutzicoside A 228–230°, IR, ^{1}H, ^{13}C	Echinocystic acid (**15**) Ara-^{4}Ara (OH-3β)	*235*
	Deutzicoside B 245–248°, IR, ^{1}H, ^{13}C	Echinocystic acid (**15**) Gal-^{4}Rha-^{4}Ara (OH-3β)	*235*
Diplazium subsinuatum (Woodriaceae)	Diplazioside I 290–291°, −17.8°, IR, ^{1}H, ^{13}C, FABMS	Aglycone (**327**) Glc ╲ ^{6} Glc (OH-3β) Ara (f) ╱ ^{2}	*236*

	Diplazioside II > 300°, +16.5°, IR, ^1H, ^{13}C, FABMS	Aglycone (**328**) Ara (f)-^{2}Glc (OH-3β)	*236*
Dianthus chinensis (Caryophyllaceae)	Dianchinenoside A 225–227°, +14.9°, ^1H, ^{13}C, FABMS	Aglycone (**41**) Ara (OH-3β) Glc (CO_2H-28)	*237*
	Dianchinenoside B 230–232°, +2.6°, ^1H, ^{13}C, 2D, FABMS	Aglycone (**41**) Xyl (OH-3β) Glc (CO_2H-28)	*237*
	Dianchinenoside C 225–227°, +12.4°, IR, ^1H, ^{13}C, FABMS	Aglycone (**41**) Glc (CO_2H-23) Glc (CO_2H-28)	*238*
	Diachinenoside D 236–238°, +3.3°, IR, ^1H, ^{13}C, FABMS	Aglycone (**41**) Glc-^{6}Glc (CO_2H-28)	*238*
	Dianchinenoside E 214–216°, +5.1°, IR, ^1H, ^{13}C, 2D, FABMS	Aglycone (**108**) Glc-^{2}Glc ╲ ^{6} Glc (CO_2H-28) Glc ╱ ^{3}	*239*
	Dianchinenoside F 215–218°, +5.0°, IR, ^1H, ^{13}C, 2D, FABMS	Aglycone (**109**) Glc-^{2}Glc ╲ ^{6} Glc (CO_2H-28) Glc ╱ ^{3}	*239*
	Dianchinenoside G 202–204°, +15.3°, IR, ^1H, ^{13}C, 2D, FABMS	Aglycone (**108**) Glc ╲ ^{6} Glc (CO_2H-28) Glc ╱ ^{3}	*239*

Table 1. *(continued)*

Source (1)	Saponin mp, $[\alpha]_D$, spectra recorded (2)	Structure (3)	Ref. (4)
	Dianchinenoside H 198–200°, +13.2°, IR, ^1H, ^{13}C, 2D, FABMS	Aglycone (**109**) Glc ╲ $^6_3$ Glc (CO_2H-28) Glc ╱	*239*
Digitalis ciliata (Scrophulariaceae)	Digitoside 268–270°, +5.1°, ^1H, ^{13}C	Oleanolic acid (**7**) Xyl-^{4}Rha (OH-3β) Glc-^{6}Glc (CO_2H-28)	*240*
Diploclisia glaucescens (Menispermaceae)	Saponin I	Aglycone (**50**) GlcA (OH-3β)	*241*
	Saponin 2	Aglycone (**98**) GlcA (OH-3β)	*241*
	Saponin 3	Aglycone (**50**) GlcA (OH-3β) Glc (CO_2H-28)	*241*
	Saponin 4	Aglycone (**50**) GlcA (OH-3β) Glc (CO_2H-28)	*241*
	Deploclisin 171–173°, +25°, IR, ^1H, ^{13}C, FABMS	Aglycone (**98**) Glc (OH-3β) Glc (CO_2H-28)	*242*
Dipsacus asper (Dipsacaceae)	Dipsacus saponin B 238–241°, −34.7°, IR, ^1H, ^{13}C, FABMS	Hederagenin (**11**) Rha ╲ $^6_4$ Glc-^{3}Rha-^{2}Ara (OH-3β) Glc ╱	*243*

	Dipsacus saponin C 256–260°, −52.7°, IR, ^{1}H, ^{13}C, FABMS	Hederagenin (**11**) Rha ╲ $_{3}^{4}$ Rha-^{2}Ara(OH-3β) Xyl-^{4}Glc-^{4}Glc ╱	*243*
	 +15°, ^{1}H, ^{13}C, FABMS	Hederagenin (**11**) (OAc-4′) Ara (OH-3β) Glc-^{6}Glc (CO_2H-28)	*244*
D. speroides	Asperosaponin F ^{1}H, ^{13}C, 2D	Hederagenin (**11**) Xyl-^{4}Glc ╲ $_{3}^{4}$ Gal-^{3}Rha-^{2}Ara (OH-3β) Rha ╱	*245*
	Asperosaponin H_1 ^{1}H, ^{13}C, 2D, FABMS	Hederagenin (**11**) Xyl-^{4}Glc ╲ $_{3}^{4}$ Gal-^{3}Rha-^{2}Ara (OH-3β) Rha ╱ Glc-^{6}Glc (CO_2H-28)	*245*
Dolichos lablab (Leguminosae)	Saponin I −92.2°, UV, ^{1}H, ^{13}C, FABMS	Aglycone (**51**) Rha-^{2}Gal-^{2}GlcA (OH-3β)	*246*
Dumasia truncata (Leguminosae)	 −10.4°, ^{1}H, ^{13}C, FABMS	Hederagenin (**11**) Rha-^{3}GlcA (OH-3β) Glc (CO_2H-28)	*247*
	 −7.1°, ^{1}H, ^{13}C, FABMS	Oleanolic acid (**7**) Rha ╲ $_{2}^{3}$ GlcA (OH-3β) Xyl ╱	*247*
Eleutherococcus *senticosus* (Araliaceae)	Saponin 1 255–258°, −45°,	Protoprimulagenin A (**193**) Glc-^{3}Gal ╲ $_{2}^{4}$ GlcA (OH-3β) Rha ╱	*248*

Table 1. *(continued)*

Source (1)	Saponin mp, $[\alpha]_D$, spectra recorded (2)	Structure (3)	Ref. (4)
	Saponin 2 230°, −15°, IR, ^{1}H, ^{13}C, FABMS	Protoprimulagenin A (**193**) Rha-^{4}Rha ⟶^{4} Glc-^{4}GlcA (OH-3β) Rha ⟶^{2}	*248*
Enterospermum pruinosum (Rubiaceae)	157–159°, 14.6°, ^{1}H, ^{13}C, FABMS	Longispinogenin (**135**) Glc (OH-3β) Glc (OH-16β)	*249*
Fagonia arabica (Zygophyllaceae)	Saponin 1 +20.4°, ^{1}H, ^{13}C, 2D, FABMS	Oleanolic acid (**7**) Glc ⟶^{3} Ara (OH-3β) Xyl ⟶^{2} Glc (CO_2H-28)	*250*
	Saponin 2 +25.4°, ^{1}H, ^{13}C, 2D, FABMS	Oleanolic acid (**7**) Glc ⟶^{3} Ara (OH-3β) Glc ⟶^{2} Glc (CO_2H-28)	*250*
	Saponin 3 +21.2°, ^{1}H, ^{13}C, 2D, FABMS	Oleanolic acid (**7**) Glc ⟶^{3} Ara (OH-3β) Xyl ⟶^{2}	*250*
	Saponin 4 +31.7°, ^{1}H, ^{13}C, 2D, FABMS	Oleanolic acid (**7**) Glc ⟶^{3} Ara (OH-3β) Glc ⟶^{2}	*250*

	Saponin 5 +16.0°, ^{1}H, ^{13}C, 2D, FABMS	27-Hydroxyoleanolic acid (**21**) Glc ─ 3 / Xyl ─ 2 Ara (OH-3β) Glc (CO_2H-28)	*250*
	Saponin 6 +14.8°, ^{1}H, ^{13}C, 2D, FABMS	Ursolic acid (**175**) Glc ─ 3 / Xyl ─ 2 Ara (OH-3β) Glc (CO_2H-28)	*250*
	Saponin 7 ^{1}H, ^{13}C, 2D, FABMS	27-Hydroxyursolic acid (**184**) Glc ─ 3 / Xyl ─ 2 Ara (OH-3β) Glc (CO_2H-28)	*250*
F. amollis	Saponin −18.5°, ^{1}H, ^{13}C,	Oleanolic acid (**7**) (Me-ester-6′) GlcA (OH-3β)	*251*
	Saponin 0°, ^{1}H, ^{13}C,	Oleanolic acid (**7**) Rha-^{3} (Me-ester-6′) GlcA (OH-3β)	*251*
	Saponin ^{1}H, ^{13}C	Oleanolic acid (**7**) Rha-^{3} (Me-ester-6′) GlcA (OH-3β) Glc (CO_2H-28)	*251*
Glycyrrhiza eurycarpa (Leguminosae)	Glyeurysaponin 288°, +22.5°,	Glycyrrhitic acid (**53**) GlcA-^{4}GlcA (OH-3β)	*252*
G. inflata	Apioglycyrrhizin 193–195°, +43°, IR, ^{1}H, ^{13}C	Glycyrrhitic acid (**53**) Apio (f)-^{2}GlcA (OH-3β)	*253*

Table 1. *(continued)*

Source (1)	Saponin mp, $[\alpha]_D$, spectra recorded (2)	Structure (3)	Ref. (4)
	Araboglycyrrhizin 237–238°, +31°, IR, ^1H, ^{13}C	Glycyrrhitic acid (**53**) Ara-^{2}GlcA (OH-3β)	*253*
G. uralensis	Licorice saponin A_3 198–199°, +69°, UV, IR, ^1H, ^{13}C	Glycyrrhitic acid (**53**) GlcA-^{2}GlcA (OH-3β) Glc (CO_2H-30)	*254*
	Licorice saponin B_2 209–210°, +54°, UV, IR, ^1H, ^{13}C	Agycone (**126**) GlcA-^{2}GlcA (OH-3β)	*254*
	Licorice saponin C_2 249–251°, −120°, UV, IR, ^1H, ^{13}C,	Aglycone (**237**) GlcA-^{2}GlcA (OH-3β)	*254*
	Licorice saponin D_3 −5.0°, UV, IR, ^1H, ^{13}C, MS	Aglycone (**54**) Rha-^{2}GlcA-^{2}GlcA(OH-3β)	*255*
	Licorice saponin E_2 218–219°, +68.0°, UV, IR, ^1H, ^{13}C	Glabrolide (**55**) GlcA-^{2}GlcA (OH-3β)	*255*
	Licorice saponin F_3 215–217°, −20°, IR, ^1H, ^{13}C	11-Deoxyglabrolide (**56**) Rha-^{2}GlcA-^{2}GlcA (OH-3β)	*256*
	Licorice saponin G_2 229–230°, +34°, UV, ^1H, ^{13}C, FABMS	24-Hydroxyglycyrrhitic acid (**57**) GlcA-^{2}GlcA (OH-3β)	*256*
	Licorice saponin H_2 209–210°, +31°, UV,	Aglycone (**58**) GlcA-^{2}GlcA (OH-3β)	*256*

	IR, ^{1}H, ^{13}C, FABMS		
	Licorice saponin J_2 263–265°, +21°, IR, ^{1}H, ^{13}C, FABMS	Aglycone (**59**) GlcA-^{2}GlcA (OH-3β)	*256*
	Licorice saponin K_2 207–209°, +28°, UV, IR, ^{1}H, ^{13}C, FABMS	Aglycone (**229**) GlcA-^{2}GlcA (OH-3β)	*256*
	Licorice saponin L_3 233–234°, +3.7°, UV, IR, ^{1}H, ^{13}C, FABMS	Aglycone (**60**) Rha-^{2}Ara-^{2}GlcA (OH-3β)	*257*
G. yunnanensis	Yunnanglysaponin A	Aglycone (**238**) GlcA-^{4}GlcA (OH-3β)	*258*
	Yunnanglysaponin B	Aglycone (**230**) GlcA-^{4}GlcA (OH-3β)	*258*
	Yunganoside A_1 +6°, ^{1}H, ^{13}C	Aglycone (**88**) Rha-^{2}GlcA-^{2}GlcA (OH-3β)	*259*
	Yunganoside B_1 −8°, ^{1}H, ^{13}C	Soyasapogenol B (**69**) Rha-^{2}GlcA-^{2}GlcA (OH-3β)	*259*
	Yunganoside C_1 −10°, ^{1}H, ^{13}C	Aglycone (**136**) Rha-^{2}GlcA-^{2}GlcA (OH-3β)	*259*
	Yunganoside D_1 +11°, ^{13}C	Aglycone (**233**) Rha-^{2}GlcA-^{2}GlcA (OH-3β)	*259*
	Yunganoside E_2 −42°, ^{13}C	Aglycone (**234**) GlcA-^{2}GlcA (OH-3β)	*259*
	Yunganoside F_2 −30°, ^{13}C	Aglycone (**235**) GlcA-^{2}GlcA (OH-3β)	*259*
	Glyyunnanprosapogenin 168–170°, UV, IR, ^{1}H, ^{13}C, FABMS	Aglycone (**230**) GlcA (OH-3β) GlcA (OH-21α)	*260*

Table 1. *(continued)*

Source (1)	Saponin mp, $[\alpha]_D$, spectra recorded (2)	Structure (3)	Ref. (4)
Guaiacum Officinale (Zygophylaceae)	Guaiacin A ^{13}C, FABMS	Aglycone (**28**) Glc-^{3}Ara (OH-3β) Glc (CO_2H-28)	*261*
	Guaiacin B ^{13}C, FABMS	Oleanolic acid (**7**) Glc-^{3}Ara (OH-3β) Glc (CO_2H-28)	*261*
	Guaianin C ^{1}H, ^{13}C, FABMS	Oleanolic acid (**7**) Rha-^{3}Glc-^{3}Ara (OH-3β) Glc (CO_2H-28)	*262*
	Guaiacin C ^{1}H, ^{13}C, FABMS	Aglycone (**28**) Glc \ ^{3} $_{2}$ Ara (OH-3β) Rha /	*263*
	Guaiacin D ^{1}H, ^{13}C, FABMS	Aglycone (**28**) Glc \ ^{3} $_{2}$ Ara (OH-3β) Rha / Glc (CO_2H-28)	*263*
	Guaiacin E ^{1}H, ^{13}C	Aglycone (**28**) Glc \ ^{3} $_{2}$ Ara (OH-3β) Glc / Glc (CO_2H-28)	*263*
		Oleanolic acid (**7**)	*263*

	Guaiacin F ^1H, ^{13}C, FABMS	Glc ╲ $^3_2$ Ara (OH-3β) Glc ╱ Glc(CO_2H-28)	
	Guaiacin H ^1H, ^{13}C, FABMS	Oleanolic acid (**7**) Rha-^{4}Glc-^{3}Ara (OH-3β)	*264*
	Guaiacin I ^1H, ^{13}C, FABMS	Oleanolic acid (**7**) Rha-^{2}Rha-^{2}Ara-^{3}GlcA (OH-3β) Glc-^{6} Glc (CO_2H-28)	*264*
	Guaiacin J ^1H, ^{13}C, FABMS	Oleanolic acid (**7**) Ara-^{3}GlcA (OH-3β) Rha-^{2}Rha-^{4}Glc-^{6}Glc (CO_2H-28)	*264*
	Guaiacin K ^1H, ^{13}C, FABMS	Oleanolic acid (**7**) GlcA (OH-3β) Glc ╲ $^3_2$ Ara-^{3}Glc (CO_2H-28) Rha-^{2}Rha ╱	*264*
	Guaiacin L ^1H, ^{13}C, FABMS	Oleanolic acid (**7**) Glc ╲ $^4_3$ GlcA (OH-3β) Ara ╱ Rha-^{6}Glc-^{6}Glc (CO_2H-28)	*264*
	Guaiacin M ^1H, ^{13}C, FABMS	Aglycone (**28**) Glc-^{2}Ara (OH-3β)	*265*
Gymnema sylvestre (Asclepiadaceae)	Gymnemaside I 159–161°, +23.7°, IR, ^1H, ^{13}C, FABMS	Aglycone (**260**) Glc (OH-3β) Glc (OH-20β)	*266*
	Gymnemaside II 212–214°, +10.5°, IR, ^1H, ^{13}C, FABMS	Aglycone (**260**) Glc-^{2}Glc (OH-3β) Glc (OH-20β)	*266*

Table 1. *(continued)*

Source (1)	Saponin mp, $[\alpha]_D$, spectra recorded (2)	Structure (3)	Ref. (4)
	Gymnemaside III 182–184°, +8.2°, IR, ^{1}H, ^{13}C, FABMS	Aglycone (**260**) Glc-^{2}Ara (OH-3β) Glc (OH-20β)	*266*
	Gymnemaside IV 256–257°, +14.5°, IR, ^{1}H, ^{13}C, FABMS	Aglycone (**260**) Glc (OH-3β) Xyl-^{6}Glc (OH-20β)	*266*
	Gymnemaside V 187–189°, +5.9°, IR, ^{1}H, ^{13}C, FABMS	Aglycone (**260**) Glc-^{2}Glc (OH-3β) Xyl-^{6}Glc (OH-20β)	*266*
	Gymnemaside VI 188–190°, +8.7°, ^{1}H, ^{13}C, FABMS	Aglycone (**262**) Rha-^{6}Glc (OH-20β)	*266*
	Gymnemaside VII 185–187°, +7.7°, ^{1}H, ^{13}C, FABMS	Aglycone (**262**) Xyl-^{6}Glc (OH-20β)	*266*
	Gymnemic acid V 202–203°, +2.2°, ^{1}H, ^{13}C, FABMS	Aglycone (**62**) GlcA (OH-3β)	*267*
	Gymnemic acid VI 225–226°, +11.7°, ^{1}H, ^{13}C, FABMS	Aglycone (**61**) Glc-^{3}GlcA (OH-3β)	*267*
	Gymnemic acid VII 222–223°, +9.6°, ^{1}H, ^{13}C, FABMS	Gymnestrogenin (**63**) GlcA (OH-3β)	*267*

Gymnemic acid XIII 218–220°, +17.4°, IR, ^{1}H, ^{13}C, FABMS	Aglycone (**64**) (Hexulo-2′) Ara-^{3}GlcA (OH-3β)	*268*
Gymnemic acid XIV 222–224°, 11.4°, IR, ^{1}H, ^{13}C, FABMS	Aglycone (**61**) (Hexulo-2′) Ara-^{3}GlcA (OH-3β)	*268*
Gymnemasin A 215–217°, +15°, IR, ^{1}H, ^{13}C, FABMS	Aglycone (**144**) Glc-^{3}GlcA (OH-3β)	*269*
Gymnemasin B 221–222°, +18.5°, IR, ^{1}H, ^{13}C, FABMS	Gymnemanol (**145**) Glc-^{3}GlcA (OH-3β)	*269*
Gymnemasin C 212–214°, +12.5°, IR, ^{1}H, ^{13}C, FABMS	Aglycone (**144**) GlcA (OH-3β)	*269*
Gymnemasin D 220–221°, +8°, IR, ^{1}H, ^{13}C, FABMS	Gymnemanol (**145**) GlcA (OH-3β)	*269*
Gymnemic acid VIII 185–187°, +21.5°, IR, ^{1}H, ^{13}C, FABMS	Aglycone (**121**) GlcA (OH-3β)	*270*
Gymnemic acid IX 194–196°, +7.6°, IR, ^{1}H, ^{13}C, FABMS	Aglycone (**122**) GlcA (OH-3β)	*270*
Gymnemic acid X 212°, +14.9°, IR, ^{1}H, ^{13}C, FABMS	Aglycone (**123**) GlcA (OH-3β)	*270*

Table 1. *(continued)*

Source (1)	Saponin mp, $[\alpha]_D$, spectra recorded (2)	Structure (3)	Ref. (4)
	Gymnemic acid XI 190–192°, +1.7°, IR, ^{1}H, ^{13}C, FABMS	Aglycone (**124**) GlcA (OH-3β)	*270*
	Gymnemic acid XII 209–211°, +11.7°, IR, ^{1}H, ^{13}C, FABMS	Aglycone (**125**) Glc-^{3}GlcA (OH-3β)	*270*
	Gymnemic acid XV +7.2°, IR, ^{1}H, ^{13}C, FABMS	Aglycone (**138**) GlcA (OH-3β)	*271*
	Gymnemic acid XVI 203–205°, −6.8°, IR, ^{1}H, ^{13}C, FABMS	Aglycone (**139**) GlcA (OH-3β)	*271*
	Gymnemic acid XVII 211-213°, +7.1°, IR, ^{1}H, ^{13}C, FABMS	Aglycone (**140**) GlcA (OH-3β)	*271*
	Gymnemic acid XVIII 201–203°, +6.4°, IR, ^{1}H, ^{13}C, FABMS	Aglycone (**141**) GlcA (OH-3β)	*271*
	Gymnemasaponin I 184–185°, +9.3°, ^{1}H, ^{13}C, FABMS	Aglycone (**137**) Glc (OH-28)	*89*
	Gymnemasaponin II 190–192°, +1.9°, ^{1}H, ^{13}C, FABMS	Aglycone (**137**) Glc (OH-23) Glc (OH-28)	*89*

	Gymnemasaponin III 203–205°, −11.6°, ^{1}H, ^{13}C, FABMS	Aglycone (**137**) Glc (OH-23) Glc-^{6}Glc (OH-28)	*89*
	Gymnemasaponin IV 201–203°, −1.1°, ^{1}H, ^{13}C, FABMS	Aglycone (**137**) Glc-^{6}Glc (OH-23) Glc (OH-28)	*89*
	Gymnemasaponin V 186–188°, −6.2°, ^{1}H, ^{13}C, FABMS	Aglycone (**137**) Glc-^{6}Glc (OH-23) Glc-^{6}Glc (OH-28)	*89*
Gynostemma compressum (Cucurbitaceae)	Gycomoside I	Aglycone (**258**) Glc-^{4}Glc (OH-20S)	*272*
G. longipes	Saponin 1	Aglycone (**273**) Xyl-^{2}Xyl (OH-3β)	*273*
	Saponin 2	Aglycone (**273**) Rha-^{2}Xyl-^{2}Xyl (OH-3β)	*273*
G. pentaphyllum	Glycoside 1 +2.5°, ^{1}H, ^{13}C, FABMS	Aglycone (**324**) Glc-^{2}Ara (OH-3β) Rha (OH-20β)	*274*
	Glycoside 2 −9.3°, ^{1}H, ^{13}C, FABMS	Aglycone (**334**) Glc-^{2}Ara (OH-3β) Rha (OH-20β)	*274*
	Glycoside 3 −8.5°, ^{1}H, ^{13}C, FABMS	Aglycone (**285**) Glc-^{2}Ara (OH-3β) Rha (OH-20β) Glc (OH-27)	*274*
	Glycoside 4 −33.6°, ^{1}H, ^{13}C, FABMS	Aglycone (**284**) Glc-^{2}Ara (OH-3β) Glc (OH-21) Rha (OH-24)	*274*

Table 1. *(continued)*

Source (1)	Saponin mp, $[\alpha]_D$, spectra recorded (2)	Structure (3)	Ref. (4)
	6″-Malonylginsenoside-Rb_1 198–200°, +7.8°, IR ^{13}C, EIMS	Aglycone (**259**) (Malonate-6′) Glc-^{2}Glc (OH-3β) Glc-^{6}Glc (OH-20β))	*275*
	6″-Malonylginsenoside-R_d 215–217°, +14.3°, IR, ^{13}C, EIMS	Aglycone (**259**) (Malonate-6′) Glc-^{2}Glc (OH-3β) Glc (OH-20β)	*275*
	6″-Malonylgypenoside-V 205–207°, +6.7°, IR, ^{13}C, EIMS	Aglycone (**259**) (Malonate-6′) Glc-^{2}Glc (OH-3β) Rha-^{6}Glc (OH-20β)	*275*
Gypsophila capillaris (Caryophyllaceae)	227–229°, +2.2°, IR, ^1H, ^{13}C, FABMS	Aglycone (**143**) Gal —(6) Glc —(3) Glc (CO_2H-28)	*276*
	212–214°, +12.5°, IR, ^1H, ^{13}C, FABMS	Gypsogenin (**13**) Glc —(6) Glc-^{2}Gal —(3) Gal (CO_2H-28)	*276*
	230–232°, +10.5°, IR, ^1H, ^{13}C, FABMS	Aglycone (**143**) Glc —(6) Glc-^{2}Gal —(3) Gal (CO_2H-28)	*276*
	225–227°, +4.5°, IR, ^1H, ^{13}C, FABMS	Aglycone (**143**) Glc (CO_2H-23) Glc —(6) Glc-^{2}Gal —(3) Gal (CO_2H-28)	*276*

G. oldhamiana	^{1}H, ^{13}C, 2D, FABMS	Aglycone (**110**) Xyl ╲ $^{3}_{2}$ GlcA (OH-3β) Glc ╱	*277*
	^{1}H, ^{13}C, 2D, FABMS	Aglycone (**152**) Xyl ╲ $^{3}_{2}$ GlcA (OH-3β) Glc ╱	*277*
	^{1}H, ^{13}C, 2D, FABMS	Quillaic acid (**46**) Xyl ╲ $^{3}_{2}$ GlcA (OH-3β) Glc ╱ Fuc ╲ $^{4}_{3}$ Rha (CO_2H-28) Glc ╱	*277*
G. paniculata	Saponin G_1 210-213°, ^{1}H, ^{13}C, 2D, FABMS	Quillaic acid (**46**) Xyl ╲ $^{3}_{2}$ GlcA (OH-3β) Gal ╱ Xyl ╲ $^{4}_{3}$ Rha-^{2}Fuc (CO_2H-28) Glc ╱	*278*
	Saponin G_2 213–215°, ^{1}H, ^{13}C, 2D, FABMS	Quillaic acid (**46**) Xyl ╲ $^{3}_{2}$ GlcA (OH-3β) Gal ╱ Ara-^{4}Ara-^{3}Xyl-^{4}Rha-^{2}Fuc (CO_2H-28)	*278*

Table 1. *(continued)*

Source (1)	Saponin mp, $[\alpha]_D$, spectra recorded (2)	Structure (3)	Ref. (4)
	Saponin G_3 207–211°, ^1H, ^{13}C, 2D, FABMS	Gypsogenin (**13**) Glc-^{2}GlcA (OH-3β) Xyl ╲^{4} Rha-^{2}Fuc (CO_2H-28) Glc ╱^{3}	*278*
	Saponin G_4 215–218°, ^1H, ^{13}C, 2D, FABMS	Gypsogenin (**13**) Xyl ╲^{3} GlcA (OH-3β) Gal ╱^{2} Xyl ╲^{4} Rha-^{2}Fuc (CO_2H-28) Glc ╱^{3}	*278*
Hedera helix (Araliaceae)	Hederasaponin E ^{13}C, FABMS	Bayogenin (**25**) Ara (OH-3β) Rha-^{4}Glc-^{6}Glc (CO_2H-28)	*279*
	Hederasaponin H ^{13}C, FABMS	Oleanolic acid (**7**) Gal-^{4}Glc (OH-3β) Rha-^{4}Glc-^{6}Glc (CO_2H-28)	*279*
	Hederasaponin I ^{13}C, FABMS	Hederagenin (**11**) Glc (OH-3β) Rha-^{4}Glc-^{6}Glc (CO_2H-28)	*279*
	Hederasaponin F ^{13}C, FABMS	Aglycone (**76**) Rha-^{4}Glc-^{6}Glc (CO_2H-28)	*279*
H. taurica	Hederoside E_1	Erythrodiol (**65**) Glc-^{2}Glc (OH-3β)	*280*

Heinsia crinata (Rubiaceae)	Saponin 1 123–124°, +14.29°, ^{1}H, ^{13}C, 2D	Heinsiagenin A (**312**) Glc-^{2}Glc ╲ $^6_2$ Glc-^{2}Glc (OH-3β) Rha ╱	*281*
	Saponin 2 105–107°, +21.43°, ^{1}H, ^{13}C, 2D	Heinsiagenin A (**312**) Rha-^{2}Glc-^{2}Glc (OH-3β)	*281*
Helianthus annuus (Compositae)	Helianthoside 1	Oleanolic acid (**7**) Xyl-^{4}Glc (OH-3β) Glc-^{4}Rha-^{2}Ara (CO_2H-28)	*282*
	Helianthoside 2	Echinocystic acid (**15**) Xyl-^{4}Glc (OH-3β) Glc-^{4}Rha-^{2}Ara (CO_2H-28)	*282*
	Helianthoside 3	Echinocystic acid (**15**) Xyl-^{4}Glc (OH-3β) Glc-^{4}Rha-^{2}Glc (CO_2H-28)	*282*
Hemsleya graciliflora (Cucurbitaceae)	Hemsloside G_1 +7.6°, ^{1}H, ^{13}C	Oleanolic acid (**7**) Ara-^{3}Glc (OH-3β) Glc-^{6}GlcA (CO_2H-28)	*283*
	Hemsloside G_2 −6.1°, ^{1}H, ^{13}C	Oleanolic acid (**7**) Glc-^{2}GlcA (OH-3β) Glc-^{6}Glc (CO_2H-28)	*283*
Herniaria glabra (Caryophylaceae)	Herniaria saponin 3 275–276°, −6.12°, ^{13}C, 2D, FABMS	Medicagenic acid (**48**) Glc (OH-3β) Glc-$^3_2$ (OAc-4′) Fuc (CO_2H-28) ╱ Rha^3-Glc	*284*
Heteropappus altaicus (Compositae)	Heteropappus saponin 5	Polygalacic acid (**27**) Glc (OH-3β) Xyl ╲ $^4_3$ Rha-^{2}Ara (CO_2H-28) Apio (f) ╱	*285*

Table 1. *(continued)*

Source (1)	Saponin mp, $[\alpha]_D$, spectra recorded (2)	Structure (3)	Ref. (4)
H. biennis	^{1}H, 2D	Arjunolic acid (**66**) Ara ╲ Xyl-3 $^{6}_{2}$ Glc (CO_2H-28) Rha ╱	*26*
Holothuria forskali (Holothuridae)	Holothurinoside A 232–233°, –0.9°, ^{1}H, ^{13}C, 2D, FABMS	Aglycone (**255**) Glc ╲ $^{4}_{2}$ Xyl (OH-3β) (3′-OMe) Glc-^{3}Glc-^{4}Quin ╱	*8*
	Holothurinoside B 230–232°, ^{1}H, ^{13}C, 2D, FABMS	Aglycone (**256**) Glc ╲ $^{4}_{2}$ Xyl (OH-3β) (3′-OMe) Glc-^{3}Glc-^{4}Quin ╱	*8*
	Holothurinoside C 223–225°, ^{1}H, ^{13}C, 2D, FABMS	Aglycone (**257**) ^{3}Glc-^{4}Quin-^{2}Xyl (OH-3β) | (3′-OMe) Glc	*8*
	Holothurinoside D 219–221°, ^{1}H, ^{13}C, 2D, FABMS	Aglycone (**255**) (3′-OMe) Glc-^{3}Glc-^{4}Quin-^{2}Xyl (OH-3β)	*8*
	Desholothurin A	Aglycone (**257**) Quin-^{2}Xyl (OH-3β)	*8*
Hovenia dulcis (Rhamnaceae)	Hoduloside I 184–186°, –19.5°, ^{13}C, 2D, FABMS	Hovenolactone (**311**) Rha-^{2}Glc (OH-3β) Glc (OH-23R)	*90*

Hoduloside II 188–190°, −14.6°, ^{13}C, 2D, FABMS	Hovenolactone (**311**) Glc–^{3}, Rha–^{2} Glc (OH-3β)	*90*
Hoduloside III 297–299°, −36.9°, ^{13}C, 2D, FABMS	Jujubogenin (**305**) Glc–^{3}, Quin–^{2} Ara (OH-3β)	*90*
Hoduloside IV 246–248°, −12.9°, ^{13}C, 2D, FABMS	Jujubogenin (**305**) Glc–^{3}, Glc–^{2} Ara (OH-3β)	*90*
Hoduloside V 215–217°, −31.4°, ^{13}C, 2D, FABMS	Jujubogenin (**305**) Glc–^{3}, Rha–^{2} Glc (OH-3β)	*90*
Hoduloside VII −52.1°, IR, ^1H, ^{13}C, 2D, FABMS	Aglycone (**286**) Rha-^{2}Ara (OH-3β) Glc (OH-30)	*286*
Hoduloside VIII −34.4°, IR, ^1H, ^{13}C, 2D, FABMS	Aglycone (**286**) Ara (OH-3β) Xyl-^{6}Glc (OH-30β)	*286*
Hoduloside IX −37.6°, IR, ^1H, ^{13}C, 2D, FABMS	Aglycone (**286**) Rha-^{2}Ara (OH-3β) Xyl-^{6}Glc (OH-30β)	*286*
Hoduloside X −35.0°, IR, ^1H, ^{13}C, 2D, FABMS	Aglycone (**286**) Glc–^{3}, Rha–^{2} Ara (OH-3β) Glc (OH-30β)	*286*

Table 1. *(continued)*

Source (1)	Saponin mp, $[\alpha]_D$, spectra recorded (2)	Structure (3)	Ref. (4)
Hydrocotyle ranunculoides (Umbelliferae)	Ranuncoside I −7.5°, IR, ^{1}H, ^{13}C, FABMS	Aglycone (**146**) Ara-^{6}Glc (OH-3β)	*287*
	Ranuncoside II −3.8°, IR, ^{1}H, ^{13}C, FABMS	Aglycone (**146**) Ara $\searrow^6$ Glc (OH-3β) Glc $\nearrow_2$	*287*
	Ranuncoside III +5.5°, IR, ^{1}H, ^{13}C, FABMS	Aglycone (**147**) Ara $\searrow^6$ Glc (OH-3β) Glc $\nearrow_2$	*287*
	Ranuncoside IV −4.2°, IR, ^{1}H, ^{13}C, FABMS	Aglycone (**148**) Ara $\searrow^6$ Glc (OH-3β) Glc $\nearrow_2$	*287*
	Ranuncoside V +4.4°, IR, ^{1}H, ^{13}C, FABMS	Aglycone (**149**) Ara $\searrow^6$ Glc (OH-3β) Glc $\nearrow_2$	*287*
	Ranuncoside VI −3.9°, IR, ^{1}H, ^{13}C, FABMS	Aglycone (**150**) Ara $\searrow^6$ Glc (OH-3β) Glc $\nearrow_2$	*287*
Ilex crenata (Aquifoliaceae)	Ilexoside III 201–203°, −8.7°, ^{1}H, ^{13}C, FABMS	Pomolic acid (**163**) Ara (OH-3β) Xyl-^{6}Glc (CO_2H-28)	*288*

Ilexoside IV 206–208°, −14.4°, ^{1}H, ^{13}C	Pomolic acid (**163**) Ara (OH-3β) Rha-^{2}Glc (CO_2H-28)	*288*
Ilexoside V 218–220°, −23.1°, ^{1}H, ^{13}C, EIMS	Pomolic acid (**163**) Ara (OH-3β) Xyl \\ $^{6}_{2}$ Glc (CO_2H-28) Rha /	*288*
Ilexoside VI 196–198°, −2.6°, ^{1}H, ^{13}C	Pomolic acid (**163**) Ara (OH-3β) Glc \\ $^{6}_{2}$ Glc (CO_2H-28) Rha /	*288*
Ilexoside VII −10.1°, ^{1}H, ^{13}C	Pomolic acid (**163**) Glc-^{3}Ara (OH-3β) Xyl-^{6}Glc (CO_2H-28)	*288*
Ilexoside VIII 202–204°, −4.5°, ^{1}H, ^{13}C, EIMS	Pomolic acid (**163**) Glc-^{3}Ara (OH-3β) Glc-^{6}Glc (CO_2H-28)	*288*
Ilexoside IX 202–204°, −23.2°, ^{1}H, ^{13}C, FABMS	Pomolic acid (**163**) Xyl \\ $^{6}_{2}$ Glc (CO_2H-28) Rha /	*289*
Ilexoside X 224–226°, −17.2°, ^{1}H, ^{13}C, FABMS	Pomolic acid (**163**) Glc-^{3}Ara (OH-3β) Xyl \\ $^{6}_{2}$ Glc (CO_2H-28) Rha /	*289*

Table 1. *(continued)*

Source (1)	Saponin mp, $[\alpha]_D$, spectra recorded (2)	Structure (3)	Ref. (4)
	Ilexoside XI 224–226°, −15.7°, ^1H, ^{13}C, FABMS	Pomolic acid (**163**) Glc-^{3}Ara (OH-3β) Glc $\searrow^6_2$ Glc (CO_2H-28) Rha $\nearrow$	*289*
	Ilexoside XII 218–220°, −16.1°, ^1H, ^{13}C, FABMS	Pomolic acid (**163**) Glc-^{3}Ara (OH-3β) Xyl $\searrow^6_2$ Glc (CO_2H-28) Rha $\nearrow$	*289*
	Ilexoside XIII 231–233°, −12.2°, ^1H, ^{13}C, FABMS	Pomolic acid (**163**) Glc $\searrow^3_2$ Ara (OH-3β) Glc $\nearrow$ Xyl $\searrow^6_2$ Glc (CO_2H-28) Rha $\nearrow$	*289*
	Ilexoside XIV 228–230°, −13.9°, ^1H, ^{13}C, FABMS	Pomolic acid (**163**) Glc $\searrow^3_2$ Ara (OH-3β) Ara $\nearrow$ Xyl $\searrow^6_2$ Glc (CO_2H-28) Rha $\nearrow$	*289*

Ilexoside XV +13.4°, ^{1}H, ^{13}C, FABMS	Siaresinolic acid (**67**) Glc-^{3}Ara (OH-3β) Glc (CO_2H-28)	*290*
Ilexoside XVI +21.8°, ^{1}H, ^{13}C, FABMS	Siaresinolic acid (**67**) Glc-^{3}(OAc-2′) Ara (OH-3β) Glc(CO_2H-28)	*290*
Ilexoside XVII +56.2°, ^{1}H, ^{13}C, FABMS	Siaresinolic acid (**67**) Glc ╲ $^{3}_{2}$ Ara (OH-3β) Glc ╱ Glc (CO_2H-28)	*290*
Ilexoside XVIII +3.9°, ^{1}H, ^{13}C, FABMS	Pomolic acid (**163**) Glc ╲ $^{3}_{2}$ Ara (OH-3β) Glc ╱ Glc (CO_2H-28)	*290*
Ilexoside XIX +10.2°, ^{1}H, ^{13}C, FABMS	Pomolic acid (**163**) Glc-^{3}(OAc-2′) Ara (OH-3β) 2 | Xyl Glc (CO_2H-28)	*290*
Ilexoside E 171.5–172.5°, −67.0°, IR, ^{1}H, ^{13}C, FABMS	Aglycone (**320**) Ara (OH-3β) Xyl ╲ $^{6}_{2}$ Glc (CO_2H-28) Rha ╱	*291*
Ilexoside F 183–185°, +37.2°, IR, ^{1}H, ^{13}C, FABMS	Aglycone (**320**) Glc-^{3}Ara (OH-3β) Xyl ╲ $^{6}_{2}$ Glc (CO_2H-28) Rha ╱	*291*

Table 1. *(continued)*

Source (1)	Saponin mp, $[\alpha]_D$, spectra recorded (2)	Structure (3)	Ref. (4)
	Ilexoside G 194–196°, −19.5°, IR, ^{1}H, ^{13}C, FABMS	Aglycone (**321**) Ara (OH-3β)	*291*
	Ilexoside H 217–219°, −11.2°, IR, ^{1}H, ^{13}C, FABMS	Aglycone (**321**) Glc-^{3}Ara (OH-3β)	*291*
	Ilexoside I 186–188°, −19.4°, IR, ^{1}H, ^{13}C, FABMS	Aglycone (**321**) Glc-^{3}Ara (OH-3β) Glc (CO_2H-28)	*291*
I. dumosa	Saponin E_3 249–252°, +34.66°, IR, ^{1}H, ^{13}C, FABMS	Oleanolic acid (**7**) Gal-^{2}Ara (OH-3β)	*292*
	Saponin E_6 262–298°, +23.84°, IR, ^{1}H, ^{13}C, FABMS	Oleanolic acid (**7**) Gal-^{2}Glc (OH-3β)	*292*
	Saponin E_7 +51.64°, IR, ^{1}H, ^{13}C, FABMS	Oleanolic acid (**7**) Gal-^{2}Ara (OH-3β) Glc (CO_2H-28)	*292*
	Saponin E_8 +16°, IR, ^{1}H, ^{13}C, FABMS	Oleanolic acid (**7**) Gal-^{2}Glc (OH-3β) Glc (CO_2H-28)	*292*
I. integra	Ilexoside XXV +14.4°, ^{1}H, ^{13}C, FABMS	Aglycone (**169**) Glc (OH-3β) Glc (CO_2H-28)	*293*

	Ilexoside XXVI −0.4°, ^{1}H, ^{13}C, FABMS	Aglycone (**169**) Glc-^{6}Glc (OH-3β) Glc (CO_2H-28)	*293*
	Ilexoside XXVII 201–202°, +16.1°, ^{1}H, ^{13}C, FABMS	Rotundic acid (**170**) Ara (OH-3β) Glc (CO_2H-28)	*293*
	Ilexoside XXVIII 218–220°, +11.9°, ^{1}H, ^{13}C, FABMS	Aglycone (**167**) Ara (OH-3β) Glc (CO_2H-28)	*293*
I. kudincha	Kudinoside D 276–279°, IR, ^{1}H, ^{13}C, FABMS	Aglycone (**182**) Glc ＼ 3 Ara (OH-3β) Rha ／ 2	*294*
	Kudinoside E 267–270°, IR, ^{1}H, ^{13}C, FABMS	Aglycone (**182**) Glc-^{2}Glc ＼ 3 Ara (OH-3β) Rha ／ 2	*294*
	Kudinoside F 270–274°, IR, ^{1}H, ^{13}C, FABMS	Aglycone (**183**) Glc ＼ 3 Ara (OH-3β) Rha ／ 2	*294*
	Kudinoside G 228–232°, IR, ^{1}H, ^{13}C, FABMS	Pomolic acid (**163**) Glc ＼ 3 Ara (OH-3β) Rha ／ 2 Glc (CO_2H-28)	*294*
	Kudinoside H 214–215°, IR, ^{1}H, ^{13}C, FABMS	Pomolic acid (**163**) Ara (OH-3β) Glc (CO_2H-28)	*294*

Table 1. *(continued)*

Source (1)	Saponin mp, $[\alpha]_D$, spectra recorded (2)	Structure (3)	Ref. (4)
I. paraguariensis	Metasaponin 2 +6.7°, ^1H, ^{13}C, FABMS	Ursolic acid (**175**) Glc ╲^{3} Ara (OH-3β) Rha ╱$_2$ Glc (CO_2H-28)	*295*
	Metasaponin 3 +4.8°, ^1H, ^{13}C, FABMS	Ursolic acid (**175**) Glc-^{3}Ara (OH-3β) Glc-^{6}Glc (CO_2H-28)	*295*
	Metasaponin 4 −8.8°, ^1H, ^{13}C, FABMS	Ursolic acid (**175**) Glc ╲^{3} Ara (OH-3β) Rha ╱$_2$ Glc-^{6}Glc (CO_2H-28)	*295*
	Metasaponin 5 +15°, ^1H, ^{13}C, FABMS	Ursolic acid (**175**) Glc ╲^{3} Ara (OH-3β) Rha ╱$_2$ Glc-^{4}Glc-^{6}Glc (CO_2H-28)	*296*
I. rotunda	Ilexoside XXIX 204–206°, +13.7°, ^1H, ^{13}C, 2D, FABMS	Aglycone (**174**) Glc (CO_2H-28)	*297*
	Ilexoside XXX 214–215°, +26.9°, IR, ^1H, ^{13}C, 2D, FABMS	Rotundioic acid (**164**) Glc (CO_2H-28)	*297*

Ilexoside XXXI −2.0°, IR, ^{1}H, ^{13}C, 2D, FABMS	Siaresinolic acid (**67**) GlcA (OH-3β) Glc (CO_2H-28)	*297*
Ilexoside XXXII 218–220°, +2.2°, IR, ^{1}H, ^{13}C, 2D, FABMS	Ilexosapogenin A (**68**) GlcA (OH-3β)	*297*
Ilexoside XLI 198–200°, +47.0°, ^{1}H, ^{13}C, FABMS	Rotundic acid (**170**) Glc-^{6}Glc (CO_2H-28)	*298*
Ilexoside XLII +30.8°, ^{1}H, ^{13}C, FABMS	23-Oxorotungenic acid (**185**) Glc (CO_2H-28)	*298*
Ilexoside XLIII +9.5°, ^{1}H, ^{13}C, FABMS	30-Hydroxyrotundic acid (**186**) Glc (CO_2H-28)	*298*
Ilexoside XLIV 225–227°, +5.8°, ^{1}H, ^{13}C, FABMS	Ilexosapogenin B (**187**) Glc (OH-3β) Glc (OH-30)	*298*
Ilexoside XLV 267–269°, +24.7°, ^{1}H, ^{13}C, FABMS	24-Hydroxyrotundioic acid (**188**) Glc (CO_2H-28)	*298*
Ilexoside XLVI −0.6°, ^{1}H, ^{13}C, 2D, FABMS	Ilexosapogenin A (**68**) GlcA (OH-3β) Glc (CO_2H-28)	*299*
Ilexoside XLVII 238–239°, −11.3°, ^{1}H, ^{13}C, 2D, FABMS	Aglycone (**156**) GlcA (OH-3β) Glc (CO_2H-28)	*299*

Table 1. *(continued)*

Source (1)	Saponin mp, $[\alpha]_D$, spectra recorded (2)	Structure (3)	Ref. (4)
	Ilexoside XLVIII 200–201°, +19.3°, ^1H, ^{13}C, 2D, FABMS	Hederagenin (**11**) GlcA (OH-3β) Glc (CO_2H-28)	*299*
	Ilexoside XLIX +18.9°, ^1H, ^{13}C, 2D, FABMS	Hederagenin (**11**) Gal-^{2}GlcA (OH-3β) Glc (CO_2H-28)	*299*
	Ilexoside L 250–252°, +5.3°, ^1H, ^{13}C, 2D, FABMS	Aglycone (**43**) GlcA (OH-3β) Glc (CO_2H-28)	*299*
	Ilexoside LI 207–209°, −1.3°, ^1H, ^{13}C, 2D, FABMS	Siaresinolic acid (**67**) Gal-^{2}GlcA (OH-3β) Glc (CO_2H-28)	*299*
Isertia haenkeana (Rubiaceae)	195–198°, −28°, ^1H, ^{13}C, FABMS	Pyrocincholic acid (**236**) (deoxy-6′) Glc (OH-3β) Glc (CO_2H-28)	*300*
Juncus effusus (Juncaceae)	Juncoside I	Aglycone (**319**) Glc-^{2}Glc-^{2}Glc (OH-3β)	*301*
Kalopanax pictus (Araliaceae)	Kalopanax saponin JLa −19.1°, ^1H, ^{13}C, FABMS	Hederagenin (**11**) Rha-^{2}Ara (OH-3β) (OAC-2′) Rha-^{4}(OAc-6′) Glc-^{6}Glc (CO_2H-28)	*302*
	Kalopanax saponin JLb −12.4°, ^1H, ^{13}C, FABMS	Hederagenin (**11**) Rha-^{2}Ara (OH-3β) (OAC-3′) Rha-^{4} (OAc-6′) Glc-^{6}Glc (CO_2H-28)	*302*

K. septemlobus	Kalopanax saponin C $-19.3°$, ^{1}H, ^{13}C, 2D	Hederagenin (**11**) Glc $\searrow^{3}$ Ara (OH-3β) Rha $\nearrow_{2}$ Rha-^{4}Glc-^{6}Glc (CO_2H-28)	*303*
	Kalopanax saponin D 235–236°, $-24.6°$, ^{1}H, ^{13}C, 2D	Oleanolic acid (**7**) Glc $\searrow^{3}$ Ara (OH-3β) Rha $\nearrow_{2}$ Rha-^{4}Glc-^{6}Glc (CO_2H-28)	*303*
	Kalopanax saponin E $+14.2°$, ^{1}H, ^{13}C, 2D	Oleanolic acid (**7**) Glc-^{3}GlcA (OH-3β)	*303*
	Kalopanax saponin F $+7.1°$, ^{1}H, ^{13}C, 2D	Oleanolic acid (**7**) Glc $\searrow^{3}$ GlcA (OH-3β) Ara $\nearrow_{2}$ Glc (CO_2H-28)	*303*
	Kalopanax saponin L_a $+40.4°$, ^{1}H, ^{13}C, FABMS	22α-Hydroxyhederagenin (**44**) Ara (OH-3β)	*304*
	Kalopanax saponin L_b $+49.3°$, ^{1}H, ^{13}C, FABMS	22α-Hydroxyhederagenin (**44**) Rha-^{2}Ara (OH-3β)	*304*
	Kalopanax saponin L_c $+45.6°$, ^{1}H, ^{13}C, FABMS	22α-Hydroxyhederagenin (**44**) Xyl-^{3}Rha-^{2}Ara (OH-3β)	*304*
Lagenaria breviflora (Cucurbitaceae)	Saponin 260–266°, IR, ^{13}C, FABMS	Oleanolic acid (**7**) Gal (OH-3β) Xyl-^{4}Rha-^{3}Xyl-^{3}Ara (CO_2H-28)	*305*
	Saponin IR, ^{13}C, FABMS	Oleanolic acid (**7**) Gal (OH-3β) Gal-^{4}Rha-^{3}Xyl-^{3}Ara (CO_2H-28)	*305*

Table 1. *(continued)*

Source (1)	Saponin mp, $[\alpha]_D$, spectra recorded (2)	Structure (3)	Ref. (4)
	Saponin 230–232°, IR, ^{1}H, ^{13}C, FABMS	Oleanolic acid (**7**) Gal (OH-3β) Ara-^{6}Gal-^{4}Rha-^{3}Xyl-^{3}Ara (CO_2H-28)	*305*
Lonicera fulvotomentosa (Caprifoliaceae)	Fulvotomentoside A 215–217°, −14.9°, IR, ^{1}H, ^{13}C	Hederagenin (**11**) Xyl-^{3}Rha-^{2}Ara (OH-3β) Glc-^{4}Glc (CO_2H-28)	*306*
L. Japonica	Loniceroside A 210–216°, −28.0°, IR, ^{1}H, ^{13}C, FABMS	Hederagenin (**11**) Ara (OH-3β) Xyl ╲ $^6_2$ Glc (CO_2H-28) Rha ╱	*307*
	Loniceroside B 218–222°, −70.3°, IR, ^{1}H, ^{13}C, FABMS	Hederagenin (**11**) Rha-^{2}Ara (OH-3β) Xyl ╲ $^6_2$ Glc (CO_2H-28) Rha ╱	*307*
Leucas nutans (Labiatae)	Leucasin 190–192°, −12°, IR, ^{1}H, ^{13}C, FABMS	Aglycone (**251**) Glc-^{2}Glc (OH-3β)	*308*
Luffa acutangula (Curcurbitaceae)	Acutoside A 265–270°, +36.5°, ^{1}H, ^{13}C, FABMS	Oleanolic acid (**7**) Glc-^{2}Glc (OH-3β)	*309*
	Acutoside B 225–250°, −18.3°, ^{1}H, ^{13}C, FABMS	Oleanolic acid (**7**) Glc-^{2} Glc (OH-3β) Xyl-^{4}Rha-^{2}Ara (CO_2H-28)	*309*

Acutoside C 220–225°, −15.5°, ^{1}H, ^{13}C, FABMS	Machaerinic acid (**45**) Glc-^{2}Glc (OH-3β) Xyl-^{4}Rha-^{2}Ara (CO_2H-28)	*309*
Acutoside D 260–265°, −21.4°, ^{1}H, ^{13}C, FABMS	Oleanolic acid (**7**) Glc-^{2}Glc (OH-3β) Xyl-^{3}Xyl-^{4}Rha-^{2}Ara (CO_2H-28)	*309*
Acutoside E 246–251°, −14.2°, ^{1}H, ^{13}C, FABMS	Oleanolic acid (**7**) Glc-^{2}Glc (OH-3β) Ara-^{3}Xyl-^{3}Rha-^{2}Ara (CO_2H-28)	*309*
Acutoside F 215–223°, −25.3°, ^{1}H, ^{13}C, FABMS	Oleanolic acid (**7**) Glc-^{2}Glc (OH-3β) Xyl ╲ $^4_3$ Rha-^{2}Ara (CO_2H-28) Xyl ╱	*309*
Acutoside G 250–252°, −22.5°, ^{1}H, ^{13}C, FABMS	Oleanolic acid (**7**) Glc-^{2}Glc (OH-3β) Ara-^{3}Xyl ╲ $^4_3$ Rha-^{2}Ara (CO_2H-28) Xyl ╱	*309*
Acutoside H 235–238°, −53.1°, ^{1}H, ^{13}C, FABMS	Oleanolic acid (**7**) Ara-^{3}GlcA (OH-3β) Xyl-^{3}Xyl ╲ $^4_3$ Rha-^{2}Ara (CO_2H-28) Xyl ╱	*310*
Acutoside I 234–237°, −28.7°, ^{1}H, ^{13}C, FABMS	Oleanolic acid (**7**) Ara-^{3}GlcA (OH-3β) Ara-^{3}Xyl ╲ $^4_3$ Rha-^{2}Ara (CO_2H-28) Xyl ╱	*310*

Table 1. *(continued)*

Source (1)	Saponin mp, $[\alpha]_D$, spectra recorded (2)	Structure (3)	Ref. (4)
L. cylindrica	Lucyoside N 268–270°, −36.1°, IR, ^{1}H, ^{13}C, FABMS	Quillaic acid (**46**) Gal-^{2}GlcA (OH-3β) Xyl \ ^{4} Rha-^{2}Ara (CO_2H-28) Glc / ^{3}	*311*
	Lucyoside P 228–230°, −12.2°, IR, ^{1}H, ^{13}C, FABMS	Gypsogenin (**13**) Gal-^{2}GlcA (OH-3β)	*311*
Lysimachia sikokiana (Primulaceae)	Lysikoianoside I −10.5°, ^{1}H, ^{13}C, FABMS	Protoprimulagenin A (**193**) Xyl-^{2}Glc \ ^{4} Ara (OH-3β) Glc / ^{2}	*312*
Madhuca butyracea (Sapotaceae)	Butyroside A 242–243°, −48.2°, ^{1}H, ^{13}C, 2D, FABMS	Protobassic acid (**37**) Glc (OH-3β) Xyl-^{4}Rha-^{2}Ara (CO_2H-28) | ^{3} Apio (f)	*313*
	Butyroside B 239–242°, −51.8°, ^{1}H, ^{13}C, 2D, FABMS	16α-hydroxyprotobassic acid (**89**) Glc (OH-3β) Apio (f)-^{3}Xyl-^{4}Rha-^{2}Ara (CO_2H-28)	*313*
	Butyroside C 216–220°, −20°, ^{1}H, ^{13}C, FABMS	Protobassic acid (**37**) GlcA (OH-3β) Rha-^{3}Xyl-^{4}Rha-^{2}Ara (CO_2H-28)	*314*
	Butyroside D 213–215°, −53°, ^{1}H, ^{13}C, FABMS	16α-hydroxyprotobassic acid (**89**) GlcA (OH-3β) Apio (f)-^{3}Xyl-^{4}Rha-^{2}Ara (CO_2H-28)	*314*

Maesa lanceolata (Myrsinaceae)	Saponin 1 ^1H, ^{13}C, FABMS	Aglycone (**211**) Rha-^{2}Glc $\searrow$ $^3_2$ GlcA (OH-3β) Glc $\nearrow$	*315*
	Saponin 2 ^1H, ^{13}C, FABMS	Aglycone (**212**) Rha-^{2}Glc $\searrow$ $^3_2$ GlcA (OH-3β) Glc $\nearrow$	*315*
	Saponin 3 ^1H, ^{13}C, FABMS	Aglycone (**213**) Rha-^{2}Glc $\searrow$ $^3_2$ GlcA (OH-3β) Glc $\nearrow$	*315*
	Saponin 4 ^1H, ^{13}C, FABMS	Aglycone (**214**) Rha-^{2}Glc $\searrow$ $^3_2$ GlcA (OH-3β) Glc $\nearrow$	*315*
	Saponin 5 ^1H, ^{13}C, FABMS	Aglycone (**215**) Rha-^{2}Glc $\searrow$ $^3_2$ GlcA (OH-3β) Glc $\nearrow$	*315*
	Saponin 6 ^1H, ^{13}C, FABMS	Aglycone (**216**) Rha-^{2}Glc $\searrow$ $^3_2$ GlcA (OH-3β) Glc $\nearrow$	*315*
Mangifera indica (Anacardiaceae)	Indicoside A 228–230°, +67.35°, IR, ^1H, ^{13}C, FABMS	Aglycone (**251**) Glc $\searrow$ $^3_2$ Ara (OH-3β) Glc $\nearrow$	*316*

Table 1. *(continued)*

Source (1)	Saponin mp, $[\alpha]_D$, spectra recorded (2)	Structure (3)	Ref. (4)
	Indicoside B 242–244°, +63.18°, IR, ^1H, ^{13}C, FABMS	Aglycone (**251**) Glc $\searrow^3$ Ara-(OH-3β) Glc-^{3}Rha $\nearrow_2$	*316*
Margyricarpus setosus (Rosaceae)	+18°, ^1H, ^{13}C, FABMS	Tormentic acid (**178**) Quin (OH-3β)	*317*
	+26°, ^1H, FABMS	Tormentic acid (**178**) Fuc (OH-3β)	*317*
	+11°, ^1H, FABMS	Tormentic acid (**178**) Rha (OH-3β)	*317*
Mazus miquelii (Scrophulariaceae)	Mazusaponin I −8.1°, ^1H, ^{13}C, FABMS	Siaresinolic acid (**67**) Ara (OH-3β) Glc-^{6}Glc (CO_2H-28)	*318*
	Mazusaponin II −8.5°, ^1H, ^{13}C, FABMS	Pomolic acid (**163**) Ara (OH-3β) Glc-^{6}Glc (CO_2H-28)	*318*
	Mazusaponin III −33.2°, ^1H, ^{13}C, FABMS	Siaresinolic acid (**67**) Rha-^{2}Ara (OH-3β) Glc-^{6}Glc (CO_2H-28)	*318*
	Mazusaponin IV −30.5°, ^1H, ^{13}C, FABMS	Pomolic acid (**163**) Rha-^{2}Ara (OH-3β) Glc-^{6}Glc (CO_2H-28)	*318*
Medicago hispida (Leguminosae)	Hispidacin 245–247°, −22.5°, ^1H, ^{13}C, FABMS	Soyasapogenol B (**69**) Rha-^{2}Glc-^{2}GlcA (OH-3β)	*319*

M. polymorpha	Medicago saponin P_1 −20.1°, IR, ^{1}H, ^{13}C, FABMS	Caulophyllogenin (**47**) Rha-^{2}Ara (OH-3β) Glc-^{6}Glc (CO_2H-28)	*320*
	Medicago saponin P_2 −4.1°, IR, ^{1}H, ^{13}C, FABMS	Caulophyllogenin (**47**) Rha-^{2}Ara (OH-3β) Glc (CO_2H-28)	*320*
M. sativa	Saponin ^{1}H, ^{13}C, 2D	Medicagenic acid (**48**) Xyl-^{4}Rha-^{2}Ara (CO_2H-28)	*7*
	Saponin ^{1}H, ^{13}C, 2D	Medicagenic acid (**48**) Glc (OH-3β) Xyl-^{4}Rha-^{2}Ara (CO_2H-28)	*7*
	Saponin ^{1}H, ^{13}C, 2D	Medicagenic acid (**48**) Glc-^{2}Glc (OH-3β) Xyl-^{4}Rha-^{2}Ara (CO_2H-28)	*7*
	Saponin ^{1}H, ^{13}C, 2D	Medicagenic acid (**48**) GlcA (OH-3β) Xyl-^{4}Rha-^{2}Ara (CO_2H-28)	*7*
	Saponin ^{1}H, ^{13}C, FABMS	Soyasapogenol B (**69**) Rha-^{2}Glc-^{2}GlcA (OH-3β)	*7*
	Medicoside L ^{1}H, FABMS	Medicagenic acid (**48**) Rha $\searrow^{3}$ Glc-^{2}Glc (OH-3β) Glc $\nearrow_{2}$ Glc (CO_2H-28)	*321*
	Azahnic acid tridesmoside ^{13}C, 2D, FABMS	Aglycone (**70**) Glc-^{2}Glc-^{2}Glc (OH-3β) Ara (CO_2H-23) Apio (f)-^{3}Xyl-^{4}Rha-^{2}Ara (CO_2H-28)	*322*

Table 1. *(continued)*

Source (1)	Saponin mp, $[\alpha]_D$, spectra recorded (2)	Structure (3)	Ref. (4)
	Medicagenic acid glycoside	Medicagenic acid (**48**) GlcA (OH-3β) Rha-^{2}Ara (CO_2H-28)	*322*
Menyanthes trifoliata (Menyanthaceae)	Menyanthoside 227–230°, −32°, IR, ^1H, ^{13}C, FABMS	Betulinic acid (**248**) Gal-^{4}GlcA (OH-3β) Apio (f)-^{6}Glc (CO_2H-28)	*323*
Mimosa tenuiflora (Mimosaceae)	Mimonoside A 243.9–245.2°, −29.2°, IR, ^1H, ^{13}C, FABMS	Oleanolic acid (**7**) Xyl —^{4} Glc (OH-3β) Ara —^{4} Xyl —^{2} Glc Rha-^{2}Glc —^{3} Xyl Rha (CO_2H-28)	*324*
	Mimonoside B 237.4–240.2°, −28.4°, IR, ^1H, ^{13}C, FABMS	Oleanolic acid (**7**) Xyl —^{4} Glc (OH-3β) Ara —^{4} Xyl —^{2} Glc Rha-^{2}Glc —^{3} Xyl	*324*
	Mimonoside C −27.2°, ^{13}C, 2D	Machaerinic acid (**45**) Xyl —^{4} Glc (OH-3β) Ara —^{4} Xyl —^{2} Glc Rha-^{2}Glc —^{3} Xyl Rha (CO_2H-28)	*325*

Mimusops elengi (Sapotaceae)	−17.9°, IR, ^{1}H, ^{13}C, 2D, FABMS	Protobassic acid (**37**) Glc (OH-3β) Rha-^{3}Xyl⟍$^{4}_{3}$ Rha-^{2}Ara (CO_2H-28) Rha⟋	*326*
	+30.7°, IR, ^{1}H, ^{13}C, 2D, FABMS	Protobassic acid (**37**) Glc-^{3}Glc (OH-3β) Rha-^{3}Xyl-^{3}Rha-^{2}Ara (CO_2H-28)	*326*
	Mimusopside A 245–248°, −20.9°, IR, ^{1}H, ^{13}C, LSIMS	Protobassic acid (**37**) Glc (OH-3β) Rha-^{2}Ara (CO_2H-28)	*327*
	Mimusopside B 214–216°, −31.8°, IR, ^{1}H, ^{13}C, LSIMS	16α-Hydroxyprotobassic acid (**89**) Glc (OH-3β) Rha-^{2}Ara (CO_2H-28)	*327*
	Saponin 4 −37.7°, ^{1}H, ^{13}C, LSIMS	Protobassic acid (**37**) GlcA (OH-3β) Rha-^{3}Xyl⟍$^{4}_{3}$ Rha-^{2}Ara (CO_2H-28) Rha⟋	*328*
M. hexandra	Saponin 5 −49.4°, ^{1}H, ^{13}C, FABMS	16α-Hydroxyprotobassic acid (**89**) GlcA (OH-3β) Rha-^{3}Xyl-^{4}Rha-^{2}Ara (CO_2H-28)	*328*
	Saponin 6 −31.5°, ^{1}H, ^{13}C, FABMS	Protobassic acid (**37**) Glc-^{3}Glc (OH-3β) Rha-^{3}Xyl-^{4}Rha-^{2}Ara (CO_2H-28)	*328*
Mussaenda pubescens (Rubiaceae)	Mussaendoside A UV, IR, ^{1}H, ^{13}C, MS	Heinsiagenin A (**312**) Glc (OH-3β)	*329*
	Mussaendoside B UV, IR, ^{13}C, MS	Heinsiagenin A (**312**) Xyl (OH-3β)	*329*
	Mussaendoside C UV, IR, ^{1}H, ^{13}C, MS	Heinsiagenin A (**312**)) Glc-^{2}Xyl (OH-3β)	*329*

Table 1. *(continued)*

Source (1)	Saponin mp, $[\alpha]_D$, spectra recorded (2)	Structure (3)	Ref. (4)
	Mussaendoside M 178°, +20.79°, UV, ^1H, ^{13}C, 2D, FABMS	Heinsiagenin A (**312**) Rha —4 Xyl (OH-3β) Rha-^{2}Glc —2	*6*
	Mussaendoside N 194°, +19.63°, UV, ^1H, ^{13}C, FABMS	Heinsiagenin A (**312**) Rha —4 Xyl (OH-3β) Glc —6 Glc —2 Rha —2	*6*
	Saponin O +2.4°, UV, ^1H, ^{13}C, 2D, FABMS	Heinsiagenin A (**312**) Rha —4 Glc (OH-3β) Rha-^{2}Glc —2	*330*
	Saponin P +7.0°, UV, ^1H, ^{13}C, 2D, FABMS	Aglycone (**313**) Rha —4 Glc (OH-3β) Rha-^{2}Glc —2	*330*
	Saponin Q +10.7°, UV, ^1H, ^{13}C, 2D, FABMS	Aglycone (**319**) Rha —4 Glc (OH-3β) Rha-^{2}Glc —2	*330*
	Mussaendoside R +4.6°, ^1H, ^{13}C, FABMS	Pomolic acid (**163**) Glc (OH-3β) Glc (CO_2H-28)	*331*

	Mussaendoside S +53.3°, ^1H, ^{13}C, FABMS	Aglycone (**151**) Glc (OH-3β) Glc (CO_2H-28)	*331*
	Mussaendoside G +13.6°, UV, IR, ^1H, ^{13}C, FABMS	Heinsiagenin A (**312**) Rha ⟍ 4 Glc ⟍ 6 / 2 Glc ⟋ 2 → 4 Glc (OH-3β) Rha ⟋ 2	*332*
	Mussaendoside K +36.0°, IR, ^1H, ^{13}C, FABMS	Aglycone (**153**) Glc (CO_2H-24) Glc (CO_2H-28)	*332*
Myrsine salicina (Myrsinaceae)	Saponin 1 268–275°, −10°, IR, ^1H, ^{13}C, 2D, FABMS	Protoprimulagenin A (**193**) Rha-^{2}Glc ⟍ 4 Ara (OH-3β) Glc ⟋ 2	*333*
	Saponin 2 272–275°, −7.5°, IR, ^1H, ^{13}C, 2D, FABMS	Protoprimulagenin A (**193**) Xyl-^{2}Glc ⟍ 4 Ara (OH-3β) Glc ⟋ 2	*333*
	Saponin 3 223–225°, −37.5°, IR, ^1H, ^{13}C, 2D, FABMS	Aglycone (**197**) Rha-^{2}Glc ⟍ 4 Ara (OH-3β) Glc ⟋ 2	*333*
	Saponin 4 229–231°, −35.6°, IR, ^1H, ^{13}C, 2D,	Aglycone (**197**) Xyl-^{2}Glc ⟍ 4 Ara (OH-3β) Glc ⟋ 2	*333*
	Saponin 7 236–239°, −13.1°, IR, ^1H, ^{13}C, 2D, FABMS	Aglycone (**189**) Rha-^{2}Glc ⟍ 4 Ara (OH-3β) Glc ⟋ 2	*333*

Table 1. *(continued)*

Source (1)	Saponin mp,$[\alpha]_D$, spectra recorded (2)	Structure (3)	Ref. (4)
	Saponin 8 >250°, −4.7°, IR, ^1H, ^{13}C, 2D, FABMS	Aglycone (**189**) Xyl-^{2}Glc $\searrow^4$ Ara (OH-3β) Glc $\nearrow_2$	*333*
Nauclea diderrichii (Rubiaceae)	Saponin 2 ^{13}C, FABMS	Quinovic acid (**171**) Rha (OH-3β) Glc (CO_2H-28)	*334*
	Saponin 3 ^{13}C, FABMS	Quinovic acid (**171**) Glc-^{2}Glc (OH-3β)	*334*
Neoalsomitra integrifoliola (Cucurbitaceae)	Neoalsoside A_2 −6.9°, ^1H, ^{13}C, FABMS	Aglycone (**271**) Glc (OH-3β)	*335*
	Neoalsoside A_3 −3.9°, ^1H, ^{13}C, FABMS	Aglycone (**271**) Rha-^{2}Glc (OH-3β)	*335*
	Neoalsoside A_4 +3.0°, ^1H, ^{13}C, FABMS	Aglycone (**271**) Rha-^{3}Glc (OH-3β)	*335*
	Neoalsoside A_5 +6.8°, ^1H, ^{13}C, FABMS	Aglycone (**271**) Glc $\searrow^3$ Glc (OH-3β) Rha $\nearrow_2$	*335*

Neoalsoside C1 $-3.8°$, ^1H, ^{13}C, FABMS	Aglycone (**287**) Rha ╲ ^{3} Glc (OH-3β) Rha ╱ $_2$	*335*
Neoalsoside C2 $-12.5°$, ^1H, ^{13}C, FABMS	Aglycone (**287**) Glc ╲ ^{3} Glc (OH-3β) Rha ╱ $_2$	*335*
Neoalsoside D1 $-18.4°$, ^1H, ^{13}C, FABMS	Aglycone (**288**) Rha ╲ ^{3} Glc (OH-3β) Rha ╱ $_2$	*335*
Neoalsoside E1 259–261°, $+14.1°$, ^1H, ^{13}C, FABMS	Aglycone (**297**) Rha ╲ ^{3} Glc (OH-3β) Rha ╱ $_2$	*335*
Neoalsoside F1 $-15.6°$, ^1H, ^{13}C, FABMS	Aglycone (**291**) Rha ╲ ^{3} Glc (OH-3β) Rha ╱ $_2$	*335*
Neoalsoside G1 $-3.9°$, ^1H, ^{13}C, FABMS	Aglycone (**289**) Rha ╲ ^{3} Glc (OH-3β) Rha ╱ $_2$	*335*
Neoalsoside H1 209–211°, $+47.6°$, ^1H, ^{13}C, FABMS	Aglycone (**290**) Rha ╲ ^{3} Glc (OH-3β) Rha ╱ $_2$	*335*

Table 1. *(continued)*

Source (1)	Saponin mp, $[\alpha]_D$, spectra recorded (2)	Structure (3)	Ref. (4)
	Neoalsoside I_1 $-19.3°$, ^1H, ^{13}C, FABMS	Aglycone (**292**) Rha—3 / Rha—2 Glc (OH-3β)	*336*
	Neoalsoside I_2 $-17.7°$, ^1H, ^{13}C, FABMS	Aglycone (**292**) Rha—3 / Rha—2 Glc (OH-3β) Glc (OH-24R)	*336*
	Neoalsoside J_1 $-13.0°$, ^1H, ^{13}C, FABMS	Aglycone (**293**) Rha—3 / Rha—2 Glc (OH-3β)	*336*
	Neoalsoside K_1 $-18.7°$, ^1H, ^{13}C, FABMS	Aglycone (**294**) Rha—3 / Rha—2 Glc (OH-3β)	*336*
	Neoalsoside L_1 $-11.4°$, ^1H, ^{13}C, FABMS	Aglycone (**295**) Rha—3 / Rha—2 Glc (OH-3β)	*336*
	Neoalsoside M_1 $0°$, ^1H, ^{13}C, FABMS	Aglycone (**296**) Rha-^{2}Glc (OH-3β)	*336*

	Neoalsoside M_2 −6.7°, ^{1}H, ^{13}C, FABMS	Aglycone (**296**) Rha ╲ $^{3}_{2}$ Glc (OH-3β) Rha ╱	*336*
	Neoalsoside M_3 −4.1°, ^{1}H, ^{13}C, FABMS	Aglycone (**296**) Rha ╲ $^{3}_{2}$ Glc (OH-3β) Rha ╱ Glc (OH-23S)	*336*
	Neoalsoside N_1 −24.2°, ^{1}H, ^{13}C, FABMS	Aglycone (**330**) Rha ╲ $^{3}_{2}$ Glc (OH-3β) Rha ╱	*336*
	Neoalsoside O_1 −22.4°, ^{1}H, ^{13}C, FABMS	Aglycone (**329**) Rha-^{2}Glc (OH-3β)	*336*
	Neoalsoside O_2 −25.6°, ^{1}H, ^{13}C, FABMS	Aglycone (**329**) Rha ╲ $^{3}_{2}$ Glc (OH-3β) Rha ╱	*336*
	Neoalsoside A 225–228°, −30.5°, ^{1}H, ^{13}C, FABMS	Aglycone (**271**) Rha ╲ $^{3}_{2}$ Glc (OH-3β) Rha ╱	*337*
Nothopanax *davidii* (Araliaceae)	Saponin I 219–224°, −6.89°, IR, ^{1}H, ^{13}C, FABMS	Aglycone (**38**) Ara-^{2}Ara (OH-3β) Rha-^{4}Glc-^{6}Glc (CO_2H-28)	*338*
	Yiyeliangwanoside IX 228–230°, −9.43°, IR, ^{1}H, ^{13}C, FABMS	Aglycone (**38**) (OAc-4′) Ara (OH-3β) Rha-^{4}Glc-^{6}Glc (CO_2H-28)	*339*

Table 1. *(continued)*

Source (1)	Saponin mp, $[\alpha]_D$, spectra recorded (2)	Structure (3)	Ref. (4)
	Yiyeliangwanoside X 210–212°, −3.9°, IR, ^1H, ^{13}C, FABMS	Aglycone (**38**) (OAc-2′) Ara (OH-3β) Rha-^{4}Glc-^{6}Glc (CO_2H-28)	*339*
	Yiyeliangwanoside X1 219–224°, −21.62°, IR, ^1H, ^{13}C, FABMS	Aglycone (**38**) Ara (OH-3β) Rha-^{4}Glc-^{6}Glc (CO_2H-28)	*339*
Opilia celtidifolia (Opiliaceae)	Saponin ^{13}C, FABMS	Hederagenin (**11**) Rha-^{3}GlcA (OH-3β) Glc (CO_2H-28)	*340*
Oxytropis bicolor (Leguminosae)	Saponin 273–275°, +3.0°, ^1H, ^{13}C, FABMS	Aglycone (**240**) Glc-^{2}Glc (OH-3β)	*341*
O. glabra	Saponin 1 230–232°, −6.0°, ^1H, ^{13}C, FABMS	Oxytrogenol (**99**) Rha-^{2}Glc-^{4}GlcA (OH-3β)	*342*
	Saponin 2 255–258°, −5.0°, ^{13}C, FABMS	Aglycone (**101**) Rha-^{2}Glc-^{4}GlcA (OH-3β)	*342*
	Saponin 3 220–223°, −15.5°, ^{13}C, FABMS	Soyasaponin E (**49**) Rha-^{2}Glc-^{4}GlcA (OH-3β)	*342*
	Saponin 4 235–238°, +3.18°, ^{13}C, FABMS	Soyasapogenol B (**69**) Rha-^{2}Glc—^{4} GlcA (OH-3β) Glc—^{2}	*342*

	Saponin I	Aglycone (**224**) Glc-^{2}GlcA (OH-3β)	*343*
	Saponin II	Aglycone (**225**) Glc-^{2}GlcA (OH-3β)	*343*
	Saponin III	Aglycone (**231**) Mann-^{2}Glc-^{4}GlcA (OH-3β)	*343*
Oxytropis species (Leguminosae)	Saponin 212–214°, +2.0°, ^{13}C, FABMS	Aglycone (**240**) Glc-^{2}Glc (OH-3β) Rha (OH-25)	*344*
Paliurus ramosissimus (Rhamnaceae)	260–262°, +120°, IR, ^{1}H, ^{13}C, FABMS	Ceanothic acid (**309**) Glc (CO_2H-28)	*345*
	270–272°, +26°, IR, ^{1}H, ^{13}C, FABMS	Isoceanothic acid (**310**) Glc (CO_2H-28)	*345*
Panax ginseng (Araliaceae)	Saponin	Aglycone (**278**) Rha-^{2}Glc (OH-6β)	*346*
	Koryoginsenoside R_1 +39.5°, ^{1}H, ^{13}C, FABMS	Aglycone (**268**) (Butenoyl-6′) Glc (OH-6α) Glc (OH-20β)	*347*
	Koryoginsenoside R_2 +12°, ^{1}H, ^{13}C, FABMS	Aglycone (**298**) Glc-^{2}Glc (OH-3β) Glc-^{6}Glc (OH-20β)	*347*
P. notoginseng	Notoginsenoside R_7	Panaxadiol (**279**) Glc (OH-3β)	*348*
	Notoginsenoside R_8 +29°, ^{1}H, ^{13}C, FABMS	Aglycone (**265**) Glc (OH-6α)	*349*

Table 1. *(continued)*

Source (1)	Saponin mp,$[\alpha]_D$, spectra recorded (2)	Structure (3)	Ref. (4)
	Notoginsenoside R_9 +27°, ^1H, ^{13}C, FABMS	Aglycone (**299**) Glc (OH-6α)	*349*
P. pseudoginseng	Chikusetsusaponin VI −10.3°, ^1H, ^{13}C, FABMS	20(S)-Protopanaxadiol (**259**) Xyl ╲ $^{6}_{2}$ Glc (OH-3β) Glc ╱ Glc-^{6}Glc (OH-20β)	*350*
	Pseudoginsenoside RI_2 197–200°, +8°, IR, ^1H, ^{13}C, EIMS	(**341**)	*351*
	Pseudoginsenoside RI_3 290–295°, +6.5°, IR, ^1H, ^{13}C, FABMS	Oleanolic acid (**7**) GlcA-^{2}GlcA-^{6}GlC (OH-3β) Xyl (CO_2H-28)	*352*
P. vietnamensis	Vina-ginsenoside R_1 −23.1°, ^1H, ^{13}C, FABMS	Aglycone (**272**) Rha-^{2}(OAc-6′) Glc (OH-6α)	*353*
	Vina-ginsenoside R_2 186–189°, −17.4°, ^1H, ^{13}C, FABMS	Aglycone (**272**) Xyl-^{2}(OAc-6′) Glc (OH-6α)	*353*
	Vina-ginsenoside R_3 254–256°, −15.0°, ^1H, ^{13}C, FABMS	Aglycone (**263**) Glc-^{2}Glc (OH-3β) Glc (OH-20S)	*354*
	Vina-ginsenoside R_4 +28.4°, ^1H, ^{13}C, FABMS	Aglycone (**268**) Glc-^{2}Glc (OH-3β) Glc (OH-20 S)	*354*

Vina-ginsenoside R_5 +38°, ^1H, ^{13}C, FABMS	Aglycone (**272**) Glc-^{4}Xyl-^{2}Glc (OH-6α)	*354*
Vina-ginsenoside R_6 +20.0°, ^1H, ^{13}C, FABMS	Aglycone (**272**) Glc $\searrow^{6}$ Glc (OH-6α) Xyl $\nearrow_{2}$	*354*
Vina-ginsenoside R_7 +17.8°, ^1H, ^{13}C, FABMS	20(S)-Protopanaxadiol (**268**) Xyl-^{2}Glc-^{2}Glc (OH-3β) Glc (OH-20S)	*354*
Vina-ginsenoside R_8 +14.0°, ^1H, ^{13}C, FABMS	Aglycone (**280**) Glc-^{2}Glc (OH-3β) Glc (OH-20S)	*354*
Vina-ginsenoside R_9 +10.5°, ^1H, ^{13}C, FABMS	Aglycone (**261**) Glc-^{2}Glc (OH-3β) Glc (OH-20S)	*354*
Vina-ginsenoside R_{10} 257–259°, +10.5°, ^1H, ^{13}C, 2D, FABMS	Aglycone (**274**) Glc (OH-6α)	*355*
Vina-ginsenoside R_{11} 251–253°, +3.8°, ^1H, ^{13}C, 2D, FABMS	Aglycone (**274**) Xyl-^{2}Glc (OH-6α)	*355*
Vina-ginsenoside R_{12} +29.3°, ^1H, ^{13}C, FABMS	Aglycone (**275**) Glc (OH-6α)	*355*
Vina-ginsenoside R_{13} +2.2°, ^1H, ^{13}C, FABMS	Aglycone (**281**) Glc-^{2}Glc (OH-3β) Glc (OH-20S)	*355*
Vina-ginsenoside R_{14} −13.7°, ^1H, ^{13}C, 2D, FABMS	Aglycone (**276**) Xyl-^{2}Glc (OH-6α)	*355*

Table 1. *(continued)*

Source (1)	Saponin mp, $[\alpha]_D$, spectra recorded (2)	Structure (3)	Ref. (4)
Periandra dulcis (Leguminosae)	Periandradulcin A 220–225°, −55.0°, ^{1}H, ^{13}C, 2D, SIMS	Aglycone (**227**) Rha-^{2}Xyl-^{2}GlcA (OH-3β)	*356*
	Periandradulcin B 225–227°, +12.0°, ^{1}H, ^{13}C, 2D, FABMS	Aglycone (**111**) Rha-^{2}Xyl-^{2}GlcA (OH-3β)	*356*
	Periandradulcin C 205–210°, −17.4°, ^{1}H, ^{13}C, EIMS	Aglycone (**232**) Rha-^{2}Glc-^{2}GlcA (OH-3β)	*356*
Petersianthus macrocarpus (Lecythidaceae)	Petersaponin I UV, IR, ^{1}H, ^{13}C, 2D, FABMS	Aglycone (**34**) Gal-$^{3}_{2}$ (Et-ester-6′) GlcA (OH-3β) Gal ⁄ [3-(3-tigloyloxynilic acid)-4-tigloyloxy] Ara (OH-21β)	*357*
	Petersaponin II UV, IR, ^{1}H, ^{13}C, 2D, EIMS	Aglycone (**93**) Gal-^{2}Gal-^{3}GlcA (OH-3β) Rha (CO_2H-28)	*357*
Phaseolus vulgaris (Leguminosae)	Phaseoluside A	Soyasapogenol B (**69**) Glc ╲ $^{4}_{3}$ Glc (OH-3β) Glc ╱	*358*
P. cocineus	Soyasaponin α_a UV, ^{1}H, ^{13}C, FABMS	Aglycone (**75**) Glc-^{2}Ara-^{2}GlcA (OH-3β)	*359*
Phytolacca acinosa (Phytolaccaceae)	Esculentoside S	Phytolaccagenin (**116**) Xyl (OH-3β) Glc (CO_2H-28)	*360*

P. bogotensis	^{1}H, ^{13}C, FABMS	Aglycone (**115**) Gal-^{3}Glc (OH-3β) Glc (CO_2H-28)	*361*
	^{1}H, ^{13}C, FABMS	Aglycone (**115**) Rha-^{2}Glc-^{2}Glc (OH-3β) Glc (CO_2H-28)	*361*
P. dodecandra	Monodesmosidic saponin 1 +19.4°, ^{1}H, ^{13}C, 2D, LSIMS	Oleanolic acid (**7**) Gal $\searrow$ $^{3}_{2}$ Glc (OH-3β) Rha $\nearrow$	*47*
	Monodesmosidic saponin 2 +12.3°, ^{1}H, ^{13}C, 2D, SIMS	Oleanolic acid (**7**) Xyl $\searrow$ $^{6}_{3}$ Glc (OH-3β) Glc $\nearrow$	*47*
	Monodesmosidic saponin I +33.5°, ^{1}H, ^{13}C, 2D, LSIMS	2β-Hydroxyoleanolic acid (**16**) Glc $\searrow$ $^{4}_{2}$ Glc (OH-3β) Glc $\nearrow$	*362*
	Monodesmosidic saponin II +10.5°, ^{1}H, ^{13}C, 2D, FABMS	2β-Hydroxyoleanolic acid (**16**) Rha-^{2}Gal-^{3}Glc (OH-3β)	*362*
	Monodesmosidic saponin III +12.2°, ^{1}H, ^{13}C, 2D, FABMS	2β-Hydroxyoleanolic acid (**16**) Gal-^{3}Glc (OH-3β)	*362*
P. esculenta	Esculentoside G 215–217°, UV, IR, ^{1}H, ^{13}C, 2D, FABMS	Aglycone (**223**) Glc-^{4}Xyl (OH-3β) Glc (CO_2H-28)	*363*
	Esculentoside I	Phytolaccagenin (**116**) Glc-^{4}Glc (OH-3β) Glc (CO_2H-28)	*364*

Table 1. *(continued)*

Source (1)	Saponin mp, $[\alpha]_D$, spectra recorded (2)	Structure (3)	Ref. (4)
	Esculentoside J	Aglycone (**157**) Glc-^{4}Xyl (OH-3β) Glc (CO_2H-28)	*365*
	Esculentoside K 212-213°, IR, ^{1}H, ^{13}C, MS	Phytolaccagenic acid (**36**) Glc-^{4}Xyl-^{4}Glc (OH-3β) Glc (CO_2H-28)	*366*
	Esculentoside L 208-210°, IR, ^{1}H, ^{13}C	Phytolaccagenic acid (**36**) Glc-^{4}Xyl (OH-3β) Glc (CO_2H-28)	*366*
	Esculentoside N	Phytolaccagenin (**116**) Glc-^{4}Xyl-^{4}Glc (OH-3β) Glc (CO_2H-28)	*364*
	Esculentoside O	Aglycone (**73**) Xyl (OH-3β)	*367*
	Esculentoside P	Aglycone (**74**) Glc (OH-3β)	*367*
	Esculentoside Q	Aglycone (**74**) Glc-^{4}Xyl-^{4}Glc (OH-3β)	*367*
P. rivinoides	Saponin ^{1}H, ^{13}C, FABMS	Aglycone (**115**) Gal —^{4} Glc (OH-3β) Glc —^{3} (to same Glc) Glc (CO_2H-28)	*361*

	Saponin ^1H, ^{13}C, FABMS	Aglycone (**115**) Glc-^{3}Gal-^{3}Glc (OH-3β) Glc (CO_2H-28)	*361*
	Saponin ^1H, ^{13}C, FABMS	Aglycone (**115**) Gal ╲ $^{4}_{2}$ Glc (OH-3β) Glc ╱	*361*
Pisum sativum (Leguminosae)	Chromosaponin I 210–212°, UV, IR, ^1H, ^{13}C, MS	Aglycone (**75**) Rha-^{2}Gal-^{2}GlcA (OH-3β)	*368*
Polemonium caeruleum (Polemoniaceae)	Polemonium saponin 1 2D, FABMS	Aglycone (**77**) Ara ╲ $^{3}_{2}$ GlcA (OH-3β) Gal ╱	*369*
	Polemonium saponin 2 2D, FABMS	Aglycone (**78**)) Ara ╲ $^{3}_{2}$ GlcA (OH-3β) Gal ╱	*369*
	Polemonium saponin 3 2D, FABMS	Aglycone (**79**) Ara ╲ $^{3}_{2}$ GlcA (OH-3β) Gal ╱	*369*
Polycarpone loeflingiae (Caryophyllaceae)	Polycarponoside A 172–174°, −28.6°, IR, ^1H, ^{13}C, FABMS	Saikogenin D (**222**) Ara (f)- $^{4}_{2}$ Ara-^{4}Glc (OH-3β) Glc ╱	*370*
Polygala japonica (Polypodiaceae)	Polygalasaponin I +25.7°, ^1H, ^{13}C, FABMS	Bayogenin (**25**) Glc (OH-3β) Glc (CO_2H-28)	*371*
	Polygalasaponin II 0°, ^1H, ^{13}C, FABMS	Bayogenin (**25**) Glc (OH-3β) Rha-^{2}Glc (CO_2H-28)	*371*

Table 1. *(continued)*

Source (1)	Saponin mp,$[\alpha]_D$, spectra recorded (2)	Structure (3)	Ref. (4)
	Polygalasaponin III $-11.5°$, ^1H, ^{13}C, FABMS	Bayogenin (**25**) Glc (OH-3β) Apio (f) ╲ $^3_2$ Glc (CO_2H-28) Rha ╱	*371*
	Polygalasaponin IV $-10.8°$, ^1H, ^{13}C, FABMS	Bayogenin (**25**) Glc (OH-3β) Apio (f) ╲ $^3_2$ Glc (CO_2H-28)) Xyl-^{4}Rha ╱	*371*
	Polygalasaponin V $-16.7°$, ^1H, ^{13}C, FABMS	Bayogenin (**25**) Glc (OH-3β) Xyl ╲ $^4_3$ Rha-^{2}Glc (CO_2H-28) Apio (f) ╱	*371*
	Polygalasaponin VI $+28.3°$, ^1H, ^{13}C, FABMS	Bayogenin (**25**) Glc-^{2}Glc (OH-3β) Glc (CO_2H-28)	*371*
	Polygalasaponin VII $+1.2°$, ^1H, ^{13}C, FABMS	Bayogenin (**25**) Glc-^{2}Glc (OH-3β) Rha-^{2}Glc (CO_2H-28)	*371*

Polygalasaponin VIII +10.6°, ^{1}H, ^{13}C, FABMS	Bayogenin (**25**) Glc-^{2}Glc (OH-3β) Apio (f) $\searrow^{3}$ $\nearrow_{2}$ Glc (CO_2H-28) Rha	*371*
Polygalasaponin IX −1.3°, ^{1}H, ^{13}C, FABMS	Bayogenin (**25**) Glc-^{2}Glc (OH-3β) Xyl-^{4}Rha-^{2}Glc (CO_2H-28)	*371*
Polygalasaponin X +17.2°, ^{1}H, ^{13}C, FABMS	Bayogenin (**25**) Glc-^{2}Glc (OH-3β) Apio (f) $\searrow^{3}$ $\nearrow_{2}$ Glc (CO_2H-28) Xyl-^{4}Rha	*371*
Polygalasaponin XI +30.6°, ^{1}H, ^{13}C, FABMS	Bayogenin (**25**) Glc-^{2}Glc (OH-3β) Glc-^{2}Glc (CO_2H-28)	*372*
Polygalasaponin XII +39.6°, ^{1}H, ^{13}C, FABMS	Aglycone (**19**) Glc (OH-3β) Glc (CO_2H-28)	*372*
Polygalasaponin XIII +69.6°, ^{1}H, ^{13}C, FABMS	Aglycone (**19**) Glc-^{2}Glc (OH-3β)	*372*
Polygalasaponin XIV +48.6°, ^{1}H, ^{13}C, FABMS	Aglycone (**19**) Glc-^{2}Glc (OH-3β) Glc (CO_2H-28)	*372*
Polygalasaponin XV +15.5°, ^{1}H, ^{13}C, FABMS	Aglycone (**19**) Glc-^{2}Glc (OH-3β) Glc-^{2}Glc (CO_2H-28)	*372*
Polygalasaponin XVI +31.9°, ^{1}H, ^{13}C, FABMS	Aglycone (**19**) Glc-^{2}Glc (OH-3β) Xyl-^{4}Glc-^{2}Glc (CO_2H-28)	*372*

Table 1. *(continued)*

Source (1)	Saponin mp, $[\alpha]_D$, spectra recorded (2)	Structure (3)	Ref. (4)
	Polygalasaponin XVII +12.0°, ^1H, ^{13}C, FABMS	Aglycone (**19**) Glc-^{2}Glc (OH-3β) Rha-^{2}Glc (CO_2H-28)	*372*
	Polygalasaponin XVIII +0.8°, ^1H, ^{13}C, FABMS	Aglycone (**19**) Glc-^{2}Glc (OH-3β) Xyl-^{4}Rha-^{2}Glc (CO_2H-28)	*372*
	Polygalasaponin XIX −10.6°, ^1H, ^{13}C, FABMS	Aglycone (**19**) Glc-^{2}Glc (OH-3β) Apio (f) ╲ $^3_2$ Glc (CO_2H-28) Xyl-^{4}Rha ╱	*372*
P. reini	Reinioside A +25.2°, ^1H, ^{13}C, FABMS	Aglycone (**154**) Glc-^{2}Glc (OH-3β)	*373*
	Reinioside B +3.6°, ^1H, ^{13}C, FABMS	Aglycone (**52**) Glc-^{2}Glc (OH-3β) Xyl-^{4}Rha-^{2} (OAc-4′) Fuc (CO_2H-28)	*373*
	Reinioside C +10.4°, ^1H, ^{13}C, FABMS	Aglycone (**52**) Glc-^{2}Glc (OH-3β) Xyl-^{4}Rha-^{2} (OAc-3′, 4′) Fuc (CO_2H-28)	*373*
	Reinioside D −5.4°, ^1H, ^{13}C, FABMS	Aglycone (**52**) Glc-^{2}Glc (OH-3β) Xyl ╲ $^4_3$ Rha-^{2} (OAc-3′, 4′) Fuc (CO_2H-28) Apio (f) ╱	*373*

	Reinioside E −4.5°, ^{1}H, ^{13}C, FABMS	Aglycone (**52**) Xyl ⟍ Glc-^{2}Glc (OH-3β) $^{4}_{3}$ Rha-^{2}(OAc-3′,4′) Fuc (CO_2H-28) (OAc-5′) Apio (f) ⟋	*373*
	Reinioside F +5.2°, ^{1}H, ^{13}C, FABMS	Aglycone (**52**) Glc-^{2}Glc (OH-3β) Gal-^{4}Xyl-^{4}Rha-^{2} (OAc-3′,4′) Fuc (CO_2H-28)	*373*
Polyscias dichroostachya (Araliaceae)	Saponin 1 220–229°, ^{13}C, CIMS	Hederagenin (**11**) Ara (OH-3β)	*374*
	Saponin 2 244–250°, ^{13}C, CIMS	Hederagenin (**11**) Rha-^{2}Ara (OH-3β)	*374*
	Saponin 3 265–271°, ^{13}C, CIMS	Hederagenin (**11**) Glc-^{2}Ara (OH-3β)	*374*
	Saponin 4 204–208°, ^{13}C, CIMS	Hederagenin (**11**) Rha-^{2}Ara (OH-3β) Glc (CO_2H-28)	*374*
P. scutellaria	Saponin C 260°, ^{1}H, ^{13}C, 2D, FABMS	Oleanolic acid (**7**) Glc-^{3}GlcA (OH-3β) Glc (CO_2H-28)	*375*
	Polysciasaponin P_2 ^{13}C, FABMS	Oleanolic acid (**7**) Glc-^{4}Glc-^{2}GlcA (OH-3β)	*376*
	Polysciasaponin P_5 ^{13}C, FABMS	Oleanolic acid (**7**) Glc-^{2}GlcA (OH-3β)	*376*
Potentilla tormentilla (Rosaceae)	Saponin 1 +7°, IR, ^{1}H, ^{13}C, FABMS	Aglycone (**177**) Glc (CO_2H-28)	*377*
	Saponin 2 +12°, IR, ^{1}H, ^{13}C, FABMS	Aglycone (**173**) Glc (CO_2H-28)	*377*

Table 1. *(continued)*

Source (1)	Saponin mp, $[\alpha]_D$, spectra recorded (2)	Structure (3)	Ref. (4)
	Saponin 3 +18°, IR, ^{1}H, ^{13}C, FABMS	Aglycone (**118**) Glc (CO_2H-28)	*377*
Primula denticulata (Primulaceae)	Denticin 201°, −4.0°, UV, IR, ^{1}H, ^{13}C, FABMS	Cyclamiretin A (**196**) Glc ╲ $^4_2$ Glc ╲ Xyl ╱ $^4_2$ Ara (OH-3β) Glc ╱	*378*
	Denticulatin 235°, +85.47°, UV, IR, ^{1}H, ^{13}C, FABMS	Cyclamiretin A (**196**) Glc-^{2}Glc ╲ $^4_2$ Glc ╲ Xyl ╱ $^4_2$ Ara (OH-3β) Glc ╱	*378*
P. macrophylla	Macrophyllicin 313–314°, −22°, IR, ^{1}H, ^{13}C, 2D, FABMS	Aglycone (**81**) Rha-^{2}Glc-^{2}Gal-^{2}GlcA (OH-3β)	*379*
	Macrophyllicinin 314–315°, −9°, IR, ^{1}H, ^{13}C, FABMS	Aglycone (**40**) Rha-^{2}Glc-^{2}Gal-^{2}GlcA (OH-3β)	*380*
P. veris	Priverosaponin B22 acetate IR, ^{1}H, ^{13}C, 2D, FABMS	Aglycone (**194**) Rha-^{2}Gal ╲ $^3_2$ GlcA (OH-3β) Glc ╱	*381*

	Priverosaponin B IR, ^{1}H, ^{13}C, 2D, FABMS	Priverogenin B (**204**) Rha-^{2}Gal—^{3} GlcA (OH-3β) Glc—^{2}	*381*
	Primulasaponin −28.4°, IR, ^{1}H, ^{13}C, FABMS	Aglycone (**195**) Rha-^{2}Gal—^{3} GlcA (OH-3β) Glc—^{2}	*381*
Pterocephalus bretschneidri (Dipsacaceae)	Bretschnoside A 216–218°, −39.64°, IR, ^{1}H, ^{13}C, FABMS	Oleanolic acid (**7**) Rha-^{3}Xyl-^{3}Rha-^{2}Xyl (OH-3β) Glc-^{6}Glc (CO_2H-28)	*382*
	Bretschnoside B 209–212°, −26.67°, IR, ^{1}H, ^{13}C, FABMS	Oleanolic acid (**7**) Glc-^{3}Xyl-^{3}Rha-^{2}Xyl (OH-3β) Glc-^{6}Glc (CO_2H-28)	*382*
P. hookeri	Hookeroside A −22.5°, ^{1}H, ^{13}C, FABMS	Oleanolic acid (**7**) Glc-^{4}Xyl-^{3}Rha-^{2}Xyl (OH-3β) Glc-^{6}Glc (CO_2H-28)	*383*
	Hookeroside B −27.14°, ^{1}H, ^{13}C, FABMS	Oleanolic acid (**7**) Xyl-^{4}Glc-^{4}Xyl-^{3}Rha-^{2}Xyl (OH-3β)) Glc-^{6}Glc (CO_2H-28)	*383*
	Hookeroside C −95.25°, ^{1}H, ^{13}C, FABMS	Oleanolic acid (**7**) Rha-^{2}Xyl-^{4}Glc-^{4}Xyl-^{3}Rha-^{2}Xyl (OH-3β) Glc-^{6}Glc (CO_2H-28)	*383*
	Hookeroside D −32.23°, ^{1}H, ^{13}C, FABMS	Oleanolic acid (**7**) Rha-^{2}Xyl-^{4}Glc-^{4}Xyl-^{3}Rha-^{2}Xyl (OH-3β))	*383*
P. koreana	Pulsatilla saponin A	Hederagenin (**11**) Rha-^{2}Ara (OH-3β)	*384*

Table 1. *(continued)*

Source (1)	Saponin mp,$[\alpha]_D$, spectra recorded (2)	Structure (3)	Ref. (4)
	Pulsatilla saponin B	Hederagenin (**11**) Glc-^{4}Ara (OH-3β)	*384*
	Pulsatilla saponin D	Hederagenin (**11**) Glc ╲ $^{4}_{2}$ Ara (OH-3β) Rha ╱	*384*
	Pulsatilla saponin F	Hederagenin (**11**) Glc ╲ $^{6}_{4}$ Glc (OH-3β) Rha ╱	*384*
	Pulsatilla saponin H	Hederagenin (**11**) Glc ╲ $^{4}_{2}$ Ara (OH-3β) Rha ╱ Rha-^{4}Glc-^{6}Glc (CO_2H-28)	*384*
Puerariae radix (Leguminosae)	Kadzusaponin SA_1 +12.1°, ^{1}H, ^{13}C, FABMS	Soyasapogenol A (**94**) Gal-^{2}GlcA (OH-3β)	*21*
	Kadzusaponin SA_2 +1.6°, ^{1}H, ^{13}C, FABMS	Soyasapogenol A (**94**) Gal-^{2}GlcA (OH-3β) Ara (OH-22β)	*21*
	Kadzusaponin SA_3 −2.8°, ^{1}H, ^{13}C, FABMS	Soyasapogenol A (**94**) Rha-^{2}Gal-^{2}GlcA (OH-3β) Ara (OH-22β)	*21*

	Kadzusaponin C_1 $-8.0°$, ^1H, ^{13}C, FABMS	Aglycone (**88**) Rha-^{2}Gal-^{2}GlcA (OH-3β) Glc (OH-21β)	*21*
Randia dumetorum (Rubiaceae)	Randianin 290–295°, +0.22°, IR, ^1H, ^{13}C, FDMS	Oleanolic acid (**7**) Glc-^{3}Glc (OH-3β)	*385*
	Saponin IR, ^1H, ^{13}C, FABMS	Oleanolic acid (**7**) Glc $\searrow^4$ $_3$ GlcA (OH-3β) Glc $\nearrow$	*386*
	Saponin IR, ^1H, ^{13}C, FABMS	Oleanolic acid (**7**) Glc $\searrow^6$ $_3$ GlcA (OH-3β) Glc $\nearrow$	*386*
Rubus coreanus (Rosaceae)	Coreanoside F_1 242–245°, +33.2°, IR, ^1H, ^{13}C, FABMS	(**337**)	*387*
Russell lupine (Leguminosae)	Lupinoside PA_1 $-15.3°$, IR, ^1H, ^{13}C, FABMS	Soyasapogenol A (**94**) Rha-^{2}Gal-^{2}GlcA (OH-3β) Xyl (OH-21β)	*388*
	Lupinoside PA_2 $-1.6°$, IR, ^1H, ^{13}C, FABMS	Soyasapogenol A (**94**) Gal-^{2}GlcA (OH-3β) Xyl (OH-21β)	*388*
	Lupinoside PA_3 $-12.9°$, IR, ^1H, ^{13}C, FABMS	Kudzusapogenol A (**71**) Gal-^{2}GlcA (OH-3β)	*388*
	Lupinoside PA_4 $-21.9°$, IR, ^1H, ^{13}C, FABMS	Soyasapogenol B (**69**) Rha-^{2}Gal-^{2}GlcA (OH-3β) Rha (OH-22β)	*388*

Table 1. *(continued)*

Source (1)	Saponin mp, $[\alpha]_D$, spectra recorded (2)	Structure (3)	Ref. (4)
	Lupinoside PA_5 +23.0°, IR, ^1H, ^{13}C, FABMS	Soyasapogenol B (**69**) Rha-^{2}Gal-^{2}GlcA (OH-3β) Glc-^{4}Rha (OH-22β)	*388*
Sanguisorba alpina (Rosaceae)	Saponin 1 277–280°, +5.0°, ^1H, ^{13}C, FABMS	Tormentic acid (**178**) Glc (OH-3β) Glc (CO_2H-28)	*389*
	Saponin 2 142–143.5°, 14.5°, ^1H, ^{13}C, FABMS	Aglycone (**180**) Glc (CO_2H-28)	*389*
	Saponin 3 247–249.5°, ^1H, ^{13}C, FABMS	Aglycone (**179**) Glc (CO_2H-28)	*389*
	Saponin 4 193–195°, +8.8°, ^1H, ^{13}C, FABMS	Tormentic acid (**178**) Gal (CO_2H-28)	*389*
Schefflera lucantha (Araliaceae)	216–220°, −67.7°, IR, ^{13}C, FABMS	Betulinic acid (**248**) Rha-^{2}Glc-^{2}GlcA (OH-3β)	*390*
	215–219°, −52.2°, IR, ^1H, ^{13}C, FABMS	Betulinic acid (**248**) Rha-^{2}Xyl-^{2}GlcA (OH-3β)	*390*
	237–242°, −15.6°, IR, ^{13}C, FABMS	Oleanolic acid (**7**) Rha-^{2}Glc-^{2}GlcA (OH-3β)	*390*
S. octophylla	Scheffoleoside A −1.4°, ^1H, ^{13}C, FABMS	Arjunolic acid (**66**) Rha-^{4}Glc-^{6}Glc (CO_2H-28)	*391*

Scheffursoside B $-4.7°$, ^{1}H, ^{13}C, FABMS	Aglycone (**168**) Rha-^{4}Glc-^{6}Glc (CO_2H-28)	*391*
Scheffoleoside B $-4.8°$, ^{1}H, ^{13}C, FABMS	Aglycone (**142**) Rha-^{4}Glc-^{6}Glc (CO_2H-28)	*391*
Scheffursoside C $+2.3°$, ^{1}H, ^{13}C, FABMS	23-Hydroxyursolic acid (**169**) Ara (OH-3β) Rha-^{4}Glc-^{6}Glc (CO_2H-28)	*391*
Scheffoleoside D $-19.4°$, ^{1}H, ^{13}C, FABMS	Aglycone (**80**) Rha-^{4}Glc-^{6}Glc (CO_2H-28)	*391*
Scheffursoside E $-28.4°$, ^{1}H, ^{13}C, FABMS	Ursolic acid (**175**) Glc-^{2}Gal-^{2}GlcA (OH-3β) Rha-^{4}Glc-^{6}Glc (CO_2H-28)	*391*
Scheffoleoside E $-16.0°$, ^{1}H, ^{13}C, FABMS	Oleanolic acid (**7**) Glc-^{2}Gal-^{2}GlcA (OH-3β) Rha-^{4}Glc-^{6}Glc (CO_2H-28)	*391*
Scheffursoside F $-5.88°$, ^{1}H, ^{13}C, FABMS	24-hydroxyursolic acid (**162**) Rha-^{4}Glc-^{6}Glc (CO_2H-28)	*391*
	Aglycone (**96**)	*391*
Scheffoleoside F $-3.08°$, ^{1}H, ^{13}C FABMS	Rha-^{4}Glc-^{6}Glc (CO_2H-28)	
232–234°, $-10°$, IR, ^{1}H, ^{13}C, FABMS	Aglycone (**181**) Rha-^{4}Glc-^{6}Glc (CO_2H-28)	*392*
>316°, $-25.8°$, ^{1}H, ^{13}C, FABMS	Aglycone (**254**) Rha-^{4}Glc-^{6}Glc (CO_2H-28)	*393*

Table 1. *(continued)*

Source (1)	Saponin mp, $[\alpha]_D$, spectra recorded (2)	Structure (3)	Ref. (4)
	138–140°, −23°, ^{1}H, ^{13}C, FABMS	3-Epibetulinic acid (**252**) Rha-^{4}Glc-^{6}Glc (CO_2H-28)	*393*
	Saponin 175–185°, +33.8°, IR, ^{1}H, ^{13}C, MS	Aglycone (**253**) Rha-^{4},Glc-^{6}Glc (CO_2H-28)	*394*
	Saponin 177–180°, −30.9°, IR, ^{1}H, ^{13}C, FABMS	3-Epibetulinic acid (**252**) Glc (OH-3α) Rha-^{4}Glc-^{6}Glc (CO_2H-28)	*395*
	Saponin 230–235°, −37°, IR, ^{1}H, ^{13}C, FABMS	3-Epibetulinic acid (**252**) (OAc-6′) Glc (OH-3α) Rha-^{4}Glc-^{6}Glc (CO_2H-28)	*396*
S. venulosa	262–264°, FABMS	Betulinic acid (**248**) Glc-^{2}Glc (OH-3β)	*397*
Scrophularia ilwensis (Scrophulariaceae)	Ilwensissaponin A	Aglycone (**205**) Rha-^{4}Glc —^{3} Fuc (OH-3β); Glc —^{2} Fuc	*398*
	Ilwensissaponin B	Aglycone (**219**) Rha-^{4}Glc —^{3} Fuc (OH-3β); Glc —^{2} Fuc	*398*
	Ilwensissaponin C	Aglycone (**83**) Rha-^{4}Glc —^{3} Fuc (OH-3β); Glc —^{2} Fuc	*398*

	Ilwensissaponin D	Aglycone (**82**) Rha-^{4}Glc ╲ $^{3}_{2}$ Fuc (OH-3β) Glc ╱	*398*
S. koelzii	Scrokoelziside A +27°, IR, ^{1}H, ^{13}C, FABMS	Aglycone (**205**)) Glc ╲ $^{2}_{3}$ Fuc-^{4}Glc (OH-3β) Rha ╱	*399*
Sesamum alatum (Pedaliaceae)	Alatoside A	Aglycone (**316**) Rha-^{2}Xyl (OH-3β)	*400*
	Alatoside B	Aglycone (**316**) Rha-^{2}Xyl (OH-3β) Glc (OH-23)	*400*
	Alatoside C	Aglycone (**316**) Rha-^{2}Glc (OH-3β) Glc (OH-23)	*400*
Sideroxylon *cubense* (Sapotaceae)	Saponin 1 −19°, ^{1}H, ^{13}C, FABMS	Protobassic acid (**37**) Glc (OH-3β) Rha-^{3}Xyl-^{4}Rha-^{2}Xyl (CO_2H-28)	*401*
	Saponin 2 +26°, ^{1}H, ^{13}C, FABMS	Protobassic acid (**37**) Glc (OH-3β)	*401*
S. foetidissimum	Sideroxyloside B −49.4°, ^{1}H, ^{13}C MS	Protobassic acid (**37**) Glc (OH-3β) Apio (f)-^{3}Xyl-^{4}Rha ╲ Apio (f)-3 $^{4}_{2}$ Ara (CO_2H-28) Rha ╱	*402*

Table 1. *(continued)*

Source (1)	Saponin mp, $[\alpha]_D$, spectra recorded (2)	Structure (3)	Ref. (4)
	Sideroxyloside C $-42.8°$, ^1H, ^{13}C, FABMS	Protobassic acid (**37**) Xyl-^{4}Rha ╲ Apio (f)-3 $^4_2$ Ara (CO_2H-28) Rha ╱	*402*
Silene janisseensis (Caryophyllaceae)	Saponin UV, IR, ^1H, ^{13}C	Quillaic acid (**46**) Gal-^{2}GlcA (OH-3β) ^{2}(4-O-trans-p-methoxy-cinnamoyl) Fuc (CO_2H-28) | Rha^2-Glc	*403*
	Saponin UV, IR, ^{13}C	Quillaic acid (**46**) Gal-^{2}GlcA (OH-3β) ^{2}(4-O-cis-p-methoxy-cinnamoyl) Fuc (CO_2H-28) | Rha^2-Glc	*403*
Solidago canadensis (Compositae)	Canadensissaponin 1 251–254°, 2D, FABMS	Bayogenin (**25**) Glc-^{3}Glc (OH-3β) Apio (f) ╲ Rha-^{3}Xyl ╲ $^3_2$ (deoxy-6′) Glc (CO_2H-28) $^4_3$ Rha ╱ Xyl ╱	*404*
	Canadensissaponin 2 234–236°, 2D, FABMS	Bayogenin (**25**) Glc-^{3}Glc (OH-3β) Apio (f) ╲ Rha-^{3}Xyl ╲ $^3_2$ Ara (CO_2H-28) $^4_3$ Rha ╱ Xyl ╱	*404*

	Canadensissaponin 3 267–269°, 2D, FABMS	Bayogenin (**25**) Glc-^{3}Glc (OH-3β) Rha ∖ $^{3}_{2}$ Quin (CO_2H-28) Rha-^{3}Xyl ∖ $^{4}_{3}$ Rha ∕ Xyl ∕	*404*
	Canadensissaponin 4 241–244°, 2D, FABMS	Bayogenin (**25**) Glc-^{3}Glc (OH-3β) Rha ∖ $^{3}_{2}$ Ara (CO_2H-28) Rha-^{3}Xyl ∖ $^{4}_{3}$ Rha ∕ Xyl ∕	*404*
S. gigantea	Saponin	Bayogenin (**25**) Apio (f)-^{3}Glc-^{3}Glc (OH-3β) Apio (f) ∖ $^{3}_{2}$ Quin (CO_2H-28) Rha-^{3}Xyl ∖ $^{4}_{3}$ Rha ∕ Xyl ∕	*405*
	Giganteasaponin 4 ^{13}C, 2D, FABMS	Bayogenin (**25**) Apio (f)-^{3}Glc-^{3}Glc (OH-3β) Rha ∖ $^{3}_{2}$ Quin (CO_2H-28) Rha-^{2}Rha-^{3}Xyl ∖ $^{4}_{3}$ Rha ∕ Xyl ∕	*406*
S. virgaurea	Virgaureasaponin 3 ^{1}H, ^{13}C, 2D	Polygalacic acid (**27**) Glc-^{3}Glc (OH-3β) ^{2}Rha-^{3}Xyl-^{4}Rha-^{2}Fuc (CO_2H-28) | Fuc	*407*

Table 1. *(continued)*

Source (1)	Saponin mp, $[\alpha]_D$, spectra recorded (2)	Structure (3)	Ref. (4)
	Solidagosaponin X −2.2°, ^{1}H, ^{13}C, FABMS	Polygalacic acid (**27**) Glc (OH-3β) ($CH_3CHOHCH_2CO$)-^{4}Ara-^{2}Glc (OH-16α) (OAc-3′,4′) Ara (CO_2H-28)	*408*
	Solidagosaponin XI −2.7°, ^{1}H, ^{13}C, FABMS	Polygalacic acid (**27**) Glc (OH-3β) ($CH_3CHOHCH_2CO$)-^{4}Ara-^{2}Glc (OH-16α) (OAc-2′,4′) Ara (CO_2H-28)	*408*
	Solidagosaponin XII −10.3°, ^{1}H, ^{13}C, FABMS	Polygalacic acid (**27**) Glc (OH-3β) Ara-^{2}Glc (OH-16α) (OAc-3′,4′) Ara (CO_2H-28)	*408*
	Solidagosaponin XIII −6.5°, ^{1}H, ^{13}C, FABMS	Polygalacic acid (**27**) Glc (OH-3β) Ara-^{2}Glc (OH-16α) (OAc-2′, 4′) Ara (CO_2H-28)	*408*
	Solidagosaponin XIV −23.6°, ^{1}H, ^{13}C, FABMS	Polygalacic acid (**27**) Glc (OH-3β) ($CH_3CHOHCH_2CO_2CH_3CHCH_2CO_2CHCH_3CH_2CO$) $^4_2$ Fuc (CO_2H-28) Rha-^{3}Xyl-^{4}Rha	*408*
	Solidagosaponin XV −33.6°, ^{1}H, ^{13}C, FABMS	Polygalacic acid (**27**) Glc (OH-3β) Rha-^{3}Xyl-^{4}Rha-^{2} (OAc-4′) Fuc (CO_2H-28)	*408*

Solidagosaponin XVI $-20.1°$, ^{1}H, ^{13}C, FABMS	Polygalacic acid (**27**) Glc (OH-3β) $(CH_3CHOHCH_2CO)$ —^{4} Fuc (CO_2H-28) Rha-^{3}Xyl-^{4}Rha —^{2}	*408*
Solidagosaponin XVII $-21.6°$, ^{1}H, ^{13}C, FABMS	Polygalacic acid (**27**) Glc (OH-3β) $(CH_3CHOCOCH_3CH_2CO)$ —^{4} Fuc (CO_2H-28) Rha-^{3}Xyl-^{4}Rha —^{2}	*408*
Solidagosaponin XVIII $-29.0°$, ^{1}H, ^{13}C, FABMS	Polygalacic acid (**27**) Glc (OH-3β) $(CH_3CHOHCH_2CO_2CHCH_3CH_2CO)$ —^{4} Fuc (CO_2H-28) Rha-^{3}Xyl-^{4}Rha —^{2}	*408*
Salidagosaponin XIX $-15.6°$, ^{1}H, ^{13}C, FABMS	Polygalacic acid (**27**) Glc (OH-3β) $(CH_3CH{=}CHCO_2CHCH_3CH_2CO)$ —^{4} Fuc (CO_2H-28) Rha-^{3}Xyl-^{4}Rha —^{2}	*408*
Solidagosaponin XX $-37.1°$, ^{1}H, ^{13}C, FABMS	Polygalacic acid (**27**) Glc (OH-3β) Rha-^{3}Xyl-^{4}Rha —^{2} (OAc-4′) Fuc (CO_2H-28) (OAc-5′) Apio (f) —^{3}	*408*
Solidagosaponin XXI $-20.1°$, ^{1}H, ^{13}C, 2D, FABMS	Polygalacic acid (**27**) Xyl-^{3}Glc (OH-3β) (Trimeric β hydroxybutyrate) —^{4} Fuc (CO_2H-28) Rha-^{3}Xyl-^{4}Rha —^{2}	*409*

Table 1. *(continued)*

Source (1)	Saponin mp,$[\alpha]_D$, spectra recorded (2)	Structure (3)	Ref. (4)
	Solidagosaponin XXII −3.8°, ^1H, ^{13}C, FABMS	Polygalacic acid (**27**) Xyl-^{3}Glc (OH-3β) (Trimeric β hydroxybutyrate) $\searrow^3$ Fuc (CO_2H-28) Rha-^{3}Xyl-^{4}Rha $\nearrow_2$	*409*
	Solidagosaponin XXIII −25.0°, ^1H, ^{13}C, FABMS	Polygalacic acid (**27**) Xyl-^{3}Glc (OH-3β) (Trimeric β hydroxybutyrate) $\searrow^4$ (OAc-5′) Apio (f)-3 Fuc (CO_2H-28) Rha-^{3}Xyl-^{4}Rha $\nearrow_2$	*409*
	Solidagosaponin XXIV −25.0°, ^1H, ^{13}C, 2D, FABMS	Polygalacic acid (**27**) Xyl-^{3}Glc (OH-3β) (Dimeric β hydroxybutyrate) $\searrow^4$ Fuc (CO_2H-28) Rha-^{3}Xyl-^{4}Rha $\nearrow_2$	*409*
	Solidagosaponin XXV −25.8°, ^1H, ^{13}C, 2D, FABMS	Polygalacic acid (**27**) Xyl-^{3}Glc (OH-3β) Rha-^{3}Xyl-^{4}Rha-^{2}Fuc (CO_2H-28)	*409*
	Solidagosaponin XXVI −22.2°, ^1H, ^{13}C, 2D, FABMS	Polygalacic acid (**27**) Glc-^{3}Glc (OH-3β) (Trimeric β hydroxybutyrate) $\searrow^4$ Fuc (CO_2H-28) Rha-^{3}Xyl-^{4}Rha $\nearrow_2$	*409*

	Solidagosaponin XXVII −31.5°, ^{1}H, ^{13}C, 2D, FABMS	Polygalacic acid (**27**) Glc-^{4}Glc (OH-3β) ^{3}Xyl-^{4}Rha-^{2}Fuc (CO_2H-28) | Rha	*409*
	Solidagosaponin XXVIII −28.6°, ^{1}H, ^{13}C, 2D, FABMS	Polygalacic acid (**27**) Glc-^{3}Glc (OH-3β) Apio (f) \ ^{3} (OAc-4′) Fuc (CO_2H-28) Rha-^{3}Xyl-^{4}Rha / ^{2}	*409*
	Solidagosaponin XXIX −28.2°, ^{1}H, ^{13}C, 2D, FABMS	Polygalacic acid (**27**) Glc-^{4}Glc (OH-3β) Apio (f) \ ^{3} (OAc-4′) Fuc (CO_2H-28) Rha-^{3}Xyl-^{4}Rha / ^{2}	*409*
Sophora flavescens (Leguminosae)	Sophoraflavoside II −17.2°, ^{1}H, ^{13}C, FABMS	Aglycone (**114**) Rha-^{2}Gal-^{2}GlcA (OH-3β)	*410*
	Sophoraflavoside III −23.0°, ^{1}H, ^{13}C, FABMS	Aglycone (**114**) Rha-^{2}Gal-^{2}GlcA (OH-3β) Ara (OH-22β)	*410*
	Sophoraflavoside IV −18.4°, ^{1}H, ^{13}C, FABMS	Aglycone (**114**) Rha-^{2}Gal-^{2}GlcA (OH-3β) Glc-^{2}Ara (OH-22β)	*410*
Spartium junceum (Leguminosae)	Junceoside UV, IR, ^{1}H, ^{13}C, 2D	Aglycone (**84**) Glc-^{2}Rha (OH-3β)	*411*
Stachys riederichamisso (Labiatae)	Stachyssaponin I −8.1°, ^{1}H, ^{13}C, FABMS	Echinocystic acid (**15**) Ara-^{6}Glc (OH-3β) Ara (CO_2H-28)	*412*

Table 1. *(continued)*

Source (1)	Saponin mp, $[\alpha]_D$, spectra recorded (2)	Structure (3)	Ref. (4)
	Stachyssaponin II −38.2°, ^{1}H, ^{13}C, FABMS	Echinocystic acid (**15**) Ara-^{6}Glc (OH-3β) Rha-^{2}Ara (CO_2H-28)	*412*
	Stachyssaponin III −15.6°, ^{1}H, ^{13}C, FABMS	Echinocystic acid (**15**) Xyl-^{6}Glc (OH-3β) Rha-^{2}Ara (CO_2H-28)	*412*
	Stachyssaponin IV −41.3°, ^{1}H, ^{13}C, FABMS	Echinocystic acid (**15**) Ara-^{6}Glc (OH-3β) Xyl-^{4}Rha-^{2}Ara (CO_2H-28)	*412*
	Stachyssaponin V −60.0°, ^{1}H, ^{13}C, FABMS	Echinocystic acid (**15**) Ara-^{6}Glc (OH-3β) (OAc-3′) Xyl-^{4}Rha-^{2}Ara (CO_2H-28)	*412*
	Stachyssaponin VI −46.3°, ^{1}H, ^{13}C, FABMS	Echinocystic acid (**15**) Ara-^{6}Glc (OH-3β) (OAc-4′) Xyl-^{4}Rha-^{2}Ara (CO_2H-28)	*412*
	Stachyssaponin VII −58.3°, ^{1}H, ^{13}C, FABMS	Echinocystic acid (**15**) Ara-^{6}Glc (OH-3β) Xyl ╲ $^4_3$ Rha-^{2}Ara (CO_2H-28) Glc ╱	*412*
	Stachyssaponin VIII −33.2°, ^{1}H, ^{13}C, FABMS	Echinocystic acid (**15**) Xyl-^{6}Glc (OH-3β) Xyl-^{4}Rha-^{2}Ara (CO_2H-28)	*412*
Stauntonia chinensis (Lardizabalaceae)	Yemuoside I 204–208°, +14.1°, IR, ^{1}H, ^{13}C, FABMS	Aglycone (**28**) Ara-^{3}Rha-^{2}Ara (OH-3β) Glc-^{6}Glc (CO_2H-28)	*413*

	Yemuoside YM_7 235–237°, +20.53°, IR, ^{1}H, ^{13}C, FABMS	Aglycone (**28**) Ara-^{3}Ara (OH-3β) Rha-^{4}Glc-^{6}Glc (CO_2H-28)	*414*
	Yemuoside YM_8 208–212°, +16.08°, IR, ^{1}H, ^{13}C, FABMS	Aglycone (**28**) Glc-^{3}Rha-^{2}Ara (OH-3β) Glc-^{6}Glc (CO_2H-28)	*415*
	Yemuoside YM_9 207–210°, +20.6°, IR, ^{1}H, ^{13}C, FABMS	Aglycone (**28**) Glc-^{3}Rha-^{2}Ara (OH-3β) Rha-^{4}Glc-^{6}Glc (CO_2H-28)	*415*
	Yemuoside YM_{10}	Aglycone (**28**) Rha-^{2}Ara (OH-3β) Rha-^{4}Glc-^{6}Glc (CO_2H-28)	*416*
	Yemuoside YM_{11} 205–208°, +38.40°, IR, ^{1}H, ^{13}C, FABMS	Aglycone (**28**) Ara (OH-3β) Glc-^{6}Glc (CO_2H-28)	*414*
	Yemuoside YM_{12}	Aglycone (**28**) Rha-^{2}Ara (OH-3β) Glc-^{6}Glc (CO_2H-28)	*416*
	Yemuoside YM_{13} 218–222°, +28.40°, IR, ^{1}H, ^{13}C, FABMS	Aglycone (**28**) Ara-^{3}Ara (OH-3β) Glc-^{6}Glc (CO_2H-28)	*414*
	Yemuoside YM_{14} 206–209°, +11.74°, IR, ^{1}H, ^{13}C, FABMS	Aglycone (**28**) Ara (OH-3β) Rha-^{4}Glc-^{6}Glc (CO_2H-28)	*414*
	Saponin III 238–242°, +21.68°, IR, ^{1}H, ^{13}C, FABMS	Aglycone (**28**) Glc-^{3}Rha-^{2}Ara (OH-3β)	*415*
S. hexaphylla	Staunoside A ^{1}H, ^{13}C, FABMS	Hederagenin (**11**) Glc (OH-3β) Glc-^{6}Glc (CO_2H-28)	*417*

Table 1. *(continued)*

Source (1)	Saponin mp, $[\alpha]_D$, spectra recorded (2)	Structure (3)	Ref. (4)
	Staunoside B ^{1}H, ^{13}C, FABMS	Hederagenin (**11**) Glc (OH-3β) Rha-^{4}Glc-^{6}Glc (CO_2H-28)	*417*
	Staunoside D 224–227°, +6.2°, IR, ^{1}H, ^{13}C, FABMS	Hederagenin (**11**) Glc ╲ $^{3}_{2}$ Glc (OH-3β) Glc ╱ Glc-^{6}Glc (CO_2H-28)	*418*
	Staunoside E 218–221°, −4.7°, IR, ^{1}H, ^{13}C, FABMS	Hederagenin (**11**) Glc ╲ $^{3}_{2}$ Glc (OH-3β) Glc ╱ Rha-^{4}Glc-^{6}Glc (CO_2H-28)	*418*
Steganotaenia araliacea (Apiaceae)	Saponin 4 −14°, ^{1}H, ^{13}C, CPDMS	Aglycone (**86**) Gal ╲ $^{3}_{2}$ GlcA (OH-3β) Gal ╱	*419*
	Saponin 5 −10°, ^{1}H, ^{13}C, CPDMS	Aglycone (**85**) Gal ╲ $^{3}_{2}$ GlcA (OH-3β) Gal ╱	*419*

	Saponin 6 $-38.6°$, ^{1}H, ^{13}C, CPDMS	Aglycone (**86**) Xyl –^{3} GlcA (OH-3β) Glc –^{2}	*419*
	Saponin 7 $-39.4°$, ^{1}H, ^{13}C,	Aglycone (**85**) Xyl –^{3} GlcA (OH-3β) Glc –^{2}	*419*
	Saponin 8 $-24°$, IR, ^{1}H, ^{13}C, CPDMS	Steganogenin (**87**) Rha –^{4} Glc (OH-3β) Glc –^{3} | ^{2} Glc	*419*
	Saponin 9 ^{1}H, ^{13}C, CPDMS	Oleanolic acid (**7**) Gal-^{2}GlcA (OH-3β) Glc (CO_2H-28)	*419*
Stephanotis lutchuensis (Asclepiadaceae)	Sitakisoside I 206–208°, $-12.4°$, UV, IR, ^{1}H, ^{13}C, 2D, FABMS	Aglycone (**158**) Xyl-^{6}Glc-^{6}Glc (OH-3β)	*420*
	Sitakisoside II 204–206°, $-8.2°$, UV, IR, ^{1}H, ^{13}C, 2D, FABMS	Aglycone (**161**) Xyl-^{6}Glc-^{6}Glc (OH-3β)	*420*
	Sitakisoside III 200–202°, $-9.5°$, UV, IR, ^{1}H, ^{13}C, 2D, FABMS	Aglycone (**159**) Xyl-^{6}Glc-^{6}Glc (OH-3β)	*420*

Table 1. *(continued)*

Source (1)	Saponin mp, $[\alpha]_D$, spectra recorded (2)	Structure (3)	Ref. (4)
	Sitakisoside IV 198–200°, −11.5°, UV, IR, ^1H, ^{13}C, 2D, FABMS	Aglycone (**158**) Glc-^{6}Glc-^{6}Glc (OH-3β)	*420*
	Sitakisoside V 202-204°, −10.0°, UV, IR, ^1H, ^{13}C	Aglycone (**160**) Xyl-^{6}Glc-^{6}Glc (OH-3β)	*420*
Symphytum officinale (Boraginaceae)	Symphytoxide A 228°, +12°, UV, IR, ^1H, ^{13}C, 2D, FABMS	Hederagenin (**11**) Glc-^{2}Glc-^{4}Ara (OH-3β)	*421*
	Symphytoxide B 192°, −2.86°, IR, ^1H, ^{13}C, FABMS	Hederagenin (**11**) Glc-^{4}Glc-^{4}Ara (OH-3β) Glc ╲^{6} Rha ╱^{4} Glc (CO_2H-28)	*422*
	+22.86°, UV, IR, ^1H, ^{13}C, FABMS	Oleanolic acid (**7**) Glc-^{4}Glc-^{4}Ara (OH-3β)	*423*
	Bisdesmosidic saponin 190°, +14.52°, UV, IR, ^1H, ^{13}C, FABMS	Hederagenin (**11**) Ara (OH-3β) Glc-^{4}Glc-^{6}Glc (CO_2H-28)	*424*
	Bisdesmosidic saponin ^1H, ^{13}C, 2D, FABMS	Hederagenin (**11**) Glc-^{4}Ara (OH-3β) Glc-^{6}Glc (CO_2H-28)	*425*

References, pp. 173–196

Terminalia bellerica (Combretaceae)	Bellericaside A 207°, −17°, IR, ^{13}C	Bellericagenin A (**119**) Glc (CO_2H-28)	*426*
	Bellericaside B 223°, +15.1°, IR, ^{13}C	Bellericagenin B (**120**) Glc (CO_2H-28)	*426*
	Bellericoside 238°, +45°, ^1H, ^{13}C	Belleric acid (**72**) Glc (CO_2H-28)	*427*
T. chebula	Chebuloside I 238–240°, +42°, IR, ^{13}C, MS	Arjunolic acid (**66**) Gal (CO_2H-28)	*428*
	Chebuloside II 215°, +25°, IR, ^{13}C, MS	Terminolic acid (**134**) Glc (CO_2H-28)	*428*
Tetrapleura tetraptera (Leguminosae)	Saponin 238–243°, UV, IR, ^1H, ^{13}C, FABMS	27-Hydroxyoleanolic acid (**21**) Glc-^{6}Glc (OH-3β)	*429*
Thalictri herba (Ranunculaceae)	Thalictoside V −16.5°, ^1H, ^{13}C, FABMS	Aglycone (**246**) Rha ⟍ ^{6} Glc (OH-3β) Rha ⟋ ^{2}	*430*
	Thalictoside IX −14°, ^{13}C, FABMS	Aglycone (**246**) Rha ⟍ ^{6} Glc (OH-3β) Rha ⟋ ^{2} Xyl-^{6}Glc (CO_2H-21)	*430*
Thalictrum foeniculaceum (Ranunculaceae)	Thalifoenoside A	Aglycone (**247**) Quin-^{2}Rha^6 (OAc-4′) Glc (OH-3β)	*431*
T. thunbergii	Thalictoside A −1.3°, ^1H, ^{13}C, FABMS	Aglycone (**247**) Quin-^{6}Glc-^{4}Fuc (OH-3β)	*432*

Table 1. *(continued)*

Source (1)	Saponin mp, $[\alpha]_D$, spectra recorded (2)	Structure (3)	Ref. (4)
	Thalictoside C −23.5°, ^1H, ^{13}C, FABMS	Aglycone (**247**) Glc ⟍ $_{2}^{6}$ Glc-^{4}Fuc (OH-3β) Rha ⟋	*432*
Thinouia coriaceae (Sapindaceae)	Saponin 1 ^{13}C, MS	Oleanolic acid (**7**) Ara (OH-3β)	*433*
	Saponin 2 ^{13}C, MS	Oleanolic acid (**7**) Rha-^{2}Ara (OH-3β)	*433*
	Saponin 3 ^{13}C, MS	Oleanolic acid (**7**) Glc-^{4}Ara (OH-3β)	*433*
	Saponin 4 ^{13}C, MS	Oleanolic acid (**7**) Glc-^{3}Rha-^{2}Ara (OH-3β)	*433*
	Saponin 5 ^{13}C, MS	Oleanolic acid (**7**) Glc ⟍ $_{2}^{4}$ Ara (OH-3β) Rha ⟋	*433*
	Saponin 6 ^{13}C, MS	Oleanolic acid (**7**) Glc ⟍ $_{2}^{4}$ Ara (OH-3β) Glc-^{3}Rha ⟋	*433*
Thladiantha dubia (Cucurbitaceae)	Dubioside A 210–215°, −31.9°, ^1H, ^{13}C, FABMS	Quillaic acid (**46**) Gla-^{2}GlcA (OH-3β) Rha-^{2}Ara (CO_2H-28)	*434*
	Dubioside B 225–226°, −26.1°, ^1H, ^{13}C, FABMS	Quillaic acid (**46**) Gal-^{2}GlcA (OH-3β) Xyl-^{4}Rha-^{2}Ara (CO_2H-28)	*434*

	Dubioside C 229–231°, −27.6°, ^{1}H, ^{13}C, FABMS	Quillaic acid (**46**) Gal-^{2}GlcA (OH-3β) Xyl-^{3}Xyl-^{4}Rha-^{2}Ara (CO_2H-28)	*434*
	Dubioside D −20.0°, ^{1}H, ^{13}C, FABMS	Quillaic acid (**46**) Glc-^{3}Gal-^{2}Glc (OH-3β) Rha-^{2}Ara (CO_2H-28)	*435*
	Dubioside E −16.9°, ^{1}H, ^{13}C, FABMS	Quillaic acid (**46**) Glc-^{3}Gal-^{2}Glc (OH-3β) ^{4}Rha-^{2}Ara (CO_2H-28) | Xyl	*435*
	Dubioside F −17.3°, ^{1}H, ^{13}C, FABMS	Quillaic acid (**46**) Glc-^{3}Gal-^{2}Glc (OH-3β) Xyl-^{3}Xyl-^{4}Rha-^{2}Ara (CO_2H-28)	*435*
T. hookeri	Thladioside H_1 +3.8°, IR, ^{1}H, ^{13}C, FDMS	Gypsogenin (**13**) Gal-^{2}GlcA (OH-3β) Xyl-^{3}Xyl-^{4}Rha-^{2}Xyl (CO_2H-28)	*436*
Tragopogon pratensis (Compositae)	Tragopogonoside A −21.9°, ^{1}H, ^{13}C, FABMS	Echinocystic acid (**15**) (Me-ester-6′) GlcA (OH-3β) Xyl (CO_2H-28)	*437*
	Tragopogonoside B −9.4°, ^{1}H, ^{13}C, FABMS	Aglycone (**113**) Gal-^{2}(Me-ester-6′) GlcA (OH-3β)	*437*
	Tragopogonoside C −16.3°, ^{1}H, ^{13}C, FABMS	Echinocystic acid (**15**) Gal-^{2}(Me-ester-6′) GlcA (OH-3β) Xyl (CO_2H-28)	*437*
	Tragopogonoside D −27.8°, ^{1}H, ^{13}C, FABMS	Echinocystic acid (**15**) (Me-ester-6′) GlcA (OH-3β) Glc-^{3}Xyl (CO_2H-28)	*437*

Table 1. *(continued)*

Source (1)	Saponin mp, $[\alpha]_D$, spectra recorded (2)	Structure (3)	Ref. (4)
	Tragopogonoside E −22.9°, ^{1}H, ^{13}C, FABMS	Echinocystic acid (**15**) Gal-^{2}(Me-ester-6′) GlcA (OH-3β) Xyl-^{3}Xyl (CO_2H-28)	*437*
	Tragopogonoside F +1.3°, UV, ^{1}H, ^{13}C, FABMS	Echinocystic acid (**15**) Gal-^{2} (Me-ester-6′) GlcA (OH-3β) Glc \\ $^3_2$ Xyl (CO_2H-28) p-coumaric acid /	*437*
	Tragopogonoside G ^{1}H, ^{13}C, FABMS	Echinocystic acid (**15**)) Gal-^{2}(Me-ester-6′) GlcA (OH-3β) p-coumaric acid-^{2}Xyl (CO_2H-28)	*437*
	Tragopogonoside H ^{1}H, ^{13}C, FABMS	Echinocystic acid (**15**) Gal-^{2} (Me-ester-6′) GlcA (OH-3β) Ferulic acid-^{2}Xyl (CO_2H-28)	*437*
	Tragopogonoside I −24.2°, ^{1}H, ^{13}C,	Acacic acid lactone (**9**) Gal-^{2} (Me-ester-6′) GlcA (OH-3β)	*437*
Tridesmostemon claessenssi (Sapotaceae)	Tridesmosaponin A −52°, ^{1}H, ^{13}C, 2D, FABMS	16α-Hydroxyprotobassic acid (**89**) Glc-^{6}Glc (OH-3β) Rha-^{3}Xyl \\ $^4_3$ Rha-^{2}Xyl (CO_2H-28) Rha /	*438*

References, pp. 173–196

	Tridesmosaponin B −71°, ^{1}H, ^{13}C, 2D, FABMS	16α-Hydroxyprotobassic acid (**89**) Rha (OH-3β) Rha —⁴ and Rha —³ Xyl-^{4}Rha-^{2}Xyl (CO_2H-28)	*438*
Trifolium alexandrinum (Leguminosae)	+28.8°, ^{1}H, ^{13}C, FABMS	Soyasapogenol B (**69**) Rha-^{2}Glc-^{2} (Me-ester-6′) GlcA (OH-3β) Glc-^{2}Glc (OH-22β)	*439*
	−14.2°, ^{1}H, ^{13}C, FABMS	Soyasapogenol E (**49**) Rha-^{2}Glc-^{2} (Me-ester-6′) GlcA (OH-3β)	*439*
T. repens	Cloversaponin I +21.1°, ^{1}H, ^{13}C, FABMS	Soyasapogenol E (**49**) (Me-ester-6′) GlcA (OH-3β)	*440*
	Cloversaponin II −10.1°, ^{1}H, ^{13}C, FABMS	Aglycone (**90**) (Me-ester-6′) GlcA (OH-3β)	*440*
	Cloversaponin III +5.0°, ^{1}H, ^{13}C, FABMS	Aglycone (**90**) Glc-^{2} (Me-ester-6′) GlcA (OH-3β)	*440*
	Cloversaponin IV +38.9°, ^{1}H, ^{13}C, FABMS	Soyasapogenol B (**69**) Xyl-^{2}(Me-ester-6′) GlcA (OH-3β)	*440*
	Cloversaponin V −6.0°, ^{1}H, ^{13}C, FABMS	Aglycone (**90**) Xyl-^{2}(Me-ester-6′) GlcA (OH-3β)	*440*
Triplostegia grandiflora (Dipsacaceae)	Triploside A 230–234°, −36.43°, ^{1}H, ^{13}C, 2D, FABMS	Oleanolic acid (**7**) Xyl-^{4}Rha-^{3}Xyl-^{3}Rha-^{2}Ara (OH-3β) 3| Xyl^{4}-Xyl	*441*

Table 1. *(continued)*

Source (1)	Saponin mp, $[\alpha]_D$, spectra recorded (2)	Structure (3)	Ref. (4)
	Triploside B 190–195°, −34.41°, ^{1}H, ^{13}C, 2D, FABMS	Oleanolic acid (**7**) Rha-^{3}Xyl-^{3}Rha-^{2}Ara (OH-3β) ^{4}| Xyl^{3}-Xyl	*441*
	Triploside C 210–215°, −26.69°, ^{1}H, ^{13}C, 2D, FABMS	Oleanolic acid (**7**) Rha-^{3}Xyl-^{3}Rha-^{2}Ara (OH-3β)	*441*
Uncaria guianensis (Rubiaceae)	^{1}H, ^{13}C, FABMS	Quinovic acid (**171**) Quin (OH-3β)	*442*
	^{1}H, ^{13}C, FABMS	Quinovic acid (**171**) Fuc (OH-3β) Glc (CO_2H-27)	*442*
Verbascum nigrum (Scrophulariaceae)	Saponin 250–254°, ^{1}H, ^{13}C, LSIMS	Aglycone (**205**) Rha-^{4}Glc-^{3}Glc-^{2}Fuc (OH-3β)	*443*
	Saponin 246–248°, ^{1}H, ^{13}C, LSIMS	Aglycone (**83**) Rha-^{4}Glc-^{3}Glc-^{2}Fuc (OH-3β)	*443*
V. songaricum	Songarosaponin A +27°, UV, ^{1}H, ^{13}C, FABMS	Aglycone (**219**) Rha-^{4}Glc-^{3}Glc-^{2}Fuc (OH-3β)	*444*
	Songarosaponin B 263–268°, +28°, ^{1}H, ^{13}C	Aglycone (**226**) Rha-^{4}Glc-^{3}Glc-^{2}Fuc (OH-3β)	*444*

	Songarosaponin C 264–270°, +26°, ^1H, ^{13}C, FABMS	Aglycone (**205**) Glc-^{4}Glc-^{3}Glc-^{2}Fuc (OH-3β)	*444*
	Songarosaponin D +20°, ^1H, ^{13}C, 2D, FABMS	Saikogenin F (**190**) Glc-^{4}Glc $\searrow^3$ Fuc (OH-3β) Glc $\nearrow^2$	*445*
	Songarosaponin E +24°, ^1H, SIMS	Aglycone (**219**) Glc-^{4}Glc $\searrow^3$ Fuc (OH-3β) Glc $\nearrow^2$	*446*
	Songarosaponin F +26°, ^1H, MS	Saikogenin A (**220**) Glc-^{4}Glc $\searrow^3$ Fuc (OH-3β) Glc $\nearrow^2$	*446*
Vicoa indica (Asteraceae)	Vicoside A 160–162°, +55°, ^1H, ^{13}C	Aglycone (**102**) Glc (OH-3β)	*447*
Wedelia calendulaceae (Asteraceae)	Saponin 224–226°, +10.3°, IR, ^1H, ^{13}C, FABMS	Oleanolic acid (**7**) Xyl-^{2}GlcA (OH-3β) Glc (CO_2H-28)	*448*
Wistaria brachybotrys (Leguminosae)	Wistariasaponin A −18.8°, IR, ^1H, ^{13}C, 2D, FABMS	Wistariasapogenol A (**103**) Rha-^{2}Xyl-^{2}GlcA (OH-3β)	*449*
	Wistariasaponin B_1 −10.9°, IR, ^1H, ^{13}C, 2D, FABMS	Wistariasapogenol B (**112**) Rha-^{2}Xyl-^{2}GlcA (OH-3β)	*449*
	Wistariasaponin B_2 −7.9°, IR, ^1H, ^{13}C, 2D	Wistariasapogenol B (**112**) Rha-^{2}Glc-^{2}GlcA (OH-3β)	*449*

Table 1. *(continued)*

Source (1)	Saponin mp, $[\alpha]_D$, spectra recorded (2)	Structure (3)	Ref. (4)
	Wistariasaponin B_3 $-2.0°$, IR, ^1H, ^{13}C, FABMS	Wistariasapogenol B (**112**) Rha-^{2}Xyl-^{2}GlcA (OH-3β) Glc (OH-30)	*450*
	Wistariasaponin C $-14.9°$, ^1H, ^{13}C, 2D	Soyasapogenol B (**69**) Rha-^{2}Xyl-^{2}GlcA (OH-3β)	*449*
	Wistariasaponin D $-11.5°$, IR, ^1H, ^{13}C	Soyasapogenol E (**49**) Rha-^{3}Xyl-^{2} (Me-ester-6′) GlcA (OH-3β)	*451*
	Wistariasaponin G $-35.6°$, IR, ^1H, ^{13}C	Aglycone (**92**) Rha-^{3}Xyl-^{2} (Me-ester-6′) GlcA (OH-3β)	*451*
	Wistariasaponin YC_1 $-50.2°$, IR, ^1H, ^{13}C, FABMS	Yunganogenin (**155**) Rha-^{2}Xyl-^{2}GlcA (OH-3β) Glc (OH-21α)	*450*
	Wistariasaponin YC_2 $-31.9°$, IR, ^1H, ^{13}C, FABMS	Yunganogenin (**155**) Rha-^{2}Gal-^{2}GlcA (OH-3β) Glc (OH-21α)	*450*
	Wistariasaponin A_2 $-12.3°$, IR, ^1H, ^{13}C, FABMS	Wistariasapogenol A (**103**) Rha-^{2}Xyl-^{2}GlcA (OH-3β) Glc (OH-30)	*450*
	Wistariasaponin A_3 $+11.2°$, IR, ^1H, ^{13}C, FABMS	Wistariasapogenol A (**103**) Gal-^{2}GlcA (OH-3β) Glc (OH-30)	*450*
Zizyphus jujuba (Rhamnaceae)	Jujubasaponin I 212–214°, $-43.3°$, ^1H, ^{13}C, FABMS	Jujubogenin (**305**) Rha-^{2}Ara (OH-3β) Rha (OH-20β)	*88*

	Jujubasaponin II 191–193°, −41.5°, ^{1}H, ^{13}C, FABMS	Jujubogenin (**305**) Rha-^{2}Ara (OH-3β) (OAc-2′) Rha (OH-20β)	*88*
	Jujubasaponin III 187–189°, −43.9°, ^{1}H, ^{13}C, FABMS	Jujubogenin (**305**) Rha-^{2}Ara (OH-3β) (OAc-3′) Rha (OH-20β)	*88*
	Jujubasaponin IV 185–187°, −3.64°, ^{1}H, ^{13}C, 2D, FABMS	Aglycone (**264**) Gal ╲^{3} Glc (OH-3β) Rha ╱^{2}	*452*
	Jujubasaponin V 210–212°, −14.2°, ^{1}H, ^{13}C, 2D, FABMS	Aglycone (**264**) Glc ╲^{3} Glc (OH-3β) Rha ╱^{2}	*452*
	Jujubasaponin VI 199–201°, −28.1°, ^{1}H, ^{13}C, 2D, FABMS	Trevoagenin D (**306**) Rha-^{2}Gal (OH-3β)	*452*
Zygophyllum album (Zygophyllaceae)	^{1}H, ^{13}C, FABMS	Ursolic acid (**175**) Quin-^{4}Quin (OH-3β)	*453*
	^{1}H, ^{13}C, FABMS	Quinovic acid (**171**) Glc-^{2}Glc (CO_2H-28)	
	^{1}H, ^{13}C, FABMS	Quinovic acid (**171**) Glc-^{2}Glc (CO_2H-27)	*453*
	^{1}H, ^{13}C, FABMS	Quinovic acid (**171**) Glc-^{2}Rha (OH-3β)	*453*
	Glycoside 2 190°, ^{1}H, ^{13}C,	Quinovic acid (**171**) Xyl-^{3}Quin (OH-3β)	*454*
	Glycoside 3 235°, ^{1}H, ^{13}C	Quinovic acid (**171**) Fuc (OH-3β) Glc (CO_2H-28)	*454*

Table 1. *(continued)*

Source (1)	Saponin mp, $[\alpha]_D$, spectra recorded (2)	Structure (3)	Ref. (4)
Z. album *Z. coccineum* *Z. dumosum* *Z. propinquum*	Zygophyloside F +23°, ^1H, ^{13}C, 2D, FABMS	Quinovic acid (**171**) (SO_3H-2′) Quin (OH-3β) Glc (CO_2H-27)	*455*
	Zygophyloside A 25.8°, −20.8°, UV, IR, ^1H, ^{13}C, FABMS	Quinovic acid (**171**) Ara-^{2}Quin (OH-3β)	*456*
	Zygophyloside B 200°, +34.09°, UV, IR, ^1H, ^{13}C, FABMS	Quinovic acid (**171**) Quin (OH-3β) Glc (OH-27)	*456*
	Zygophyloside C ^1H, ^{13}C, FABMS	Quinovic acid (**171**) Ara-^{2}Quin (OH-3β) Glc (OH-27)	*457*
	Zygophyloside D ^1H, ^{13}C, FABMS	Quinovic acid (**171**) (SO_3Na-2′)) Quin (OH-3β)	*458*
	Zygophyloside E ^1H, ^{13}C, FABMS	Quinovic acid (**171**) (SO_3Na-2′) Quin (OH-3β) Glc (CO_2H-28)	*458*

Abbreviations: Glc = β-D-glucopyranosyl; GlcA = β-D-glucuronic acid pyranosyl; Gal = β-D-galactopyranosyl; Ara = α-L-arabinopyranosyl; Xyl = β-D-xylopyranosyl; Rha = α-L-rhamnopyranosyl; Ara(f) = α-L-arabinofuranosyl; Apio(f) = β-D-apiofuranosyl; Fuc = β-D-fucopyranosyl; Quin = β-D-quinovopyranosyl; Mann = β-D-manno-pyranosyl

Acknowledgements

Financial support from CSIR, New Delhi in the form of ES to SBM and in the form of SRF to SG is gratefully acknowledged.

References

1. MAHATO, S.B., S.K. SARKAR, and G. PODDAR: Triterpenoid Saponins. Phytochemistry, **27**, 3037 (1988).
2. MAHATO, S.B., and A.K. NANDY: Triterpenoid Saponins Discovered between 1987 and 1989. Phytochemistry, **30**, 1357 (1991).
3. BADER, G., and K. HILLER: Triterpenoid Saponins. Pharmazie, **42**, 577 (1987).
4. SCHÖPKE, T., and K. HILLER: Triterpenoid Saponins. Pharmazie, **45**, 313 (1990).
5. JIA, Z., K. KOIKE, T. OHMOTO, and M. NI: Triterpenoid Saponins from *Ardisia crenata*. Phytochemistry, **37**, 1389 (1994).
6. ZHAO, W., J. XU, G. QUIN, and R. ZU: New Triterpenoid Saponins from *Mussaenda pubescens*. J. Nat. Prod., **57**, 1613 (1994).
7. MASSIOT, G., C. LAVAUD, V. BESSON, L.L. MEN-OLIVIER, and G.V. BINST: Saponins from Aerial Parts of Alfalfa (*Medicago sativa*). J. Agric. Food Chem., **39**, 78 (1991).
8. RODRIGUEZ, J., R. CASTRO, and R. RIGUERA: Holothurinosides: New Antitumour Non Sulphated Triterpenoid Glycosides from the Sea Cucumber *Holothuria forskalii*. Tetrahedron, **47**, 4753 (1991).
9. MAHATO, S.B.: Saponins with Antifilarial Activity from *Acacia auriculiformis*. In: Saponins: Chemistry and Biological Activity (G.R. WALLER, K. Yamasaki, eds.). New York: Plenum, 1996.
10. WILLIAMS, D.H., C. BRADLEY, G. BOJESEN, S. SANTIKARN, and L.C.E. TAYLOR: Fast Atom Bombardment Mass Spectrometry: A Powerful Technique for the Study of Polar Molecules. J. Am. Chem. Soc., **103**, 5700 (1981).
11. FENSELAU, C.: Fast Atom Bombardment and Middle Molecule Mass Spectrometry. J. Nat. Prod., **47**, 215 (1984).
12. LATTIMER, R.P., and H.R. SCHULTEN: Field Ionization and field Desorption Mass Spectrometry: Past, Present and Future. Anal. Chem., **61**, 1201A (1989).
13. SUNDVIST, B., P. ROEPSTORFF, J. FOHLMAN, A. HEDIN, P. HAKANSSON, I. KAMENSKY, M. LINDBERG, M. SALEHPOOUR, and G. SAVE: Molecular Weight Determination of Proteins by Californium Plasma Desorption Mass Spectrometry. Science, **226**, 696 (1984).
14. KARAS, M., and F. HILLENKAMP: Laser Desorption Ionization of Proteins with Molecular Masses Exceeding 10,000 Daltons. Anal. Chem., **60**, 2299 (1988).
15. COVEY, T.R., E.C. HUANG, and J.D. HENION: Structural Characterization of Protein Tryptic Peptides via Liquid Chromatography/Mass Spectrometry and Collision-Induced Dissociation of their Doubly Charged Molecular Ions. Anal. Chem., **63**, 1193 (1991).
16. IMOTO, T., F.M. YAMAMOTO, A. MIYASAKA, and H. HATANO: High Performance Liquid Chromatography-Atmospheric Pressure Ionization Mass Spectrometry of Gymnemic Acids. J. Chromatography, **557**, 383 (1991).
17. CASTELLO, C.E.: Application of Tandem Mass Spectral Approaches to Structural Determination of Saponins. Symposium on Saponins: Chemistry and Biological Activity, 210th ACS National Meeting, Chicago, Illinois, August 20–24, 1995.

18. LEE, M.R., J.S. LEE, J.C. WANG, and G.R. WALLER: Structural Determination of Saponins from Mung Bean Sprouts by Tandem Mass Spectrometry. Symposium on Saponins: Chemistry and Biological Activity, 210th ACS National Meeting, Chicago, Illinois, August 20–24, 1995.
19. LEE, M.K., Y.C. LING, M. JURZYSTA, and G.R. WALLER: Saponins from Alfalfa, Clover, and Mungbeans Analyzed by Electrospray Iionization Mass Spectrometry Compared with Positive and Negative FAB Mass Spectrometry. Symposium on Saponins: Chemistry and Biological Activity, 210th ACS National Metting, Chicago, Illinois, August 20–24, 1995.
20. WEST, P.R., P.W. GENO, G.R. WALLER, W. OLESZEK, and M. JURZYSTA: Structure Confirmation of Alfalfa Saponins by LS/MS and B/E LS/MS/MS. Symposium on Saponins: Chemistry and Biological Activity, 210th ACS National Meeting, Chicago, Illinois, August 20–24, 1995.
21. ARAO, T., J. KINJO, T. NOHARA, and R. ISOBE: Oleanene-type Triterpene Glycosides from *Puerariae Radix* II. Isolation of Saponins and the Application of Tandem Mass Spectrometry to their Structure Determination. Chem. Pharm. Bull., **43**, 1176 (1995).
22. SCHÖPKE, T., C. AL-TAWAHA, V. WRAY, M. NIMTZ, A. MEYER, and K. HILLER: Triterpenoid Saponins from the Aerial Parts of *Aster bellidiastrum*. Phytochemistry, **40**, 1489 (1995).
23. PERERA, P., R. ANDERSSON, L. BOHLIN, C. ANDERSSON, D. LI, N.L. OWEN, R. DUNKEL, C.L. MAYNE, R.J. RUGMIRE, D.M. GRANT, and P.A. COX: Structure Determination of a New Saponin from the Plant *Alphitonia zizyphoides* by NMR Spectroscopy. Magn. Res. Chem., **31**, 472 (1993).
24. GARAI, S., S.B. MAHATO, K. OHTANI, and K. YAMASAKI: Dammarane-Type Triterpenoid Saponins from *Bacopa monniera*. Phytochemistry, **42**, 815 (1996).
25. WALTHO, J.P., D.H. WILLIAMS, S.B. MAHATO, B.C. PAL, and J.C.J. BARNA: Structure Elucidation of Two Triterpenoid Tetrasaccharides from *Androsace saxifragifolia*. J. Chem. Soc. Perkin Trans. 1, 1517 (1986).
26. BADER, G., T. DANZANDARJAA, K. HILLER, G. RENICEK, J. JURENITSCH, M. GOLLY, H. SCHRÖDER, M. SCHUBERT-ZSILAVEEZ, and E. HASLINGER: A New Triterpene Saponin from *Heteropappus biennis* with an Unusual Carbohydrate Chain. Helv. Chim. Acta, **77**, 1861 (1994).
27. SHAO, Y., B. ZHOU, L. LIN, and G.A. CORDELL: Triterpenoid Saponins from *Aster batangensis*. Phytochemistry, **38**, 927 (1995).
28. BADER, G., Y. KULHANEK, and H. ZIEGLER-BOEHME: Antifungal Action of Glycosides of Polygalacic Acid. Pharmazie, **45**, 618 (1990).
29. TAKECHI, M., and Y. TANAKA: Structure-Activity Relationships of the Saponin α-Hederin. Phytochemistry, **29**, 451 (1990).
30. SHIMOYAMADA, M., M. SUZUKI, H. SONTA, M. MURUAMA, and K. OKUBO: Antifungal Activity of the Saponin Fraction from *Asparagus officinalis* L. and its Active Principle. Agric. Biol. Chem., **54**, 2553 (1990).
31. LALITHA, T., and L.V. VENKATARAMAN: Antifungal Activity and Mode of Action of Saponins from *Madhuca butyraceae* Macb. Indian J. Chem., **29**, 558 (1991).
32. ZEHAVI, U., O. ZIV-FECHT, M. LEVY, M. NAIM, R. EVRON, and I. POLACHEK: Synthesis and Antifungal Activity of Medicagenic Acid Saponins on Plant Pathogens: Modification of the Saccharide Moiety and the 23α-Substitution. Carbohydr. Res., **244**, 161 (1993).
33. BOWER, P., B.R. CLARKE, P. LUNNES, M.J. DANIELS, and A.E. OSBOURN: Host Range of a Plant Pathogenic Fungus Determined by Saponin Detoxifying Enzyme. Science, **267**, 371 (1995).

34. FAVEL, A., M.D. STEINMETZ, P. REGLI, E. VIDAL-OLLIVIER, R. ELIAS, and G. BALANSARD: In Vitro Antifungal Activity of Triterpenoid Saponins. Planta Med., **60**, 50 (1994).
35. DALSGAARD, K.: A Study of the Isolation and Characterization of the Saponin Quil A: Evaluation of its Adjuvant Activity with Special Reference to the Application in the Vaccination of Cattle Against Foot and Mouth Disease. Acta Vei. Scand., **695**, 1 (1978).
36. WU, J.Y., B.H. GARDNER, C.I. MURPHY, J.R. SEALS, C.R. KENSIL, J. RECCHIA, G.A. BELTZ, G.W. NEWMAN, and M.J. NEWMAN: Saponins Adjuvant Enhancement of Antigen-Specific Immune Responses to an Experimental HIV-1 Vaccine. J. Immunol., **148**, 1519 (1992).
37. WHITE, A.C., P. CLOUTIER, and R.T. COUGHLIN: A Purified Saponin Acts as an Adjuvant for a T-Independent Antigen. Adv. Exp. Med. Biol., **303**, 207 (1991).
38. NEWMAN, M.J., J.Y. WU, B.H. GARDNER, and K.J. MUNROE: Saponin Adjuvant Induction of Ovalbumin-Specific CD^{8+} Cytotoxic T Lymphocyte Responses. J. Immunol., **148**, 2357 (1992).
39. KIRKLEY, J.E., P.H. NAYLOR, D.J. MARCIANI, C.R. KENSIL, M. NEWMAN, and A.L. GOLDSTEIN: QS-21 Augments the Antibody Response to a Synthetic Peptide Vaccine Compared to Alum. Comb. Ther. [Proc. Int. Symp.] 1st 1991 (A.L. GOLDSTEIN, GARACI, eds.). New York: Plenum. 1992.
40. BOMFORD, R., M. STAPLETON, S. WINSOR, J.E. BEESLEY, E.A. JSSUP, K.R. PRICE, and G.R. FENWICK: Adjuvanticity and ISCOM Formation by Structurally Diverse Saponins. Vaccine, **10**, 572 (1992).
41. NEWMAN, M.J., J.Y. WU, R.T. COUGHLIN, C.I. MURPHY, J.R. SEALS, M.S. WYAND, and C.R. KENSIL: Immunogenicity and Toxicity Testing of an Experimental HIV-1 Vaccine in Nonhuman Primates. AIDS Res. Hum. Retroviruses, **8**, 1413 (1992).
42. NEWMAN, M.J., K.J. MUNROE, C.A. ANDERSON, C.I. MURPHY, D.L. PANICALI, J.R. SEALS, J.Y. WU, M.S. WYANO, and C.R. KENSIL: Induction of Antigen-Specific Killer T Lymphocyte Responses Using Subunit SIV mac 251 gag and enV Vaccines Containing QS-21 Saponin Adjuvant – Immunochemistry. AIDS Res. Hum. Retroviruses, **10**, 853 (1994).
43. YANG, C., and G. YANG: Immunologic Effect of *Panaxquinquefolium* Saponin in vitro and in vivo. Zhongguo Mianyixue Zazhi, **8**, 188 (1992).
44. HE, G., J. ZHANG, X. XUE, and J. WANT: Immune Function Stimulatory and Regulatory Action of Ginseng Saponins on Chronic Pulmonary Heart Diseases. Baiqiuen Yike Daxue Xuebao, **19**, 50 (1993).
45. RONNBERG, B., M. FEKADLI, and B. MOREIN: Adjuvant Activity of Nontoxic Quillaja Saponaria molina Components for Use in ISCOM Matrix. Vaccine, **13**, 1375 (1995).
46. TAKEDA, O., S. TANAKA, K. YAMASAKI, H. KHODA, Y. YWANAGA, and M. TSUJI: Screening for Molluscicidal Activity in Crude Drugs. Chem. Pharm. Bull., **37**, 1090 (1989).
47. THIILBORG, S.T., S.B. CHRISTENSEEN, C. CORNETT, C.E. OLSEN, and E. LEMMICH: Molluscicidal Saponins from *Phytolacca dodecandra*. Phytochemistry, **32**, 1167 (1993).
48. NDAMA, J., E. LEMMICH, and P. MOELGAARD: Release of Molluscicidal Saponins from *Phytolacca dodecandra* Aqueous Berry Extracts as Influenced by the Male Plant and the Extraction Procedure. Biochem. Syst. Ecol., **22**, 249 (1994).
49. HAMBURGER, M., I. SLACANIN, K. HOSTETTMANN, W. DYATMIKO, and S. JADI: Acetylated Saponins with Molluscicidal Activity from *Sapindus rarak*: Unambiguous Structure Determination by PMR and Quantitative Analysis. Phytochem. Anal., **3**, 231 (1992).
50. LEMMICH, E., C. CORNETT, P. FURU, C.L. JORSTIAN, A.D. KNUDSEN, and C.E. OLSEN: Molluscicidal Saponins from *Catunaregam nilotica*. Phytochemistry, **39**, 63 (1995).

51. Pakrashi, A., H. Ray, B.C. Pal, and S.B. Mahato: Sperm Immobilizing Effect of Triterpene Saponins from *Acacia auriculiformis*. Contraception, **43**, 475 (1991).
52. Garg, S., G. Dancel, S. Chabra, S.N. Upadhyay, and G.P. Talwar: Synergistic Spermicidal Activity of Neem Seed Extract Reetha Saponins and Quinine Hydrochloride. Contraception, **50**, 185 (1994).
53. Rajasekaran, M., A.G.R. Nair, W.J.G. Hellstron, and S.C. Sikka: Spermicidal Activity of an Antifungal Saponin Obtained from the Tropical Herb *Mollugo pentaphylla.* Contraception, **47**, 401 (1993).
54. Mahato, S.B.: Bioactive Gymnemic Acids and Congeners from *Gymnema sylvestre.* In: Studies in Natural Products Chemistry (A. Rahman, ed.), Vol. 18, pp. 649–676. Amsterdam: Elsevier 1996.
55. Kang, J.S., H. Koh, and T.K. Suh: Effect of Gymnemic Acid on the Elevation of Blood Glucose Concentration Induced with Oral Sucrose in Strepto-zotocin-Diabetic Rats. Hanyang Vidae Haksulchi, **10**, 587 (1990).
56. Harms, S.D., Z. Lu, S. Li, and Y. Cai: Hypoglycemic Effect of Saponin Isolated from Leaves of *Acanthopanax senticosus* (Rupr. et Maxim.). Zhongguo Zhongyao Zazhi, **19**, 683 (1994).
57. Jod, C.N., and J.Y. Kim: Hypoglycemic Action of Ginseng Saponins on Streptozotocin-Induced Diabetes in Rats (II). Koryo Insam Hakhoechi, **16**, 198 (1992).
58. Wang, D., H. Yu, S. Hu, P. Qiu, and H.W. Leng: Hypoglycemic Effect of Total Saponins of *Aralia decaisneana* in Rat and Mice Models. Zhongguo Yaoxue Zazhi, **30**, 407 (1995).
59. Tokuda, H., T. Konoshima, M. Kuzuka, and T. Kimura: Inhibition of 12-O-Tetradecanoylphorbol-13-acetate-promoted Mouse Skin Papilloma by Saponins. Oncology, **48**, 77 (1991).
60. Kim, W., and N.P. Jung: Effects of a Ginseng Saponin Fraction on the Tumoricidal Activity of Murine Macrophage against K562 Cells. Koryo Insam Hakhoechi, **13**, 24 (1989).
61. Hesegawa, H., S. Matsumiya, M. Uchiyama, T. Kurokawa, Y. Inouye, R. Kasai, S. Ishibashi, and K. Yamasaki: Inhibitory Effect of Some Triterpenoid Saponins on Glucose Transport in Tumor Cells and its Application to in vitro Cytotoxic and Antiviral Activities. Planta Med., **60**, 240 (1994).
62. Rao, A.V., and M.K. Sung: Saponins as Anticarcinogens. J. Nutr., **125**, 7175 (1995).
63. Sung, M.K., C.W.C. Kendall, M.M. Koo, and A.V. Rao: Effect of Soybean Saponins and Gypsophila Saponins on Growth and Viability of Colon Carcinoma Cells in Culture. Nutr. Cancer, **23**, 259 (1995).
64. Amarowicz, R., M. Shimoyamad, and K. Okubo: Hypocholesterolemic Effects of Saponins. Rocz. Panstw. Zakl. Hig., **45**, 125 (1994).
65. Xu, Q., Y. Zhao, and G. Cheng: Blood Lipid-decreasing Action of Total Saponins of *Panax notoginseng* (Bark). F.H. Chen. Zhongguo Zhongyao Zazhi, **18**, 367 (1993).
66. Li, J., Z. Lu, and Y. Lu: Effects of *Panax quinquefolium* Saponin on Serum Lipoprotein-Cholesterol and Its Anti-peroxidation in Hyperlipidemic Rats. Zhongguo Yaoxue Zazhi, **28**, 355 (1993).
67. Yamamoto, M., and S. Kameda (MARUZEN SEIYAKU KK): Feed Additives Containing Quillaja Saponins and/or Yucca Saponins for Poultry Egg Production with the Saponins and Low-Cholesterol Eggs Produced by the Poultry. Japanese patent JP 0507, 462 (1993).
68. Wu, C., Q. Yu, W. Liu, Y. Guo, and G. Zhang: A Study on the Antiaging Effect of Ginseng Stem and Leaf Saponin in Terms of the Free Radical Theory of Aging. Shenyang Yaoxueyuan Xuebao, **9**, 37 (1992).

69. PAN, H., L. HUANG, Z. ZHOU, G. LIANG, X. HONG, and Z. HUAN: Experimental Studies on Anti-Aging Effect of Total Saponin from Stalk and Leaf of Sanchi (*Panax notoginseng*). Zhongcaoyao, **24**, 581 (1993).
70. MA, C., Z. LU, and Y. JIANG: Antioxidative Effect of *Panax quinquefolium* Saponin on the Myocardium Injury Induced by Dexorubicin in Rats. Zhongguo Yaolixue Yu Dulixue Zazhi, **7**, 267 (1993).
71. YANG, S., J. QU, G. ZHONG, and W. ZHANG: Protective Effects of PQS on Oxidative Damage of Cultured Rat Cardiac Cells. Zhongguo Zhongyao Zazhi, **17**, 555 (1992).
72. WANG, S., Y. JIANG, G. ZHANG, and X. SUN: Anti-Free-Radical Damaging Action of PDS, PTS on the Myocardial Contractibility of the Isolated Rat Working Heart. Zhongguo Zhongyao Zazhi, **18**, 113 (1993).
73. WANG, S., R. SUN, S. LI, Y. WANG, X. XIAO, L. HAO, and X. MA: Anti-Aging Effects of Rhodosin and Ginseng Stem-Leaf Saponins. Baiqien Yike Daxue Xuebao, **20**, 329 (1994).
74. LI, H., and L. SHI: Effects of Total Saponinsfrom *Panax notogenseng* on Calcium Influx into Myocardial Cells. Zhongguo Yaote Xuebao, **11**, 213 (1990).
75. ZHANG, K., Y. ZOU, and C. ZHOU: Preventive Effects of Crude Saponins of Glabrous Greenbrier (*Smilax glabra*) on the cholesterol-Fed Atherosclerosis of Quails. Zhoncaoyao, **22**, 411 (1991).
76. HUANG, C., T. GU, Y. GUEAN, and J. WANG: Effects of *Panax notoginseng* Saponins on Myocardial Ischemia and Reprefusion Injury in Conscious Rabbit. Zhongguo Yaolixue Tongbao, **7**, 190 (1991).
77. LI, L., Z. WANG, Z. HUANG, Y. ZHANG, and J. WEI: Effects of *Panax notoginseng* Saponins on Acute Cerebral Ischemia. Zhonguo Yoolixue Tonbao, **1**, 56 (1991).
78. GAO, B.Y., X.J. LI, L. LIU, and B.H. ZHANG: Effect of Panaxatriol Saponins (PTS) Isolated from *Panax notoginseng* on Myocardial Ischemic Arrhythmia in Mice and Rats. Yaoxue Xuebao, **27**, 641 (1992).
79. WU, J., and J. SUN: Comparative Effects of *Panax notoginseng* Saponins, Verapamil, and Norepinephrine on Cerebral Circulation in Anesthetized Rats and Rabbits. Zhongguo Yaoli Xuebao, **13**, 520 (1992).
80. ZHONG, S.J., H. LI, W.J. ZHEG, C.W. SUN, X.K. MA, and L. WANG: Block Effect of PQS on Calcium Channel on Culture Rat Myocardiocyte. Zhongguo Yaoxue Zazhi, **29**, 525 (1994).
81. HANG, H., X. LIN, Q. WANG, J. WEI, and Y. CHEN: Protective Effects of *Panax notoginseng* saponins (PNS), *Panax Japonicus* Saponins (PJS) and Gypenosides (GP) on Myocardial Ischemia and Reperfusion Injury. Zhongguo Bingli Shengli Zazhi, **10**, 128 (1994).
82. LU, C., H. JIN, W. LU, and Y. CHEN: Effects of *Astragalus* saponins on Dog Cardiac Functions in vivo. Baiquen Yike Daxue Xuebao, **20**, 326 (1994).
83. JIANG, Y., G. ZHONG, C. SHAO, and G. YUE: Calcium Channel Blocking Effect of Panaxadiol Saponins and Panaxatriol Saponins on Cultured Cardiac Cells. Zhongguo Zhongyao Zazhi, **17**, 172 (1992).
84. SUI, D., Z. LU, L. MA, and Z. FAN: Effects of Saponins of *Acanthopanax senticosus* (Rupr. et. Maxim) Harms. On Myocardial Infarct Size in Acute Ischemic Dogs. Zhongguo Zhongyao Zazhi, **19**, 746 (1994).
85. LI, J., X. WANG, and H. ZHANG: Protective Effect of Saponin of Chinese and American inseng Stem and Leaf on Cells Infected by HSV-1. Zhongcaoyao, **23**, 249 (1992).
86. MARIA, L., and C. PIZZA: Structure and in vitro Antiviral Activity of Triterpenoid Saponins from *Calendula arvensis*. Planta Med., **57**, 250 (1991).

87. LIN, X., Y. CAI, and P. XIAO: Inhibition of SIV in vitro by *Cimicifuga dahurica* Saponin and Its Action Mechanism. Huaxi Yaoxue Zazhi, **9**, 221 (1994).
88. YOSHIKAWA, K., N. SHIMONO, and S. ARIHARA: Antisweet Substances, Jujuba saponins I–III from *Zizyphus jujuba*. Revised Structure of Ziziphin. Tetrahedron Lett., **32**, 7059 (1991).
89. YOSHIKAWA, K., S. ARIHARA, and K. MATSUURA: A New Type of Antisweet Principles occurring in *Gymnema sylvestre*. Tetrahedron Lett., **32**, 789 (1991).
90. YOSHIKAWA, K., S. TUMURA, K. YAMADA, and S. ARIHARA: Antisweet Natural Products. VII Hodulosides I, II, III, IV and V from the Leaves of *Hovenia dulcis* Thunb. Chem. Pharm. Bull., **40**, 2287 (1992).
91. IMOTO, T.: Sugar Discriminations and Plant Originated Glycosides and Gymnemic Acids. Seibatsi Butsuri, **30**, 146 (1990).
92. KIM, H.S., S.H. KIM, M.K. LEE, K.J. CHOI, and S.C. KIM: Effects of Ginseng Leaf Saponins on the Development of Morphine Tolerance and Dependence in Mice. Koryo Insam Hakoechi, **13**, 8 (1989).
93. KIM, H.S., J.G. KANG, H.M. RHEU, D.H. CHO, and K.W. OH: Blockade by Ginseng Total Saponin of the Development of Methamphetamine Reverse Tolerance and Dopamine Receptor Supersensitivity in Mice. Planta Med., **61**, 22 (1995).
94. KIM, H.S., J.G. KANG, Y.H. SEONG, K.Y. NAM, and K.W. OH: Blockade by Ginseng Total Saponin of the Development of Cocaine Induced Reverse Tolerance and Dopamine Receptor Supersensitivity in Mice. Pharmacol. Biochem. Behav., **50**, 23 (1995).
95. KIM, H.S., K.W. OH, and Y.H. SEONG: The Role of Ginseng Total Saponins in the Inhibition of the Development of Analgesic Tolerance to Morphine. Koryo Insam Hokhoechi, **15**, 179 (1991).
96. MAJESTER-SAVORNIN, B., R. ELIAS, A.M. DIAZ-LANZA, G. BALANSARD, M. GASQUET, and F. DELMAS: Saponins of Ivy Plant, *Hedera helix*, and their Leishmanicidic Activity. Planta Med., **57**, 260 (1991).
97. GHOSH, M., S.P. SINHA BABU, N.C. SUKUL, and S.B. MAHATO: Antifilarial Effect of Two Triterpenoid Saponins Isolated from *Acacia auriculiformis*. Indian J. Exp. Biol., **31**, 604 (1993).
98. WALLER, G.R., M. JURZYSTA, and R.L.Z. THORNE: Allelopathic Activity of Root Saponin from Alfalfa (*Medicago sativa* L.) on Weeds and Wheat. Bot. Bull. Acad. Sin., **34**, 1 (1993).
99. OLESZEK, W.: Allelopathic Potentials of Alfalfa (*Medicago sativa*) Saponins: Their Relation to Antifungal and Hemolytic Activities. J. Chem. Ecol., **19**, 1063 (1993).
100. HIJI, Y.: Gymnemic Acid for Prevention of Dental Caries. US Patent, US4, 912, 089, 27 Mar 1990.
101. KATO, T., N. NAGATA, and S. WATABE (Rohto Pharmaceutical Co. Ltd.): Ethanol-Absorption Inhibitors Containing Saponins. Japanese patent JP 04,145,028, 19 May 1992.
102. KENNELLY, E.J., L. CAI, N.C. KIM, and A.D. KINGHORN: Abrusoside E, a Further Sweet-Tasting Cycloartane glycoside from the Leaves of *Abrus precatorius*. Phytochemistry, **41**, 1381 (1996).
103. PAUTHE-DAYDE, D., M. ROCHD, and M. HENRY: Triterpenoid Saponins Production in Callus and Multiple Shoot Cultures of *Gypsophila* spp. Phytochemistry, **29**, 483 (1990).
104. BESSON, V., C. LAVAUD, G. MASSIOT, L.E. MEN-OLIVIER, S. BIANCHI, P. FLAMENT, and Y. DATTEE: Medicagenic Acid and Saponins in Callus and Suspension Cultures of Alfalfa. Phytochemistry, **28**, 1379 (1989).

105. Yu, X.: Preliminary Study on Saponin Content in Callus of Five Leaf Gynostemma (*Gynostemma pentaphyllum*). Zhongcaoyao, **22**, 468 (1991).
106. Katakura, M., and T. Kimura: Production of Saikosaponins by Tissue Culture of *Bupleurum falcatum* L. Endo. I. Eng., **7**, 97 (1991).
107. Kubo, Y., and M. Murakami: Tissue Culture of *Panax japonicus*. III. Saponin Production in Adventitious Root Cultures. Toyama-ken Yakuji Kenkyusho Nenpo **18**, 60 (1990).
108. Zhou, L.G., G.Z. Zheng, F.Y. Gan, S.L. Wang, C.R. Yang, and C. Xu: Separation and Identification of Main Medicinal Saponin Components from Cell Cultures of *Panax notoginseng*. F.H. Chem. Yaoxue Xuebao, **26**, 876 (1991).
109. Ionkova, I.: Producing Triterpene Saponins by Conventional and Genetically-Transformed Cultures of *Astragalus hamosus* L. Probl. Farmakol. Farm., **5**, 32 (1991).
110. Choi, K.T., and J.C. Park: Production of Ginseng Saponin in Tissue Culture of Ginseng (*Panax ginseng* C.A. Meyer). Fiziol. Rast., **41**, 891 (1994).
111. Zhou, Y., M. Hirotani, H. Rui, and T. Furuya: Two Triglycosidic Triterpene Astragalosides from Hairy Root Cultures of *Astragalus membranaceus*. Phytochemistry, **38**, 1407 (1995).
112. Mahato, S.B.: Glycosides, Saponins and Sapogenins. In: Second Supplement to the 2nd Edition of Rodd's Chemistry of Carbon Compounds, Vol. II B (Partial) and E (M. Sainsbury, ed.) pp. 509–554. Amsterdam: Elsevier, 1994.
113. Sakai, Y., T. Takeshita, J. Kinjo, Y. Ito, and T. Nohara: Leguminous Plants 17. Two New Triterpenoid Sapogenals and a New Saponin from *Abrus cantoniensis* (II). Chem. Pharm. Bull., **38**, 824 (1990).
114. Choi, Y.H., R.A. Hussain, J.M. Pezzuto, A.D. Kinghorn, and J.F. Morton: Abrusosides A-D, Four Novel Sweet-tasting Triterpene Glycosides from the Leaves of *Abrus precatorius*. J. Nat. Prod., **52**, 1118 (1989).
115. Mahato, S.B., B.C. Pal, and A.K. Nandy: Structure Elucidation of Two Acylated Triterpenoid Bisglycosides from *Acacia auriculiformis Cunn.* Tetrahedron, **48**, 6717 (1992).
116. Kohda, H., S. Tanaka, and Y. Yamaoko: Saponins from Leaves of *Acantho-panax hypoleucus Makino*. Chem. Pharm. Bull., **38**, 3380 (1990).
117. Sawada, H., M. Miyakoshi, S. Isoda, Y. Ida, and J. Shoji: Saponins from Leaves of *Acanthopanax siebodianus*. Phytochemistry, **34**, 1117 (1993).
118. Miyakoshi, M., Y. Ida, S. Isoda, and J. Shoji: 3α-Hydroxy-oleanane-type Triterpene Glycosyl Esters from Leaves of *Acanthopanax spinosus. Phytochemistry*, **34**, 1599 (1993).
119. Miyakoshi, M., Y. Ida, S. Isoda, and J. Shoji: 3 *Epi*-oleanane Type Triterpene Glycosyl Esters from Leaves of *Acanthopanax spinosus*. Phytochemistry, **33**, 891 (1993).
120. Lacaille-Dubois, M.A., B. Hanquet, A. Rustaiyan, and H. Wagner: Squarroside A, a Biologically Active Triterpene Saponin from *Acanthophyllum squarrosum*. Phytochemistry, **34**, 489 (1993).
121. Ida, Y., Y. Satoh, M. Katoh, M. Katsumata, M. Nagasao, K. Yamaguchi, H. Kamei, and J. Shoji: Achyranthosides A and B, Novel Cytotoxic Saponins from *Achyranthes fauriei* Root. Tetrahedron Lett., **35**, 6887 (1994).
122. Ida, Y., Y. Satoh, M. Katsumata, M. Nagasao, and J. Shoji: Achyranthosides C and D, Novel Glucuronide Saponins from *Achyranthes fauriei* Root. Chem. Pharm. Bull., **43**, 896 (1995).
123. Faini, F., M. Castillo, R. Torres, G. Delle Monache, and E. Gacs-Baitz: Malabaricane Triterpene Glucosides from *Adesmia aconcaquensis*. Phytochemistry, **40**, 885 (1995).

124. FANG, S., Z. HE, J. GAO, and P. WANG: Triterpenoid Glycosides from *Adina rubella*. Phytochemistry, **39**, 1241 (1995).
125. IKUTA, A., and H. ITOKAWA: 30-Noroleanane Saponins from Callus Tissues of *Akebia quinata*. Phytochemistry, **28**, 2663 (1989).
126. IKUTA, A.: Saponins and Triterpenes from Callus Tissues of *Akebia trifoliata* and Comparison with the Constituents of Other Lardizabalaceous Callus Tissues. J. Nat. Prod., **58**, 1378 (1995).
127. PAL, B.C., B. ACHARI, K. YOSHIKAWA, and S. ARIHARA: Saponins from *Albizia lebbeck*. Phytochemistry, **38**, 1287 (1995).
128. ORSINI, F., F. PELIZZONI, and L. VEROTTA: Saponins from *Albizzia lucida*. Phytochemistry, **30**, 4111 (1991).
129. LI, D., N.L. OWEN, P. PERERA, C. ANDERSSON, L. BOHLIN, P.A. COX, R.J. PUGMIRE, C.L. MAYNE, and D.M. GRANT: Structure Elucidation of Three Triterpenoid Saponins from *Alphitonia zizypleoides* Using 2D NMR Techniques. J. Nat. Prod., **59**, 218 (1994).
130. RASTRELLI, L., C. PIZZA, P. SATURNINO, O. SCHETTINO, and A. DINI: Studies on the Constituents of *Amaranthus caudatus* (Kiwicha) Se eds. Isolation and Characterization of Seven New Triterpene Saponins. J. Agric. Food Chem., **43**, 904 (1995).
131. KOHDA, H., S. TANAKA, Y. YAMOKA, and Y. OHHARA: Saponins from *Amaranthus hypochodriacus*. Chem. Pharm. Bull., **39**, 2609 (1991).
132. BRANDAO, M.G.L., M.A. LACAILEE-DUBOIS, M.A. TEIXEIRA, and H. WAGNER: Triterpene Saponins from the Roots of *Ampelozizphus amaronicus*. Phytochemistry, **31**, 352 (1992).
133. BRANDAO, M.G.L., M.A. LACAILLE-DUBOIS, M.A. TEIXEIRA, and H. WAGNER: A Dammarane-Type Saponin from the Roots of *Ampelozizyphus amaronicus*. Phytochemistry, **34**, 1123 (1993).
134. SHOJI, N., A. UMEYAMA, K. YOSHIKAWA, and S. ARIHARA: Structures of Anagallosaponins I-V and Their Companion Substances from *Anagallis arvensis* L. Chem. Pharm. Bull., **42**, 1874 (1994).
135. SHOJI, N., A. UMEYAMA, K. YOSHIKAWA, and S. ARIHARA: Triterpenoid Glycosides from *Anagallis arvensis*. Phytochemistry, **37**, 1397 (1994).
136. MAHATO, S.B., N.P. SAHU, S.K. ROY, and S. SEN: Structure Elucidation of Four New Triterpenoid Oligoglycosides from *Anagallis arvensis*. Tetrahedron, **47**, 5215 (1991).
137. LI, X., C. YANG, Y. LIU, R. KASAI, K. OHTANI, and K. YAMASAKI: Triterpenoid Glycosides from *Anemocleama glaucifolium*. Phytochemistry, **39**, 1175 (1995).
138. WANG, M., F. WU, and Y. CHEN: Triterpenoid Saponins from *Anemone hupehensis*. Phytochemistry, **34**, 1395 (1993).
139. DIJOUX, M.G., C. LAVAUD, G. MASSIOT, L. LEMENOLIVER, and D.M. SHEELEY: A Saponin from Leaves of *Aphloia madagascariensis*. Phytochemistry, **34**, 497 (1993).
140. HU, M., K. OGAWA, Y. SASHIDA, and P. XIAO: Triterpenoid Glucurono-saponins from Root Bark of *Aralia armata*. Phytochemistry, **39**, 179 (1995).
141. MIYASE, T., N. SUTOH, D.M. ZHANG, and A. UENO: Araliasaponins XII-XVIII, Triterpene Saponins from the Roots of *Aralia chinensis*. Phytochemistry, **42**, 1123 (1996).
142. KAWAI, H., M. NISHIDA, Y. TASHIRO, M. KUROYANAGI, A. UENO, and M. SATAKE: Studies on the Structures of Udosaponins A, B, C, D, E and F from *Aralia cordates* Thumb. Chem. Pharm. Bull., **37**, 2318 (1989).
143. MIYASE, T., K. SHIOKAWA, D.M. ZHANG, and A. UENO: Araliasaponins I-XI, Triterpene Saponins from the Roots of *Aralia decaisneana*. Phytochemistry, **41**, 1411 (1996).

144. SAITO, S., S. SUMITA, N. TAMURA, Y. NAGAMURA, K. NISHIDA, M. ITO, and I. ISHIGURO: Saponins from the Leaves of *Aralia elata Seem* (Araliaceae). Chem. Pharm. Bull., **38**, 411 (1990).

145. JIANG, Y., S. XU, X. YAO, and Y. CHEN: Chemical Constituents and Pharmacological Activities of *Aralia elata.* Shenyang Yaoxueynan Xuebao, **8**, 290 (1991).

146. SAKAI, S., M. KATSUMATO, Y. SATOH, M. NAGASAO, M. MIYAKOSHI, Y. IDA, and J. SHOJI: Oleanolic Acid Saponins from Root Bark of *Aralia elata.* Phytochemistry, **35**, 1319 (1994).

147. SATOH, Y., S. SAKAI, M. KATSUMATO, M. NAGASAO, M. MIYAKOSHI, Y. IDA, and J. SHOJI: Oleanolic acid Saponins from Root Bark of *Aralia elata.* Phytochemistry, **36**, 147 (1994).

148. YU, S., D. YU, and X. LIANG: Triterpenoid Saponins from the Roots of *Aralia spinifolia.* J. Nat. Prod., **57**, 978 (1994).

149. JIA, Z., K. KOIKE, T. NIKAIDO, T. OHMOTO, and M. NI: Triterpenoid Saponins from *Ardisia crenata* and their Inhibitory Activity on CAMP Phosphodiesterase. Chem. Pharm. Bull., **42**, 2309 (1994).

150. JIA, Z., K. KOIKE, T. NIKAIDO, and T. OHMOTO: Two Novel Triterpenoid Pentasaccharides with an Unusual Glycosylglycerol Side Chain from *Ardisia crenata*, Tetrahedron, **50**, 11853 (1994).

151. WANG, M., X. GUAN, X. HAN, and S. HONG: A New Triterpenoid Saponin from *Ardisia crenata.* Planta Med., **58**, 205 (1992).

152. DE TOMMASI, N., S. PIACENTE, F. DESIMONE, C. PIZZA, and Z.L. ZHOU: Characterization of Three New Triterpenoid Saponins from *Ardisia japonica.* J. Nat. Prod., **56**, 1669 (1993).

153. CHARROUF, Z., J.M. WIERUSZESKI, S. FKIH-TETOUANI, Y. LEROY, M. CHARROUF, and B. FOURNET: Triterpenoid Saponnins from *Argania spinosa.* Phytochemistry, **31**, 2079 (1992).

154. SHAO, Y., B. ZHOU, K. MA, and H. WU: New Triterpenoid Saponins, Asterbatanoside D and E from *Aster batangensis.* Planta Med., **61**, 246 (1995).

155. SHAO, Y., B. ZHOU, K. MA, H. WU, L. LIN, and G. A. CORDELL: Medicagenic Acid Saponins from *Aster batangensis.* Phytochemistry, **39**, 875 (1995).

156. NAGAO, T., R. TANAKA, Y. IWASE, and H. OKABE: Studies on the Constituents of *Aster scaber* Thunb. IV. Structure of Four Echinocystic acid Glycosides Isolated from the Herb. Chem. Pharm. Bull., **41**, 659 (1993).

157. NAGAO, T., Y. IWASE, and H. OKABE: Studies on the Constituents of *Aster scaber* Thumb. V. Structures of Six New Echinocystic Acid Glycosides Isolated from the Herb. Chem. Pharm. Bull., **41**, 1562 (1993).

158. TANAKA, R., T. NAGAO, H. OKABE, and T. YAMAUCHI: Studies on the Constituents of *Aster tataricus* L.f.I.V. Structures of Aster Saponins Isolated from the Herb. Chem. Pharm. Bull., **38**, 1153 (1992).

159. CHENG, D., and Y. SHAO: Terpenoid Glycosides from the Roots of *Aster tataricus.* Phytochemistry, **35**, 173 (1994).

160. NAGAO, T., H. OKABE, and T. YAMAUCHI: Studies on the Constituents of *Aster tataricus* L.f.III. Structures of Aster Saponins E and F Isolated from the Root. Chem. Pharm. Bull., **38**, 783 (1990).

161. SHAO, Y., B. ZHOU, K. MA, and H. WU: Echinocystic Acid Saponins from *Aster Yunnanensis.* J. Nat. Prod., **58**, 837 (1995).

162. SHAO, Y., B. ZHOU, L. LIN, and G.A. CORDELL: Triterpene Saponins from *Aster yunnanensis.* Phytochemistry, **38**, 1487 (1995).

163. SHAO, Y., B. ZHOU, J. GAO, L. LIN, and G.A. CORDELL: Glycosides from *Aster yunnanensis*. Phytochemistry, **38**, 675 (1995).

164. ESPADA, A., C. JIMENEZ, J. RODRIGUEZ, P. CREWS, and R. RIGUERA: Sarasinosides D-G Four New Triterpenoid Saponins from the Sponge *Asteropus sarasinosum*. Tetrahedron, **48**, 8685 (1992).

165. ORSINI, F., L. VEROTTA, L. BARBONI, N.A. EL-SEBAKHY, A.M. ASAAD, R.M. ABDALLAH, and S.M. TOAIMA: Cycloaratane Triterpene Glycosides from *Astragalus alexandrinus*. Phytochemistry, **35**, 745 (1994).

166. WANG, H.K., K. HE, L. JI, Y. TEZUKA, and K.I. TOHRU: Asernestioside C, a New Saponin from the Roots of *Astragalus ernestii Comb*; First Example of Negative Nuclear Overhauser Effect in the Saponins. Chem. Pharm. Bull., **37**, 2041 (1989).

167. ZHU, Y.Z., S. LU, Y. OKADA, M. TAKATA, and T. OKUYAMA: Two New Cycloartane-type Glycosides Mongholicoside I and II from Aerial Part of *Astragalus mongholicus* Bungi. Chem. Pharm. Bull., **40**, 2230 (1992).

168. EL-SEBAKHY, N.A., F.M. HADRAZ, R.M. ABDALLAH, A.M. ASAAD, F. ORSINI, F. PELIZZONI, G. SELLO, and L. VEROTTA: Cycloartane Triterpene Glycosides from Egyptian *Astragalus* Species. Phytochemistry, **29**, 3271 (1990).

169. GARIBOLDI, P., F. PELIZZONI, M. TATO, L. VEROTTA, N. EL-SEBAKHY, A.M. ASAAD, R.M. ABDALLAH, and S.N. TOAIMA: Cycloartane Triterpene Glycosides from *Astragalus trigonus*. Phytochemistry, **40**, 1755 (1995).

170. HILLER, K., M. LESKA, E. GRUENDEMANN, G. DUBE, A. KARWATZKI, and P. FRANKE: Structure of Bisdesmosidic Saponins from *Astrantia major* L. Pharmazie, **45**, 615 (1990).

171. JAIN, P., and D.K. KULSHRESHTHA: Bacoside A_1, a Minor Saponin from *Bacopa monniera*. Phytochemistry, **33**, 449 (1993).

172. RASTOGI, S., R. PAL, and D.K. KULSHRESHTHA: Bacoside A_3, a Triterpenoid Saponin from *Bacopa monniera*. Phytochemistry, **36**, 133 (1994).

173. GARAI, S., S.B. MAHATO, K. OHTANI, and K. YAMASAKI: Bacopasaponin D-A Pseudojujubogenin Glycosides from *Bacopa monniera*. Phytochemistry, **43**, 447 (1996).

174. PAL, B.C., T. CHAUDHURI, K. YOSHIKAWA, and S. ARIHARA: Saponins from *Barringtonia acutangula*. Phytochemistry, **35**, 1315 (1994).

175. PAL, B.C., B. ACHARI, and K.R. PRICE: A Triterpenoid Glucoside from *Barringtonia acutangula*. Phytochemistry, **30**, 4177 (1991).

176. SCHÖPKE, T., V. WRAY, B. RZAZEWSKA, and K. HILLER: Bellissaponins BA_1 and BA_2, Acylated Saponins from *Bellis perennis*. Phytochemistry, **30**, 627 (1991).

177. SCHÖPKE, T., V. WRAY, A. KUNATH, and K. HILLER: Bayogenin and Asterogenic Acid Glycosides from *Bellis perennis*. Phytochemistry, **31**, 2555 (1992).

178. SCHÖPKE, T., K. HILLER, V. WRAY, K.D. KOPPEL, K. YAMASAKI, and R. KASAI: Triterpenoid Saponins from *Bellis sylvestris*, I. Structures of the Major Deacyl Saponins. J. Nat. Prod., **57**, 1279 (1994).

179. SCHÖPKE, T., M.I.H. AGHA, V. WRAY, and K. HILLER: Triterpenoid Saponins from *Bellium bellidioides*. Phytochemistry, **36**, 449 (1994).

180. SCHÖPKE, T., V. WRAY, M. NIMTZ, and K. HILLER: Triterpenoid saponins from *Bellium bellidioides*. Structure of the Minor Deacylsaponins. Phytochemistry, **41**, 1399 (1996).

181. RIDOUT, C.L. K.R. PRICE, G. PARKIN, M.G. DIJOUX, and C. LAVAUD: Saponins from Sugar Beet and the *Floc Problem*. J. Agric. Food. Chem., **42**, 279 (1994).

182. FUCHINO, H., T. SATOH, and N. TANAKA: Chemical Evaluation of *Betula* Species in Japan. I. Constituents of *Betula ermanii*. Chem. Pharm. Bull., **43**, 1937 (1995).

183. OHASHI, K., H. KOJIMA, T. TANIKAWA, Y. OKUMURA, K. KAWAZOE, N. TATARA, H. SHIBUYA, and I. KITAGAWA: Indonesian Medicinal Plants. IX. Chemical Structures of Gongganosides A, B and C, Three New Quinovic Acid Glycosides from the Bark of *Bhesa paniculata* (Celastraceae). Chem. Pharm. Bull., **42**, 1596 (1994).

184. ESPADA, A., J. RODRIGUEZ, M.C. VILLAVERDE, and R. RIGUERA: Boussingoside D_1, a New Triterpenoid Saponin from *Boussingaultia baselloides.* Liebigs Ann. Chem., **3**, 291 (1991).

185. ESPADA, A., J. RODRIGUEZ, M.C. VILLAVERDE, and R. RIGUERA: Hypoglycemic Triterpenoid Saponins from *Boussingaultia baselloides.* Can. J. Chem., **68**, 2039 (1990).

186. OOBYYASHI, K., K. YOSHIKAWA, and S. ARIHARA: Structural Revision of Bryonoside and Structure Elucidation of Minor Saponins from *Bryonia dioica.* Phytochemistry, **31**, 943 (1992).

187. EBATA, N., K. NAKAJIMA, H. TAGUCHI, and H. MITSUHASHI: Isolation of New Saponins from the Root of *Bupleurum falcatum* L. Chem. Pharm. Bull., **38**, 1432 (1990).

188. PISTELLI, L., A. CAMMILLI, A. MANUNTA, A. MARSILI, and I. MORELLI: Triterpenoid Saponins and Flavonoid Glycosides from *Bupleurum falcatum* subsp. *cirnuum.* Phytochemistry, **33**, 1537 (1993).

189. EBATA, N., K. NAKAJIMA, K. HAYASHI, M. OKADA, and M. MARUNO: Saponins from the Root of *Bupleurum falcatum.* Phytochemistry, **41**, 895 (1996).

190. GUINEA, M.C., J. PARELLADA, M.A. LACAILLE-DUBOIS, and H. WAGNER: Biologically Active Triterpene Saponins from *Bupleurum fruticosum.* Planta Med., **60**, 163 (1994).

191. PISTELLI, L., A.R. BILIA, A. MARSILI, N. DE TOMMASI, and A. MANUNTA: Triterpenoid Saponins from *Bupleurum fruticosum.* J. Nat. Prod., **56**, 233 (1993).

192. ZHANG, R.Y., X.K. CHEN, X.B. YANG, W.Y. HE, and L. TAN: Structures of Saikosaponin M and N from *Bupleurum smithi.* Yaoxue Xuebao, **29**, 684 (1994).

193. LUO, H.S., Y.Y. ZHAO, L.B. MA, Y.Z. JIN, J.R. PENG, and R.Y. ZHANG: Isolation and Identification of Saikogenin Q and Saikosaponin Q from *Bupleurum smithi.* Yaoxue Xuebao, **30**, 435 (1995).

194. TAKEDA, T., T. NAKAMURA, S. TAKASHIMA, O. YANO, and Y. OGIHARA: Studies on the Constituents of *Calliandra anomala* (Kunth) Macbr. I. Structure Elucidation of Two Acylated Triterpenoidal Saponins. Chem. Pharm. Bull., **41**, 2132 (1993).

195. NAKAMURA, T., T. TAKEDA, and Y. OGIHARA: Studies on the Constituents of *Calliandra anomala* (Kunth) Macbr. II. Structure Elucidation of Four Acylated Triterpenoidal Saponins. Chem. Pharm. Bull., **42**, 1111 (1994).

196. YOSHIKAWA, M., E. HARADA, T. MURAKAMI, H. MATSUDO, J. YAMAHARA, and N. MURAKAMI: Camellia Saponins B_1, B_2, C_1 and C_2, New Type Inhibitors of Ethanol Absorption in Rats from the Seeds of *Camellia japonica* L. Chem. Pharm. Bull., **42**, 742 (1994).

197. AHMAD, W., K. USMANGHANI, I. AHMAD, V.U. AHMAD, and T. MIYASE: Isolation and Characterization of Saponins from *Castanospermum australe* Cunn. et. Fraster. Chem. Pharm. Bull., **42**, 314 (1994).

198. AHMAD, V.U., W. AHMAD, and K. USMANGHANI: Triterpenoid Saponins from Leaves of *Castanospermum australe.* Phytochemistry, **31**, 2805 (1992).

199. SAHU, N.P., S.K. ROY, and S.B. MAHATO: Spectroscopic Determination of Structures of Triterpenoid Trisaccharides from *Centella asiatica.* Phytochemistry, **28**, 2852 (1989).

200. GUPTA, D., and J. SINGH: Triterpenoid Saponins from *Centipeda minima.* Phytochemistry, **28**, 1197 (1989).

201. CALISKAN, O.A., H. ANIL, and G.R. STEPHENSON: Cephalaria Saponin A, a New Bidesmosidic Triterpene Saponin from *Cephalaria transylvanica.* J. Nat. Prod., **57**, 1001 (1994).

202. ALANKUS-CALISKAN, O., and H. ANIL: A Bidesmosidic Triterpene Saponin from *Cephalaria transylvanica.* Phytochemistry, **38**, 1493 (1995).

203. KIRMIZIGUL, S., and H. ANIL: An Acidic Triterpene Glycoside from *Cephalaria transsylyvanica.* Phytochemistry, **36**, 1555 (1994).

204. KIRMIZIGUL, S., H. ANIL, and M.E. ROSE: Triterpenoid Glycosides from *Cephalaria transsylvanica.* Phytochemistry, **39**, 1171 (1995).

205. MIZUI, F., R. KASAI, K. OHTANI, and O. TANAKA: Saponins from Bran of Quinoa, *Chenopodium quinoa* Wild. II. Chem Pharm. Bull., **38**, 375 (1990).

206. KUSANO, A., K. SHIMIZU, M. IDOJI, M. SHIBANO, K. MINOURA, and G. KUSANO: Studies on the Constituents of *Cimicifuga* species. XVI. Three New Cycloartane Xylosides from the Aerial Parts of *Cimicifuga simplex.* Chem. Pharm. Bull., **43**, 279 (1995).

207. SHAO, B., G. QIN, R. XU, H. WU, and K. MA: Triterpenoid Saponins from *Cleamatis chinensis.* Phytochemistry, **38**, 1473 (1995).

208. BAHUGUNA, R.P., J.S. JANGWAN, T. KAIYA, and J. SAKAKIBARA: Clemontanoside A, a Bisglycoside from *Clematis montana.* Phytochemistry **28**, 2511 (1989).

209. THAPLIYAL, R.P. and R.P. BAHUGUNA: Clemontanoside E, an New Saponin from *Clematis montana.* Int. J. Pharmacogn., **32**, 373 (1994).

210. WHANG, W.: Aromatic Compounds and Triterpenoidal Saponins from *Clematis koreana* var. *umbrosa.* Arch. Pharmacol. Res., **17**, 5 (1994).

211. KUNTZE, O., S.R. XUE, J.Q. LIU, G. WANG, J.Q. SHI, Q.J. WU, and S.Z. HU: Triterpenoid Saponin of *Clinopodium chinense* (Benth.) O. Yaoxue Xuebao, **27**, 207 (1992).

212. LIU, Z., Z. JIA, R.G. CATES, D. LI, and N.L. OWEN: Triterpenoid Saponins from *Clinopodium chinensis.* J. Nat. Prod., **58**, 184 (1995).

213. LIU, Z., D. LI, N.L. OWEN, D.M. GRANT, R.G. CATES, and Z. JIA: Oleanane Triterpene Saponins from the Chinese Medicinal herb *Clinopodium chinensis.* J. Nat. Prod., **58**, 1600 (1995).

214. MORI, F., T. MIYASE, and A. UENO: Oleanane-Triterpene Saponins from *Clinopodium chinese* var. *Parviflovrum* Phytochemistry, **36**, 1485 (1994).

215. YAMAMOTO, A., T. MIYASE, A. UENO, and T. MAEDA: Clinoposaponins I-V, New Oleanane Triterpene Saponins from *Clinopodium gracile.* O. Kuntze Chem. Pharm. Bull., **41**, 1270 (1993).

216. YAMAMOTO, A., H. SUZUKI, T. MIYASE, A. UENO, and T. MAEDA: Clinoposaponins VI and VIII, Two Oleanane-Triterpene Saponins from *Clinopodium micranthum.* Phytochemistry, **34**, 485 (1993).

217. XUE, S., J. LIU, and G. WANG: Triterpenoid Saponins from *Clinopodium polycephalum.* Phytochemistry, **31**, 1049 (1992).

218. JOSHI, B.S., K.M. MOORE, S.W. PELLETIER, M.S. PUAR, and B.N. PRAMANIK: Saponins from *Collinsonia canadensis.* J. Nat. Prod., **55**, 1468 (1992).

219. SEAFORTH, C.E., S. MOHAMMED, A. MAXWELL, W.F. TINTO, and W.F. REYNOLDS: Mabioside A, New Saponin from *Colubrina elliptica.* Tetrahedron Lett., **33**, 4111 (1992).

220. TINTO, W., W.F. REYNOLDS, C.E. SEAFORTH, S. MOHAMMED, and A. MAXWELL: New Bitter Saponins from the Bark of *Colubrina elliptica*: Proton and Carbon-13 Assignments by 2D NMR Spectroscopy. Magn. Reson. Chem., **31**, 859 (1993).

221. OULAD-ALI, A., D. GUILLAUME, B. WENIGER, Y. JIANG, and R. ANTON: Mabiosides C-E: Triterpenoid Saponins from the Bark of *Colubrina elliptica.* Phytochemistry, **36**, 445 (1994).

222. ABDUL QUADER, M., A.I. GRAY, P.G. WATERMAN, C. LAVAUD, G. MASSIOT, C.M. HASAN, and M.D. AHMAD: Capsugenin-25-30-O-β-D-diglucopyranoside: a New Glycoside from the Leaves of *Corchorus capsularies.* J. Nat. Prod., **53**, 527 (1990).

223. ASADA, Y., T. UEOKA, and T. FURUYA: Novel Acylated Saponins from Montbretia (*Crocosmia crocosmiiflora*). Isolation of Saponins and the Structures of Crocosmiosides A, B and H. Chem. Pharm. Bull., **37**, 2139 (1989).
224. ASADA, Y, T. UEOKA, and T. FURUYA: Novel Acylated Saponins from Montbretica (*Crocosmia crocosmiiflora*). II. The Structures of Crocosmiosides C, D, E, F, G and I. Chem. Pharm. Bull., **38**, 142 (1990).
225. ASADA, Y., M. IKENO, and T. FURUYA: Acylated Saponins, Masonosides A-C, from the Corms of *Crocosmia masoniorum*. Phytochemistry, **35**, 757 (1994).
226. ASADA, Y., M. IKENO, T. UEOKA, T. FURUYA: Desacylsaponins, Desacylmasonosides 1, 2 and 3 from the Corns of *Crocosmia masonorum*. Chem. Pharm. Bull., **37**, 2747 (1989).
227. ASADA, Y., M. IKENO, and T. FURUYA: Minor Desacylsaponins from the Corms of *Crocosmia masoniorum*. Phytochemistry, **36**, 455 (1994).
228. GARIBOLDI, P., L. VEROTTA, and B. GOBETTA: Saponins from *Crossopteryx febrifuga*. Phytochemistry, **29**, 2629 (1990).
229. DING, Y., J. KINJO, C. YANG, and T. NOHARA: Studies of the Constituents of Leguminous Plants. Part XX. Oleanane Glycosides from *Crotalaria albida*. Chem. Pharm. Bull., **39**, 496 (1991).
230. REZNICEK, G., J. JORENITSCH, W. ROBIEN, and W. KUBELKA: Saponins in Cyclamen Species: Configuration of Cyclamiretin C and Structure of Isocyclamin. Phytochemistry, **28**, 825 (1989).
231. PAMBOU, T.H., B. KOUDOGBO, J.C. TABET, and E. CASADEVALL: A Triterpene Saponin from *Cylicodiscus gabunensis*. Phytochemistry, **29**, 2723 (1990).
232. TCHIVOUNDA, H.P., B. KOUDOGBO, Y. BESACE, and E. CASADEVALL: Triterpene Saponins from *Cylicodiscus gabunensis*. Phytochemistry, **30**, 2711 (1991).
233. KONG, J., X. LI, B. WEI, and C. YANG: Triterpenoid Glycosides from *Decaisnea fargesii*. Phytochemistry, **33**, 427 (1993).
234. KUBO, T., S. HAMADA, T. NOHARA, Z.H. WANG, H. HIRAYAMA, K. IKEGAMI, K. YASUKAWA, and M. TAKIDO: Leguminous Plants. XIV. Study on the Constituents of *Desmodium styracifolium*. Chem. Pharm. Bull., **37**, 2229 (1989).
235. MALAVIYA, N., R. PAL, and N.M. KHANNA: Saponins from *Deutzia corymbosa*. Phytochemistry, **30**, 2798 (1991).
236. TAGAWA, N.T., Y. INATOMI, M. NISHI, H. MURATA, A. INADA, and S. AIBARA: Two New Hopane-Triterpene Glycosides from a Fern *Diplazium subsinuatum* (Wall. ex. Hook et Grev.). Chem. Pharm. Bull., **43**, 2256 (1995).
237. LI, H.Y., K. KOIKE, T. OHMOTO, and K. IKEDA: Dianchinenosides A and B, two New Saponins from *Dianthus chinensis*. J. Nat. Prod., **56**, 1065 (1993).
238. LI, H.Y., K. KOIKE, and T. OHMOTO: Triterpenoid Saponins from *Dianthus chinensis*. Phytochemistry, **35**, 751 (1994).
239. KOIKE, K. H. LI, Z. JIA, H. MURAOKA, S. FUKUI, M. INOUE, and T. OHMOTO: Triterpenoid Saponins from *Dianthus chinensis*. Tetrahedron, **50**, 1281 (1994).
240. KOMERTELIDZE, E.P., L.N. GVAZAVA, M.D. ALANIYA, V.S. KIKOLADZE, and V.G. TSITSISHVILI: Digitoside, a Novel Triterpene Glycoside from *Digitalis ciliata*. J. Nat. Prod., **55**, 217 (1992).
241. JAYASINGHE, U.L.B., G.P. WANNIGAMA, and J.K. MACLEOD: Saponins of *Diploclisia glaucescens*. Nat. Prod. Lett., **2**, 249 (1993).
242. BANDARA, B.M.R., L. JAYASINGHE, V. KARUNARATNE, G.P. WANNIGAMA, W. KRAUS, M. BOKEL, and S. SOTHEESWARAN: Diploclisin, a Bidesmosicid Triterpenoid Saponin from *Diploclisia glaucescens*. Phytochemistry, **28**, 2783 (1989).
243. JUNG, K.Y., J.C. DO, and K.H. SON: Triterpenoid Glycosides from the Roots of *Dipsacus asper.* J. Nat. Prod., **56**, 1912 (1993).

244. KOUNO, I., A. TSUBOI, M. NANRI and N. KAWANO: Acylated Triterpene Glycoside from Roots of *Dipsacus asper.* Phytochemistry, **29**, 338 (1990).
245. WEI, F., Z.C. LOU, Y.M. LIU, and Z.C. MIAO: Application of New Techniques of NMR in the Structure Elucidation of Two New Saponins, Asperosaponin F and H_1, and Spectral Rules Involved. Yaoxue Xuebao, **29**, 511 (1994).
246. YOSHIKI, Y., J.H. KIM, K. OKUBO, I. NAGOYA, T. SAKABE, and N. TAMURA: A Saponin Conjugated with 2,3,-Dihydro-2,5-dihydroxy-6-methyl-4H-pyran-4-one from *Dolichos lablab.* Phytochemistry **38**, 229 (1995).
247. KINJO, J., K. SUYAMA, and T. NOHARA: Triterpenoidal Saponins from *Dumasia truncata.* Phytochemistry, **40**, 1765 (1995).
248. SEGIET-KUJAWA, E., and M. KALOGA: Triterpenoid Saponins of *Eleuthero-coccus senticosus* Roots. J. Nat. Prod., **54**, 1044 (1991).
249. RASOANAIVO, P., G. MULTAR, E. FEDERICI, and C. GALEFFI: Triterpenoid Diglucoside of *Enterospermum pruinosum.* Phytochemistry, **39**, 251 (1995).
250. MIYASE, T., F.R. MELEK, O.D. EL-GINDI, S.M. ABDEL-KHALIK, M.R. EL-GINDI, M.Y. HAGGAG and S.H. HILAL: Saponins from *Fagonia arabica.* Phytochemistry **41**, 1175 (1996).
251. MELEK, F.R., T. MIYASE, O.D. EL-GINDI, S.M. ABDEL-KHALIK, and M.Y. HAGGAG: Saponins from *Fagonia Mollis.* Phytochemistry **42**, 1405 (1996).
252. CAI, L.N., R.V. ZHANG, Z.L. ZHANG, B. WANG, L. QIAO, L.R. HUANG, and J.R. CHENG: The Structure of Glyeurysaponin. Yaoxue Xuebao, **29**, 447 (1991).
253. KITAGAWA, I., K. HORI, M. SAKAGAMI, F. HASHIUCHI, M. YOSHIKAWA, and J. REN: Saponin and Sapogenol. XLIX. On The constituents of the Roots of *Glycyrrhiza inflata* Batalin from Xinjiang, China; Characterization of Two Sweet Oleanane-type Triterpene Oligoglycosides, Apioglycyrrhizin And Arabogycyrrhizin. Chem. Pharm. Bull., **41**, 1350 (1993).
254. KITAGAWA, I., K. HORI, T. TANIYAMA, J.L. ZHOU, and M. YOSHIKAWA: Saponin and Sapogenol. XLVII. On the Constituents of the Roots of *Glycyrrhiza uralensis* Fischer From Northeastern China. (1) Licorice-Saponins A_3, B_2, and C_2. Chem. Pharm. Bull., **41**, 43 (1993).
255. KITAGAWA, I., K. HORI, M. SAKAGAMI, J.L. ZHOU, and M. YOSHIKAWA: Saponin and Sapogenol. XLVIII. On the Constituents of the Roots of *Glycyrrhiza uralensis* Fischer from Northeastern China. (2) Licorice-Saponins D_3, E_2, F_3, G_2, H_2, J_2 and K_2. Chem. Pharm. Bull., **41**, 1337 (1993).
256. KITAGAWA, I., J. ZHOU, M. SAKAGAMI, E. UCHIDA, and M. YOSHIKAWA: Licorice-Saponins F_3, G_2, H_2, J_2 and K_2, Five New Oleanane-Triterpene Oligoglycosides fromthe Root of *Glycyrrhiza uralensis.* Chem. Pharm. Bull., **39**, 244 (1991).
257. KITAGAWA, I., K. HORI, E. UCHIDA, W.Z. CHEN, M. YOSHIKAWA, and J. REN: Saponin and Sapogenol. L. On the Constituents of the Roots of *Glycyrrhiza uralensis* Fischer from Xinjiang, China. Chemical Structures of Licorice-Saponin L_3 and Isoliquiritin Apioside. Chem. Pharm. Bull., **41**, 1567 (1993).
258. GAO, D., R. ZHANG, and B. YAO: Structure Identification of Yunnangly-saponin A and B. J. Chin. Pharm. Sci., **3**, 1 (1994).
259. OHTANI, K., K. OGAWA, R. KASAI, C. YANG, K. YAMASAKI, J. ZHOU, and O. TANAKA: Studies on the Constituents of *Glycyrrhiza yunnanensis.* Part 1. Oleanane Glycosides from *Glycyrrhiza yunnanensis* Roots. Phytochemistry **31**, 1747 (1992).
260. ZANG, L., R.Y. ZHANG, D. WANG, J.H. PANG, Z.L. ZHANG, C.Y. GAO, and Z.C. LOU: Glyyunnanprosapogenin and Glyyunnsapogenin from the Roots of *Glycyrrhiza yunnanensis.* Yaoxue Xuebao, **25**, 750 (1991).

261. AHMAD, V.U., S. PERVEEN, and S. BANO: Guaiacin A and B from the Leaves of *Guaiacum officinale*. Planta Med., **55**, 307 (1989).
262. AHMAD, V.U., N. BANO, S. UDDIN, S. PERVEEN, and I. FATIMA: Structure of Guaianin C from *Guaiacum officinale*. Fitoterapia, **60**, 255 (1989).
263. AHMAD, V.U., S. PERVEEN, and S. BANO: Saponins from the Leaves of *Guaiacum officinale*. Phytochemistry, **29**, 3287 (1990).
264. AHMAD, V.U., S. UDDIN, and S. BANO: Isolation and Structure Elucidation of Saponins from the Fruit of *Guaiacum officinale*. J. Nat. Prod., **53**, 1168 (1990).
265. AHMAD, V.U., N. SABA and S. PERVEEN: Structure of Guaianin M from *Guaiacum officinale*. Fitoterapia, **63**, 226 (1992).
266. YOSHIKAWA, K., S. ARIHARA, K. MATSUURA, and T. MIYASE: Dammarane Saponins from *Gymnema sylvestre*. Phytochemistry, **31**, 237 (1992).
267. YOSHIKAWA, K., K. AMIMOTO, S. ARIHARA, and K. MATSUURA: Gymnemic Acid V, VI and VII from Gur-Ma, the Leaves of *Gymnema sylvestre* R. Br. Chem. Pharm. Bull., **37**, 852 (1989).
268. KIUCHI, F., H.M. LIU, and Y. TSUDA: Two New Gymnemic acid Congeners Containing a Hexulopyranoside Moiety. Chem. Pharm. Bull., **38**, 2326 (1990).
269. SAHU, N.P., S.B. MAHATO, S.K. SARKAR, and G. PODDAR: Triterpenoid Saponins from *Gymnema sylvestre*. Phytochemistry, **41**, 1181 (1996).
270. YOSHIKAWA, K., M. NAKAGAWA, R. YAMAMOTO, S. ARIHARA, and K. MATSUURA: Antisweet Natural Products. V. Structures of Gymnemic Acids VIII-XII from *Gymnema sylvestre* R. Br. Chem. Pharm. Bull., **40**, 1779 (1992).
271. YOSHIKAWA, K., Y. KONDO, S. ARIHARA, and K. MATSUURA: Antisweet Natural Products. IX. Structures of Gymnemic Acids XV-XVIII from *Gymnema sylvestre* R.Br.V. Chem. Pharm. Bull., **41**, 1730 (1993).
272. DING, S., and Z. ZHU: Gycomoside I: A New Dammarane Saponin from *Gynostemma compressum*. Planta Med., **59**, 373 (1993).
273. ZHANG, X., H. ZHANG, W. SUN, and Z. SHAN: Studies on the Saponin Constituents of *Gynostemma longipes*. Chin. Chem. Lett., **2**, 221 (1991).
274. PIACENTE, S., C. PIZZA, T.N. DE, and S.F. DE: New Dammarane-Type Glycosides from *Gynostemma pentaphyllum*. J. Nat. Prod., **58**, 512 (1995).
275. KUWAHARA, M., F. KAWANISHI, T. KOMIYA, and H. OSHIO: Dammarane Saponins of *Gynostemma pentaphyllum* Makino and Isolation of Malonylginsenosides-Rb_1, -Rd, and Malonylgypenoside V. Chem. Pharm. Bull., **37**, 135 (1989).
276. ELGAMAL, M.H.A., H.S.M. SOLIMAN, M.S. KARAWYA, B. MIKHOVA, and H. DUDDECK: Isolation of Triterpene Saponins from *Gypsophila capillaris*. Phytochemistry, **38**, 1481 (1995).
277. LIU, Z., D. LI, N. OWEN, D.M. GRANT, R.G. CATES, and Z. JIA: Triterpenoid Saponins from *Gypsophila oldhamiana*. J. Nat. Prod., **58**, 1632 (1995).
278. FRECHET, D., B. CHRIST, B.M. DUSORBIER, H. FISCHER, and M. VUILHORGNE: Four Triterpenoid Saponins from Dried Roots of *Gypsophila* Species. Phytochemistry **30**, 927 (1991).
279. ELIAS, R., A.M. DIAZ-LANZA, E. VIDAI-OLLIVIER, G. BALANSARD, R. FAURE, and A. BABADJAMIAN: Triterpenoid Saponins from the Leaves of *Hedera helix*. J. Nat. Prod., **54**, 98 (1991).
280. GRISHKOVETS, V.I., A.A. LOLOIKO, A.S. SHASHKOV, and V.Y. CHIRVA: Triterpene Glycosides of *Hedera taurica*. V. Structure of Hederosides C and E_1 from Crimean Ivy Berries. Khim. Prir. Soedin., 230 (1990).
281. BABADY-BILA, W.C., S. TOPPET, A. KILONDA, and G. HOORNAERT: Two Triterpenoid Saponins from *Heinsia crinata*. Phytochemistry, **36**, 1489 (1994).

282. BADER, G., M. ZIESCHANG, K. WAGNER, E. GRUENDEMANN, and K. HILLER: New Triterpenoid Saponins from *Helianthus annuus*. Planta. Med., **57**, 471 (1991).
283. KASAI, R., T. TANAKA, R.L. NIE, M. MIYAKOSHI, J. ZHOU, and O. TANAKA: Saponins from Chinese Medicinal plant, *Hemsleya graciliflora* (Cucurbitaceae). Chem. Pharm. Bull., **38**, 1320 (1990).
284. SCHROEDER, H., M. SCHUBERT-ZSILAVECZ, G. REZNICEK, J. CART, J. JURENITSCH, and E. HASLINGER: A Triterpene Saponin from *Herniaria glabra*. Phytochemistry, **34**, 1609 (1993).
285. TAMAMSCH, B.G., D. TUJA, V. WRAY, and K. HILLER: New Triterpenoid Saponins from *Heteropappus altaicus* (Willd.) Novopokr. and *Heteropappus biennis* (Ldb.) Tam. Pharmazie, **49**, 209 (1994).
286. YOSHIKAWA, K., Y. NAGAI, M. YOSHIDA, and S. ARIHARA: Antisweet Natural Products. VIII. Structures of Hodulosides VII-X from *Hovenia dulcis* Thunb. Chem. Pharm. Bull., **41**, 1722 (1993).
287. DELLAGRECA, M., A. FIORENTINO, P. MONACO, and L. PREVITERA: Oleanane Glycosides from *Hydrocotyle ranunculoids*. Phytochemistry, **36**, 1479 (1994).
288. KAKUNO, T., K. YOSHIKAWA, and S. ARIHARA: Saponin from Ilex spp. Part 3. Triterpene Saponins from Fruit of *Ilex crenata*. Phytochemistry, **31**, 2809 (1992).
289. KAKUNO, T., K. YOSHIKAWA, and S. ARIHARA: Triterpenoid Saponins from *Ilex crenata*. Phytochemistry **31**, 3553 (1992).
290. HATA, C., M. KAKUNO, K. YOSHIKAWA, and S. ARIHARA: Triterpenoid Saponins of Aquifoliaceous Plants. V. Ilexosides XV-XIX from the Bark of *Ilex crenata*. Chem. Pharm. Bull., **40**, 1990 (1992).
291. KAKUNO, T., K. YOSHIKAWA, S. ARIHARA, M. TAKEI, and K. ENDO: Ilexo-sides E, F, G, H and I, Novel 18,19-Seco-ursane Glycosides from Fruit of *Ilex crenata*. Tetrahedron, **47**, 7219 (1991).
292. HEINZMANN, B.M., and E.P. SCHENKEL: Saponins from *Ilex dumosa*. J. Nat. Prod., **58**, 1419 (1995).
293. YANO, I., C. NISHIZUMI, K. YOSHIKAWA, and S. ARIHARA: Triterpenoid Saponins of Aquifoliaceous Plants. Part VII. Triterpenoid Saponins from *Ilex integra*. Phytochemistry, **32**, 417 (1993).
294. OUYANG, M., C.R. YANG., Z. CHEN, and H. WANG: Triterpenes and Triterpenoid Glycosides from the Leaves of *Ilex kudincha*. Phytochemistry, **41**, 871 (1996).
295. GOSMANN, G., D. GUILLIAUME, A.T.C. TAKETA, E.P. SCHENKEL: Triterpenoid Saponins from *Ilex paraguariensis*. J. Nat. Prod., **58**, 438 (1995).
296. KRAEMER, K.H., A.T.C. TAKETA, E.P. SCHENKEL, G. GOSMANN, and D. GUILLIAUME: Matesaponin 5, a Highly Polar Saponin from *Ilex paraguariensis*. Phytochemistry, **42**, 1119 (1996).
297. AMIMOTO, K., K. YOSHIKAWA, and S. ARIHARA: Triterpenoid Saponins of Aquifoliaceous plants. VIII. Ilexosides XXIX–XXXII from the Leaves of *Ilex rotunda* Thunb. Chem. Pharm. Bull., **40**, 3138 (1992).
298. AMIMOTO, K., K. YOSHIKAWA, and S. ARIHARA: Triterpenoid Saponins of Aquifoliaceous Plants. XI. Ilexosides XLI-XLV from the Leaves of *Ilex rotunda* Thunb. Chem. Pharm. Bull., **41**, 39 (1993).
299. AMIMOTO, K., K. YOSHIKAWA, and S. ARIHARA: Triterpenoid Saponins of Aquifoliaceous Plants. XII. Ilexosides XLVI-LI from the Leaves of *Ilex rotunda* Thunb. Chem. Pharm. Bull., **41**, 77 (1993).
300. RUMBERO-SANCHEZ, A., and P. VAZQUEZ: A Nor-triterpene Glycoside from *Isertia haenkeana* and A–^{13}C NMR study of Cincholic Acid. Phytochemistry, **30**, 623 (1991).

301. DELLA, G.M., A.J. FIORENTINO, P. MONACO, and L. PREVITERA: Juncoside I, a New Cycloartane Lactone Glycoside from *Juncus effusus*. Nat. Prod. Lett., **4**, 183 (1994).

302. SHAO, C., R. KASAI, K. OHTANI, O. TANAKA, and H. KOHDA: Saponins from Leaves of *Kalopanax pictus* (Thunb.). NAKAI, Harigiri: Structures of Kalopanaxsaponins JL_a and JL_b. Chem. Pharm. Bull., **38**, 1087 (1990).

303. SHAO, C., R. KASAI, J.D. XU, and O. TANAKA: Saponins from Roots of *Kalopanax septemlobus* (Thunb.). Koidz: Structures of Kalopanax saponins C, D, E and F. Chem. Pharm. Bull., **37**, 311 (1989).

304. SHAO, C., R. KASAI, K. OHTANI, J. SU, O. TANAKA: Saponins from leaves of *Kalopanax septemlobus* (Thunb.) Koidz. Structures of Kalopanax Saponins L_a, L_b and L_c. Chem. Pharm. Bull., **37**, 3251 (1989).

305. ELUJOBA, A., F. FELL, P.A. LINLEY, and D.J. MAITLAND: Triterpenoid Saponins from Fruits of *Legenaria breviflora*. Phytochemistry, **29**, 3281 (1990).

306. MAO, Q., and X.S. JIA: Chemical Constituents of *Lonicera fulvotomentosa* Hsu et S.C. Cheng. Yaoxue Xuebao, **24**, 269 (1989).

307. SON, K.H., K.Y. JUNG, H.W. CHANG, H.P. KIM, and S.S. KANG: Triterpenoid Saponins from the Aerial Parts of *Lonicera japonica*. Phytochemistry, **35**, 1005 (1994).

308. HASAN, M., D.K. BURDI, and V.U. AHMAD: Leucasin, a Triterpene Saponin from *Leucas nutans*. Phytochemistry, **30**, 4181 (1991).

309. NAGAO, T., R. TANAKA, Y. IWASE, H. HANAZONO, and H. OKABE: Studies on the Constituents of *Luffa acutangula* Roxb.I. Structures of Acutosides A-G, Oleanane-type Triterpene Saponins Isolated from the Herb. Chem. Pharm. Bull., **39**, 599 (1991).

310. NAGAO, T., R. TANAKA, and H. OKABE: Studies on the Constituents of *Luffa acutangula* Roxb.II. Structures of Acutosides H and I, Oleanolic Acid Saponins isolated from the Seed. Chem. Pharm. Bull., **39**, 889 (1991).

311. YOSHIKAWA, K., S. ARIHARA, J. WANG, T. NARUI, and T. OKUYAMO: Structures of Two New Fibrinolytic Saponins from the Seed of *Luffa cylindrica* Roem. Chem. Pharm. Bull., **39**, 1185 (1991).

312. KOHDA, H., O. TAKEDA, and S. TANAKA: Molluscicidal Triterpenoidal Saponin from *Lysimachia sikokiana*. Chem. Pharm. Bull., **37**, 3304 (1989).

313. NIGAM, S.K., X. LI, D. WANG, G. MISRA, and C. YANG: Triterpenoid Saponins from *Madhuca butyracea*. Phytochemistry, **31**, 3169 (1992).

314. LI, X., Y. LIU, D. WANG, C. YANG, S.K. NIGAM, and G. MISRA: Triterpenoid Saponins from *Madhuca butyracea*. Phytochemistry, **37**, 827 (1994).

315. SINDAMBIWE, J.B., A.M. BALDE, O. BRUYNE, L. PIETERS, H. VAN DEN HEUVEL, M. CLAEYS, D.A. VAN DEN BERGHE, and A.J. VLIETINCK: Triterpenoid Saponins from *Maesa lanceolata*. Phytochemistry **41**, 269 (1996).

316. KHAN, M.N.I., S.S. NIZAMI, M.A. KHAN, and Z. AHMAD: New Saponins from *Mangifera indica*. J. Nat. Prod., **56**, 767 (1993).

317. DE TOMMASI, N.D., L. RASTRELLI, J. CUMANDA, G. SPERANZA, and C. PIZZA: Aryl and Triterpenic Glycosides from *Margyricarpus setosus*. Phytochemistry, **42**, 163 (1996).

318. YAGUCHI, E., T. MIYASE, and A. UENO: Mazusaponins I-IV, Triterpene Saponins from *Mazus miquelii*. Phytochemistry, **39**, 185 (1995).

319. MAHATO, S.B.: Triterpenoid Saponins from *Medicago hispida*. Phytochemistry, **30**, 3389 (1991).

320. KINJO, J., H. UEMURA, M. NAKAMURA, and T. NOHARA: Leguminous Plants. XL. Two New Triterpenoidal Glycosides from *Medicago polymorphe* L. Chem. Pharm. Bull., **42**, 1339 (1994).

321. TIMBEKOVA, A.E., A.L. VERESHCHAGIN, A.A. SEMENOV, N.K. ABUBAKIROV: Triterpene Glycosides of Alfalfa. VI. Medicoside L. Khim. Prir. Soedin. 221 (1990).

322. OLESZEK, W., M. JURZYSTA, M. PLOSZYNSKI, I.J. COLQUHOUN, K.R. PRICE, and G.R. FENWICK: Zahnic Acid Tridesmoside and Other Dominant Saponins from Alfalfa (*Medicago sativa* L.) Aerial Parts. J. Agric. Food. Chem., **40**, 191 (1992).
323. JANECZKO, Z., J. SENDRA, K. KMIEC, and C.H. BRIESKORN: A Triterpenoid Glycoside from *Menyanthes trifoliata*. Phytochemistry, **29**, 3885 (1990).
324. JIANG, Y., G. MASSIOT, C. LAVAUD, J.-M. TEULON, C. GUÉCHOT, M. HAAG-BERRURIER, and R. ANTON: Triterpenoid Glycosides from the Bark of *Mimosa tenuiflora*. Phytochemisty, **30**, 2357 (1991).
325. JIANG, Y., M. HAAG-BERRURIER, R. ANTON, G. MASSIOT, C. LAVAUD, J.M. TEULON, and C. GUÉCHOT: Structure of a New saponin from the Bark of *Mimosa tenuiflora*. J. Nat. Prod., **54**, 1247 (1991).
326. SAHU, N.P., K. KOIKE, Z. JIA, and T. NIKAIDO: Novel Triterpenoid Saponins from *Mimusops elengi*. Tetrahedron, **51**, 13435 (1995).
327. SAHU, N.P.: Triterpenoid Saponins of *Mimusops elengi*. Phytochemistry, **41**, 883 (1996).
328. LAVAUD, C., G. MASSIOT, M. BECCHI, G. MISRA, and S.K. NIGAM: Saponins from Three Species of *Mimusops*. Phytochemistry, **41**, 887 (1996).
329. XU, J., R. XU, Z. LUO, and J. DONG: New Triterpenoidal Saponins from Folk Contraceptive Medicine, *Mussaenda pubescens*. Huaxue Xuebao, **49**, 621 (1991).
330. ZHAO, W., J. XU, G. QIN, R. XU, H. WU, and G. WENG: New Triterpenoid Saponins from *Mussaenda pubescens*. J. Nat. Prod., **57**, 1613 (1994).
331. ZHAO, W., J. XU, G. QIN, and R. XU: Saponins from *Mussaenda pubescens*. Phytochemistry, **39**, 191 (1995).
332. ZHAO, W., R. XU, G. QIN, T. VAISAR, and M.S. LEE: Saponins from *Mussaenda pubescens*. Phytochemistry **42**, 1131 (1996).
333. BLOOR, S.J., and L. QI: Cytotoxic Saponins from New Zealand Myrsine Species. J. Nat. Prod., **57**, 1354 (1994).
334. LAMIDI, M., E. OLLIVIER, R. FAURE, L. DEBRAUWER, L. NZE-EKEKANG, and G. BALANSARD: Quinovic Acid Glycosides from *Nauclea diderrichii*. Phytochemistry, **38**, 209 (1995).
335. FUJITA, S., R. KASAI, K. OHTANI, K. YAMASAKI, M. CHIU, R. NIE, and O. TANAKA: Dammarane Glycosides from Aerial Parts of *Neoalsomitra integrifoliola*. Phytochemistry, **38**, 465 (1995).
336. FUJITA, S., R. KASAI, K. OHTANI, K. YAMASAKI, M. CHIU, R. NIE, and O. TANAKA: Dammarane Glycosides from Aerial Part of *Neoalsomitra integrifoliola*. Phytochemistry, **39**, 591 (1995).
337. CHIU, M., R. NIE, H. NAGASAWA, A. ISOGAI, J. ZHOU, and A. SUZUKI: A Dammarane Saponin from *Neoalsomitra integrifoliola*. Phytochemistry, **31**, 2451 (1992).
338. YU, S., Z. XIAO, P. CAI, T. JIANG, and J.K. SNYDER: A new Triterpenoid Saponin from the Chinese Traditional Medicine *Nothopanax davidii* Harms (Araliaceae). Tetrahedron, **50**, 11601 (1994).
339. YU, S.S., D.Q. YU, and X.T. LIANG: Triterpenoid Saponins from the Bark of *Nothopanax davidii*. Phytochemistry, **38**, 695 (1995).
340. CRESPIN, F., E. OLLIVIER, C. LAVAUD, A. BABADJAMIAN, R. FAURE, L. DEBRAUWER, G. BALANSARD, and G. BOUDON: Triterpenoid Saponins from *Opilia celtidifolia*. Phytochemistry **33**, 657 (1993).
341. SUN, R., and Z. JIA: A Triterpenoid Saponin from *Oxytropis bicolor*. Phytochemistry, **30**, 3480 (1991).
342. SUN, R., and Z. JIA: Saponins from *Oxytropis glabra*. Phytochemistry, **29**, 2032 (1990).

343. SUN, R., Z. JIA, and Z. ZHU: Chemical Constituents of *Oxytropis glabra*. Chin. Sci. Bull., **34**, 609 (1989).
344. SUN, R., Z. JIA, and D. CHENG: Three Saponins from *Oxytropis* Species. Phytochemistry, **30**, 2707 (1991).
345. LEE, S.S., W.C. SU, and K.C. LIU: Two New Triterpene Glucosides from *Paliurus ramosissimus*. J. Nat. Prod., **54**, 615 (1991).
346. ZHAO, Y.Q., L.J. WU, C.L. YUAN, X. LI, and Y.J. CHEN: A New Triterpene Saponin from the Stems and Leaves of *Panax ginseng*. Chin. Chem. Lett., **3**, 887 (1992).
347. KIM, D., Y. CHENG, U. ZEDK, P. ZHAO, Y. LIU, and C. YANG: Dammarane Saponins from *Panax ginseng*. Phytochemistry, **40**, 1493 (1995).
348. ZHAO, P., Y. LIU, and C. YANG: Minor Constituents from the Roots of *Panax notoginseng*. Yunnan Zhiwu Yanjiu, **15**, 409 (1993).
349. ZHAO, P., Y.Q. LIU, and C. YANG: Minor Dammarane saponins from *Panax notoginseng*. Phytochemistry, **41**, 1419 (1996).
350. KOHDA, H., S. TANAKA, Y. YAMAOKA, H. IZUMI, M. NUNO, S. ISODA, K. GOTOH, T. WATANABE, S. KATSUKI, and M. SATAKE: Chikusetsusaponin. VI. A New Saponin from the Rhizome of *Panax pseudo-ginseng* var. *angustatus*. Chem. Pharm. Bull., **39**, 1588 (1991).
351. SHUKLA, Y.N., and R.S. THAKUR: Studies on Indian ginseng. Part 9. A Triterpenoid Saponin from *Panax pseudoginseng* subsp. *hemalaicus* var. *angustifolium*. Phytochemistry, **29**, 239 (1990).
352. SHUKLA, Y.N., R.S. THAKUR, and P. PACHALY: Studies on Indian Ginseng. Part 13. A Bidesmosidic Oleanolic Acid Saponin from *Panax pseudo-ginseng*. Phytochemistry, **31**, 1046 (1992).
353. NGUYEN, M.D., T.N. NGUYEN, R. KASAI, A. ITO, K. YAMASAKI, and O. TANAKA: Saponins from Vietnamese Ginseng *Panax vietnamensis* Ha et Grushv. Collected in Central Vietnam. I. Chem. Pharm. Bull., **41**, 2010 (1993).
354. NGUYEN, M.D., R. KASAI, K. OHTANI, A. ITO, T.N. NGUYEN, K. YAMASAKI, and O. TANAKA: Saponins from Vietnamese Ginseng *Panax vietnamensis* Ha et Grushv. Collected in Central Vietnam. II. Chem. Pharm. Bull., **42**, 115 (1994).
355. NGUYEN, M.D., R. KASAI, K. OHTANI, A. ITO, T. NGUYEN, K. YAMASAKI, and O. TANAKA: Saponins from Vietanamese Ginseng *Panax vietnamensis* Ha et Grushv. Collected in Central Vietnam. III. Chem. Pharm. Bull., **42**, 634 (1994).
356. IKEDA, Y., M. SUGIURA, C. FUKAYA, K. YOKOYAMA, Y. HASHIMOTO, K. KAWA-NISHI, and M. MORIYASU: Poriandradulcins A,B and C: Phosphodiesterase Inhibitors from *Periandra dulcis* Mart. Chem. Pharm. Bull., **39**, 566 (1991).
357. MASSIOT, G., X.-F. CHEN, C. LAVAUD, L. LE MEN-OLIVIER, C. DELAUDE, A. VIARI, P. VIGNY, and J. DUVAL: Saponins from Stem Bark of *Petersianthus macrocarpus*. Phytochemistry, **31**, 3571 (1992).
358. JAIN, D.C., P.K. AGRAWAL, and R.S. THAKUR: Phaseoluside-A, a New Soyasapogenol B Triglucoside from *Phaseolus vulgaris* Seeds. Planta Med., **57**, 94 (1991).
359. YOSHIKI, Y., J. KIM, and K. OKUBO: Saponins Conjugated with 2,3,-Dihydro-2,5-dihydroxy-6-methyl-4H-pyran-4-one from *Phaseolus cocineus*. Phytochemistry, **36**, 1009 (1994).
360. SPENGEL, S., and W. SCHAFFNER: Esculentoside S, a New Saponin from the Leaves of *Phytolacca acinosa*. Nat. Prod. Lett., **2**, 243 (1993).
361. NIELSEN, S.E., U. ANTHONI, C. CHRISTOPHERSEN, and C. CORNETT: Triterpenoid Saponins from *Phytolacca rivinoides* and *Phytolacca bogotensis*: Phytochemistry, **39**, 625 (1995).

362. THIIBORG, S.T., S. BROEGGER CHRISTENSEN, C. CORNETT, C.E. OLSEN, and E. LEMMICH: Molluscicidal Saponins from a Zimbabwean Strain of *Phytolacca dodecandra.* Phytochemistry **36**, 753 (1994).
363. YI, Y.: A Triterpenoid Saponin from *Phytolacca esculenta.* Phytochemistry, **31**, 2552 (1992).
364. YI, Y.: Two New Saponins from the Roots of *Phytolacca esculenta.* Planta Med., **58**, 99 (1992).
365. YI, Y., and F. DAI: A New Triterpenoid and Its Glycoside from *Phytolacca esculenta.* Planta. Med., **57**, 162 (1991).
366. YI, Y.: Esculentoside L and K: Two New Saponins from *Phytolacca esculenta.* Planta. Med., **56**, 301 (1990).
367. YI, Y.H., and X. HUANG: Isolation and Identification of Three New Saponins from Shanglu (*Phytolacca esculenta*). Yaoxue Xuebao, **25**, 745 (1991).
368. SEIJI, T., T. TAKAGI, and T. HASHIMOTO: A γ-Pyronyl-triterpenoid Saponin from *Pisum sativum.* Phytochemistry, **31**, 2435 (1992).
369. REZNICEK, G., H. SCHROEDER, M. SCHUBERT-ZSILAVECZ, T.H. SCHÖPKE, S. LEHRKINDER, E. HASLINGER, K. HILLER, J. JURENITSCH, and W. KUBELKA: *Polemonium caeruleum* L.; the Main Saponins of Lower Polarity. Pharmazie, **49**, 58 (1994).
370. BHANDARI, S.P.S., P.K. AGARWAL, and H.S. GARG: A Triterpenoid Saponin from *Polycarpone Loeflingiae.* Phytochemistry, **29**, 3889 (1990).
371. ZHANG, D., T. MIYASE, M. KUROYANAGI, K. UMEHARA, and A. UENO: Studies on the Constituents of *Polygala japonica* Houtt. I. Structures of Polygalasaponins I-X. Chem. Pharm. Bull., **43**, 115 (1995).
372. ZHANG, D., T. MIYASE, M. KUROYANAGO, K. UMEHARA, and A. UENO: Studies on the Constituents of *Polygala japonica.* Houtt.II. Structures of Polygalasaponins XI-XIX. Chem. Pharm. Bull., **43**, 966 (1995).
373. MIYASE, T., H. SAITOH, K. SHIOKAWA, and A. UENO: Six New Presenegenin Glycosides, Reiniosides A-F, from *Polygala reini* Root. Chem. Pharm. Bull., **43**, 466 (1995).
374. GOPAL SAMY, N., J. GUEHO, H.R. JULIEN, A.W. OWADALLY, and K. HOSTETTMANN: Molluscicidal Saponins of *Polyscias dichroostachya.* Phytochemistry, **29**, 793 (1990).
375. PAPHASSARANG, S., J. RAYNAUD, M. LUSSIGNOL, and M. BECCHI: Triterpenic Glycosides from *Polyscias scutellaria.* Phytochemistry **28**, 1539 (1989).
376. PAPHASSARANG, S., J. RAYNAUD, M. LUSSIGNOL, and M. BECCHI: Triterpenoid Saponins from *Polyscias scutellaria.* J. Nat. Prod., **52**, 239 (1989).
377. BILIA, A.R., E. PALME, S. CATALANO, G. FLAMINE, and I. MORELLI: New Triterpenoid Saponins from the Roots of *Potentilla tormentilla.* J. Nat. Prod., **57**, 333 (1994).
378. AHMAD, V.U., V. SULTANA, and Q.N. SAQIB: Triterpenoid Saponins from *Primula denticulata.* Planta. Med., **56**, 94 (1990).
379. AHMAD, V. U., M.G. SHAH, F.V. MOHAMMAD, and F.T. BAQAI: Macrophyllicin, a Saponin from *Primula macrophylla.* Phytochemistry, **32**, 1543 (1993).
380. AHMAD, V.U., and M.G. SHAH: Macrophyllicinin, a Saponin from *Primula macrophylla.* J. Nat. Prod., **56**, 1580 (1993).
381. CALIS, I., A. YURUKER, H. RUEGGER, A.D. WRIGHT, and O. STICHER: Triterpene Saponins from *Primula veris subsp. macrocalyx* and *Primula elatior subsp. meyeri.* J. Nat. Prod., **55**, 1299 (1992).
382. TIAN, J., F. WU, M. QIU, and R. NIE: Two Triterpenoid Saponins from *Pterocephalus bretschneidri.* Phytochemistry, **32**, 1539 (1993).
383. TIAN, J., F. WU, M. QIU, and R. NIE: Triterpenoid Saponins from *Pterocephalus hookeri.* Phytochemistry, **32**, 1535 (1993).

384. Kang, S.S.: Saponin from the Roots of *Pulsatilla koreana*. Arch. Pharmacol. Res., **12**, 42 (1989).

385. Sotheesawaran, S., M. Bokel, and W. Kraus: A Hemolytic Saponin, Randianin, from *Randia dumetorum*. Phytochemistry, **28**, 1544 (1989).

386. Dubois, M.A., S. Benze, and H. Wagner: New Biologically Active Triterpene Saponins from *Randia dumetorum*. Planta. Med., **56**, 451 (1990).

387. Ohtani, K., C. Miyajima, T. Takahasi, R. Kasai, O. Tanaka, D.R. Hahn, and N. Naruhashi: A Dimeric Triterpene Glycoside from *Rubus coreanus*. Phytochemistry, **29**, 3275 (1990).

388. Kinjo, J., F. Kishida, K. Watanabe, F. Hashimoto, and T. Nohara. Leguminous plants. XLII. Five New Triterpene Glycosides from Russell Lupine. Chem. Pharm. Bull., **42**, 1874 (1994).

389. Jia, Z., X. Liu, and Z. Liu: Triterpenoids from *Sanguisorba alpina*. Phytochemistry, **32**, 155 (1993).

390. Pancharoen, O., P. Tuntiwachwuttikul, W.C. Taylor, and K. Picker: Triterpenoid Glycosides from *Schefflera lucantha*. Phytochemistry, **35**, 987 (1994).

391. Maeda, C., K. Ohtani, R. Kasai, K. Yamasaki, N.M. Duc, N.T. Nham, and G.K.Q. Cu: Oleanane and Ursane Glycosides from *Schefflera octophylla*. Phytochemistry, **37**, 1131 (1994).

392. Sung, T.V., C. Lavaud, A. Porzel, W. Steglich, and G. Adam: Triterpenoids and their Glycosides from the Bark of *Schefflera octophylla*. Phytochemistry, **31**, 227 (1992).

393. Sung, T.V., W. Steglich, and G. Adam: Triterpene Glycosides from *Schefflera octophylla*. Phytochemistry, **30**, 2349 (1991).

394. Sung, T.V., and G. Adam: A Sulfated Triterpenoid Saponin from *Schefflera octophylla*. Phytochemistry, **30**, 2717 (1991).

395. Sung, T.V., J. Peter-Katalinic, and G. Adam: A Bidesmosidic Triterpenoid Saponin from *Schefflera octophylla*. Phytochemistry, **30**, 3717 (1991).

396. Tran, V.S., and G. Adam: An Acetylated Bidesmosidic Saponin from *Schefflera octophylla*. J. Nat. Prod., **55**, 503 (1992).

397. Purohit, M.C., G. Pant, and M.S.M. Rawat: A Betulinic Acid Glycoside from *Schefflera venulosa*. Phytochemistry, **30**, 2419 (1991).

398. Calis, I., M. Zor, A.A. Basaran, A.D. Wright, and O. Sticher: Ilwensisaponins A, B, C and D: Triterpene Saponins from *Scrophularia ilwensis*. Helv. Chim. Acta., **76**, 1352 (1993).

399. Bhandari, S.P.S., R. Roy, P.K. Agrawal, and H.S. Garg: Scrokoelziside A, a Triterpene Glycoside from *Scrophularia koelzii*. Phytochemistry, **41**, 879 (1996).

400. Potterat, O., K. Hostettmann, H. Stoeckli-Evang, and M. Saadou: Saponins with an Unusual Secoursene Skeleton from *Sesamum alatum* Thonn. Helv. Chim. Acta, **75**, 883 (1992).

401. Jiang, Y., A.O. Ali, D. Guillaume, B. Weniger, and R. Anton: Triterpenoid Saponins from the Root of *Sideroxylon cubense*. Phytochemistry, **35**, 1013 (1994).

402. Nicolas, G., A. Ouland-Ali, D. Guillaume, A. Lobstein, B. Weniger, and R. Anton: Triterpenoid Saponins from the Root of *Sideroxylon foetidissimum*. Phytochemistry, **38**, 225 (1995).

403. Lacille-Dubois, M., B. Hanquet, Z. Cui, Z. Lou, and H. Wagner: Acylated Triterpene Saponins from *Silene jenisseensis*. Phytochemistry, **40**, 509 (1995).

404. Reznicek, G., J. Jurenitsch, M. Plasun, S. Korhammer, E. Haslinger, K. Hiller, and W. Kubelka: Four Major Saponins from *Solidago canadensis*. Phytochemistry, **30**, 1629 (1991).

405. Reznicek, G., W. Kubelka, J. Jurenitsch, G. Michl, and E. Haslinger: Structure elucidation of a Saponin from *Solidago gigantea.* Bull. Magn. Reson., **11**, 402 (1989).

406. Reznicek, G., J. Jurenitsch, G. Michl, and E. Haslinger: The First Structurally Confirmed Saponin from *Solidago gigantea.* Structure Elucidation by Modern NMR Techniques. Tetrahedron Lett., **30**, 4097 (1989).

407. Bader, G., V. Wray, and K. Hiller: Virgaureasaponin 3, a 3,28-Bisdesmosidic Triterpenoid Saponin from *Salidago virgaurea.* Phytochemistry, **31**, 621 (1992).

408. Inose, Y., T. Miyase, and A. Ueno: Studies on the Constituents of *Solidago virgaaurea* L. II. Structures of Solidagosaponins X-XX. Chem. Pharm. Bull., **40**, 946 (1992).

409. Miyase, T., Y. Inose, and A. Ueno: Studies on the Constituents of *Solidago Virgaurea* L.III. Structures of Solidagosaponins XXI-XXIX. Chem. Pharm. Bull., **42**, 617 (1994).

410. Ding, Y., R. Tian, J. Kinjo, T. Nohara, and I. Kitagawa: Constituents of Leguminous Plants. XXXIII. Three New Oleanene Glycosides from *Sophora flavescens.* Chem. Pharm. Bull., **40**, 2990 (1992).

411. Bilia, A.R., F. Flamini, G. Flamini, I. Morelli, and A. Marsili: Flavonoids and a Saponin from *Apartium junceum.* Phytochemistry, **34**, 847 (1993).

412. Yamamoto, R., T. Miyase, and A. Ueno: Stachyssaponins I-VIII, New Oleanane-type Triterpene Saponins from *Stachys riederichamisso.* Chem. Pharm. Bull., **42**, 1291 (1994).

413. Wang, H.B., D.Q. Yu, and X. Liang: Yemuoside I, a New Nor-triterpenoid glycoside from *Stauntonia chinensis.* J. Nat. Prod., **54**, 1097 (1991).

414. Wang, H., D. Yu, X. Liang, N. Watanabe, M. Tamai, and S. Omura: Yemuoside YM_7, YM_{11}, YM_{13} and YM_{14}: Four Nortriterpenoid Saponins from *Stauntonia chinensis.* Planta. Med., **55**, 303 (1989).

415. Wang, H., D. Yu, X. Liang, N. Watanabe, M. Tamai, and S. Omura: Structures of Two Nortriterpenoid Saponins from *Stauntonia chinensis.* J. Nat. Prod., **53**, 313 (1990).

416. Wang, H.B., D.Q. Yu, X.T. Liang, N. Watanabe, M. Tamai, and S. Omura: The Structures of Yamuoside YM_{10} and YM_{12} from *Stauntonia chinensis.* Yaoxue Xuebao, **24**, 444 (1989).

417. Wang, H.B., R. Mayer, and G. Ruecker: Triterpenoid Glycosides from *Stauntonia hexaphylla.* Phytochemistry, **33**, 1496 (1993).

418. Wang, H.B., R. Mayer, and G. Ruecker: Triterpenoid Glycosides from *Stauntonia hexaphylla.* Phytochemistry, **34**, 1389 (1993).

419. Lavaud, C., G. Massiot, L. Le Men-Olivier, A. Viari, P. Vigny, and C. Delaude: Saponins from *Steganotaenia araliacea.* Phytochemistry, **31**, 3177 (1992).

420. Yoshikawa, K., H. Taninaka, Y. Kan, and S. Arihara: Antisweet Natural Products. X. Structures of Sitakisoside I-V from *Stephanotis lutchuensis* Koidz, var. *japonica.* Chem. Pharm. Bull., **42**, 2023 (1994).

421. Ahmad, V.U., M. Noorwala, F.V. Mohammad, B. Sener, A.H. Gilani, and K. Aftab: Symphytoxide A, a Triterpenoid Saponin from the Roots of *Symphytum officinale.* Phytochemistry, **32**, 10003 (1993).

422. Mohammad, F.V., M. Noorwala, V.U. Ahmad, and B. Sener: A Bidesmosidic Hederagenin Hexasaccharide from the Roots of *Symphytum officinale.* Phytochemistry, **40**, 213 (1995).

423. Ahmad, V.U., M. Noorwala, F.V. Mohammad, and B. Sener: A New Triterpene Glycoside from the Roots of *Symphytum officinale.* J. Nat. Prod., **56**, 329 (1993).

424. Noorwala, M., F.V. Mohammad, V.U. Ahmad, and B. Sener: A Bidesmosidic Triterpene Glycoside from the Roots of *Symphytum officinale.* Phytochemistry, **36**, 439 (1994).

425. AHMAD, V.U., F.V. MOHAMMAD, M. NOORWALA, and B. SENER: A New Bidesmosidic Triterpenoidal Saponin from the Roots of *Symphytum officinale*. Planta Med., **59**, 461 (1993).
426. MAHATO, S.B., A.K. NANDY, and A.P. KUNDU: Pentacyclic Triterpenoid Sapogenols and Their Glycosides from *Terminalia bellerica*. Tetrahedron, **48**, 2483 (1992).
427. NANDY, A.K., G. PODDER, N.P. SAHU, and S.B. MAHATO: Triterpenoids and Their Glucosides from *Terminalia bellerica*. Phytochemistry, **28**, 2769 (1989).
428. KUNDU, A.P., and S.B. MAHATO: Triterpenoids and Their Glycosides from *Terminalia chebula*. Phytochemistry, **32**, 999 (1993).
429. MAILLARD, M., C.O. ADEWUNMI, and K. HOSTETTMANN: A Triterpene Glycoside from the Fruits of *Tetrapleura tetraptera*. Phytochemistry, **31**, 1321 (1992).
430. YOSHIMITSU, H., K. HAYASHI, M. KUMABE, and T. NOHARA: Cycloartane Type Glycosides from *Thalictri herba*. Phytochemistry, **38**, 939 (1995).
431. YI, Y., and Z. WU: A New Triterpene Glycoside of *Thalictrum foeniculaceum*. Zhongguo Yaoxue Daxue Xuebao, **22**, 270 (1991).
432. YOSHIMITSU, H., K. HAYASHI, K. SHINGU, J. KINJO, S. YAHARA, K. NAKANO, K. MURAKAMI, T. TOMIMATSU, and T. NOHARA: Two New Cycloartane Glycosides, Thalictoside A and C from *Thalictrum thunbergii* D.C. Chem. Pharm. Bull., **40**, 2465 (1992).
433. SCHENKEL, E.P., W. WERNER, and K.E. SCHULTE: Saponins from *Thinouia coriaceae*. Planta. Med., **57**, 463 (1991).
434. NAGAO, T., H. OKABE, K. MIHASHI, and T. YAMAUCHI: Studies on the Constituents of *Thladiantha dubia* Bunge. I. The Structures of Dubiosides A, B and C, the Quillaic Acid Glucuronide Saponins Isolated From the Tuber. Chem. Pharm. Bull., **37**, 925 (1989).
435. NAGAO, T., R. TANAKA, H. OKABE, and T. YAMAUCHI: Studies on the Constituents of *Thladiantha dubia* Bunge. II. Structures of Dubiosides D, E, and F, Neutral Saponins of Quillaic acid Isolated from the Tuber. Chem. Pharm. Bull., **38**, 378 (1990).
436. NIE, R., T. TANAKA, M. MIYAKOSHI, R. KASAI, T. MORITA, J. ZHOU, and O. TANAKA: A Triterpenoid Saponin from *Thladiantha hookeri* var. *pentadactyla*. Phytochemistry, **28**, 1711 (1989).
437. MIYASE, T., H. KOHSAKA, and A. UENO: Tragopogonosides A-I, Oleanane Saponins from *Tragopogon pratensis*. Phytochemistry, **31**, 2087 (1992).
438. MASSIOT, G., C. LAVAUD, C. DELAUDE, G. VAN BINST, S.P.F. MILLER, and H.M. FALES: Saponins from *Tridesmostemon claessenssi*. Phytochemistry, **29**, 3291 (1990).
439. MOHAMED, K.M., K. OHTANI, R. KASAI, and K. YAMASAKI: Oleanene Glycosides from Seeds of *Trifolium alexandrium*. Phytochemistry, **40**, 1237 (1995).
440. SAKAMOTO, S., S. KOFUJI, M. KUROYANAGI, A. UENO, and S. SEKITA: Saponins from *Trifolium repens*. Phytochemistry, **31**, 1773 (1992).
441. MA, W., D. WANG, Y. ZENG, and C. YANG: Three Triterpenoid Saponins from *Triplostegia grandiflora*. Phytochemistry, **30**, 3401 (1991).
442. YEPEZ P.A.M., O.L. DE UGAZ, C.M. ALVAREZ A, V. DE FEO, R. AQUINO, F. DE SIMONE, and C. PIZZA: Quinovic Acid Glycosides from *Uncaria guianensis*. Phytochemistry, **30**, 1635 (1991).
443. KLIMEK, B., C. LAVAUD, and G. MASSIOT: Saponins from *Verbascum nigrum*. Phytochemistry, **31**, 4368 (1992).
444. SEIFERI, K., A. PREISS, S. JOHNE, N.T. JUERGEN LIEN, C. LAVAUD, and G. MASSIOT: Triterpene Saponins from *Verbascum songaricum*. Phytochemistry, **30**, 3395 (1991).
445. HARTLEB, I., and K. SEIFERT: Songarosaponin D – a Triterpenoid Saponin from *Verbascum songaricum*. Phytochemistry, **35**, 1009 (1994).

446. HARTLEB, I., and K. SEIFERT: Triterpenoid Saponins from *Verbascum songaricum.* Phytochemistry, **38**, 221 (1995).

447. VASANTH, S., A.B. KUNDU, S.K. PANDA, and A. PATRA: Vicoside A, a 28-Nortriterpenoid Glucoside from *Vicoa indica.* Phytochemistry, **30**, 3053 (1991).

448. GOVINDACHARI, T.R., and M.S. PREMILA: Triterpenoid Saponins from *Wedelia calendulaceae* Less. Indian J. Chem. Sect. B., **30B**, 466 (1991).

449. KONOSHIMA, T., M. KOZUKA, M. HARUNA, K. ITO, T. KIMURA, and H. TOKUDA: Studies on the Constituents of Leguminous Plants XII. The Structures of New Triterpenoid Saponins from *Wistaria brachybotrys* Sieb. *et* Zucc. Chem. Pharm. Bull., **37**, 2731 (1989).

450. KINJO, J., Y. FUJISHIMA, K. SAINO, R. TIAN, and T. NOHARA: Leguminous Plants. XLV. Five New Triterpene Glycosides from *Wistaria brachybotrys* (Leguminosae). Chem. Pharm. Bull., **43**, 636 (1995).

451. KONOSHIMA, T., M. KOZUKA, M. HARUNA, and K. ITO: Constituents of Leguminous Plants, XIII. New Triterpenoid Saponins from *Wistaria brachybotrys.* J. Nat. Prod., **54**, 830 (1991).

452. YOSHIKAWA, K., N. SHIMONO, and S. ARIHARA: Antisweet Natural Products, VI. Jujubasaponins IV, V and VI from *Zizyphus jujuba* Mill. Chem. Pharm. Bull., **40**, 2275 (1992).

453. HASSANEAN, H.A., M.M.A. EL-HAMOULY, S.A. EL-MOGHAZY, and D.W. BISHAY: 14-Decarboxyquinovic and Quinovic Acid Glycosides from *Zygophyllum album.* Phytochemistry, **33**, 667 (1993).

454. HASSANEAN, H.H., E.K. DESOKY, and M.M.A. EL-HAMOULY: Quinovic Acid Glycosides from *Zygophyllum album.* Phytochemistry, **33**, 663 (1993).

455. ELGAMAL, M.H.A., K.H. SHAKER, K. POELLMANN, and K. SEIFERT: Triterpenoid Saponins from *Zygophyllum* Species. Phytochemistry, **40**, 1233 (1995).

456. AHMAD, V.U., U.S. GHAZALA, and S. BANO: Saponins from *Zygophyllum propinquum.* J. Nat. Prod., **53**, 1193 (1990).

457. AHMAD, V.U., and U.S. GHAZALA: A Triterpenoid Saponin from *Zygophyllum propinquum.* Phytochemistry, **31**, 1051 (1992).

458. AHMAD, V.U., U.S. GHAZALA, and S.M. AKI: Saponins from *Zygophyllum propinquum.* Phytochemistry, **33**, 453 (1993).

(Received May 7, 1997)

Synthesis of 6-Deoxyamino Sugars

L. A. OTSOMAA and A. M. P. KOSKINEN
Department of Chemistry, University of Oulu,
Linnanmaa, Oulu, Finland

Contents

1. Introduction

1.1. General

All cells of higher organisms are covered with surface carbohydrates, which are linked to peptides or fatty acids to form glycoconjugates (*1*). These cell surface glycoconjugates (glycoproteins, proteoglycans, glycosphingolipids, and glycosyl phosphatidyl inositols) play an important role in biological recognition, carrying encoded biological information that is recognized by other cells, viruses, bacteria, and toxins (*2*). This is another example of the lock and key mechanism, which was first used by Emil Fischer in 1897 to explain the interactions between enzymes and substrates. The recognition event is important for the regulation of cell-substratum adhesion and cell proliferation, for the binding and uptake of extracellular components, and for the regulation of extracellular matrix formation (*3*).

Intracellular carbohydrates are thus important for signaling and activation. Owing to their structural diversity, oligosaccharides are particularly good information carriers. For example, two identical monosaccharides can form 11 different disaccharides, while two identical amino acids form only one dipeptide. Glycoconjugates on eukaryotic cells contain glucose, mannose, galactose, the deoxy sugar fucose, the amino sugars *N*-acetylgalactosamine and *N*-acetylglucosamine, and a few acidic sugars like *N*-acetylneuraminic acid (*1*).

Along with several other types of amino sugars 6-deoxyaminohexoses have been identified as important structural components in antibiotics, including enediynes, macrolides, and anthracyclines (*4*, *5*, *6*), where they function as recognition elements and contribute to the high selectivity of the antibiotics. In several DNA-interacting antibiotics the sugar units are suspected to function as minor groove binders. 6-Deoxyaminohexoses are also found on some bacterial cell walls (*7*). Owing to their varied biological actions, 6-deoxyamino sugars are synthetic targets of great interest for potential pharmaceutical use. Some of them inhibit protein glycosylation. The 6-deoxyaminohexoses were first isolated from nature and synthesized in the late sixties.

Carbohydrates have challenged chemists for a long time, but only recently has their biological function become understood. The relative configurations we know today were laboriously worked out by EMIL FISCHER, who pioneered the field of sugar chemistry one hundred years ago (*8*).

A very limited number of carbohydrate drugs are in commercial use (*e.g.* heparin). Two significant problems in the development of carbohydrate drugs are the chemical instability of carbohydrates and the inefficient methods for synthesizing them. These problems might be overcome by the development of glycomimics which are more robust towards hydrolysis and other degradation reactions. Such development requires a deeper understanding of the biological phenomena and of the structural behavior of glycomimics.

1.2. Antibiotics

Many antibiotics contain carbohydrate units to render the usually poorly water-soluble active ingredient more water-soluble and thereby more effective. Sugar units are also believed to play a role in recognition (*9*). Often these carbohydrate units contain novel amino sugars including 6-deoxyaminohexoses. Classification of antibiotics is based on the aglycone (the non-carbohydrate unit): anthracycline, macrolide, azalide, enediyne, *etc.*

The first anthracycline antibiotic, β-rhodomycin II, was isolated in 1950 by BROCKMANN and BAUER (*6*, *10*). The earliest of the anthracyclines displayed potent antibacterial activity in cell culture. Probably the most familiar antibiotic in this class is daunomycin, also known as daunorubicin or rubidomycin, which contains the novel 6-deoxyamino-hexose daunosamine (*6*, *11*). Daunomycin was isolated in the early 1960s and was the first antibiotic of this type to show activity against acute leukemia (*6*, *12*, *13*).

Many anthracycline antibiotics exhibit anticancer activity due to their ability to intercalate into double helical DNA (*6*). The aglycones formed in metabolic transformations in liver through reductive scission of the glycosidic bond are of pharmacological and clinical concern because they do not appear to contribute to cytotoxic or anticancer activity, but rather they appear to contribute to general side-effect toxicities (*6*).

During the 1950s BROCKMANN and HENKEL isolated the first 14-membered macrolide antibiotic, picromycin, from an *Actinomyces* culture (*14*). Picromycin contains desosamine (see Table 6), a 6-deoxyaminohexose, as the carbohydrate unit. Shortly after the isolation of picromycin, several other macrolide antibiotics were isolated from natural sources: 14-membered erythromycin and megalomicin, 12-membered methymycin, and 16-membered mycinamicin (*15–19*). Their

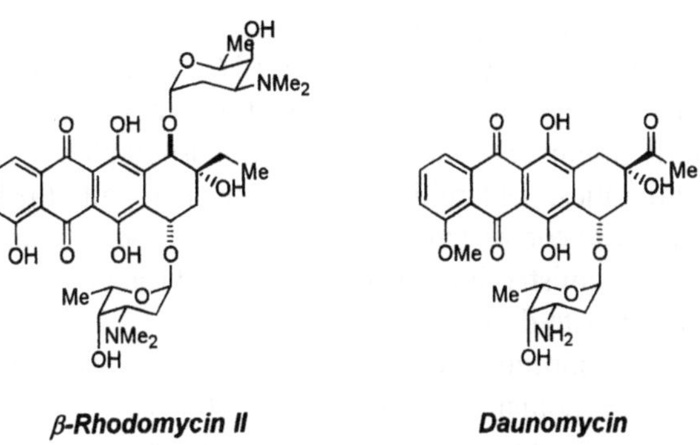

Fig. 1. Two anthracycline antibiotics

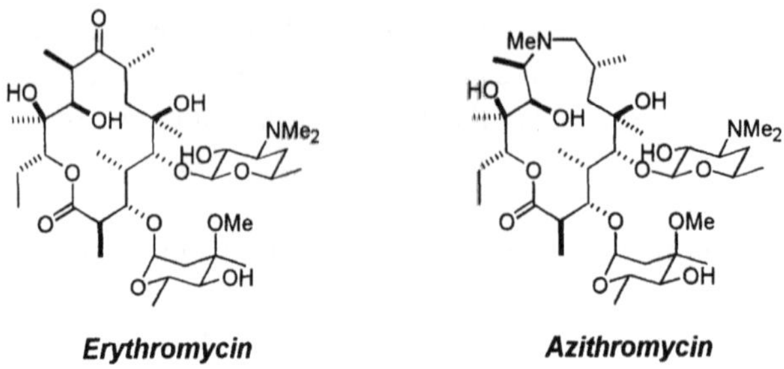

Fig. 2. A 6-deoxyaminohexose as a part of erythromycin A and azithromycin

biosyntheses involve the polyketide pathway. These compounds usually exhibit activity against Gram-positive bacteria and only weak activity against Gram-negative bacteria.

Azithromycin, the first member of yet another class of antibiotics known as azalides, is an effective therapeutic agent for oral treatment of sexually transmitted diseases, upper and lower respiratory tract infections, and skin infections (*20*). Azithromycin differs structurally from erythromycin A in the insertion of a methyl-substituted nitrogen in the lactone ring so as to create a 15-membered macrocycle (*21*). This modification produces enhanced potency against bacteria, superior stability in acid environment, as well as much longer half-lives and much higher tissue concentrations compared to erythromycin A 2 (*22*).

Enediyne anticancer antibiotics have attracted growing synthetic interest since the structure of the neocarzinostatin (NCS) chromophore was reported by EDO *et al.* in 1985 (*23*). Several structurally related species have been discovered, including the calicheamicins, esperamicins, and dynemicins (Fig. 3). Successful synthetic routes have been developed to many of these compounds (*24*). Kedarcidin is a recently discovered member of this exciting class of natural products (*4*). Its core enediyne unit resembles that of the NCS chromophore. Like typical enediyne antibiotics, kedarcidin is glycosylated with the unusual 6-deoxyaminohexose component kedarosamine, whose structure, including absolute stereochemistry, has been established through X-ray analysis of the *p*-bromobenzoate derivative (*25*).

The enediyne portion is responsible for the DNA cleaving action and the carbohydrate domain is suggested to function as minor groove binder, thus contributing to the high sequence selectivity observed within this class of DNA cleavers (*26, 27*).

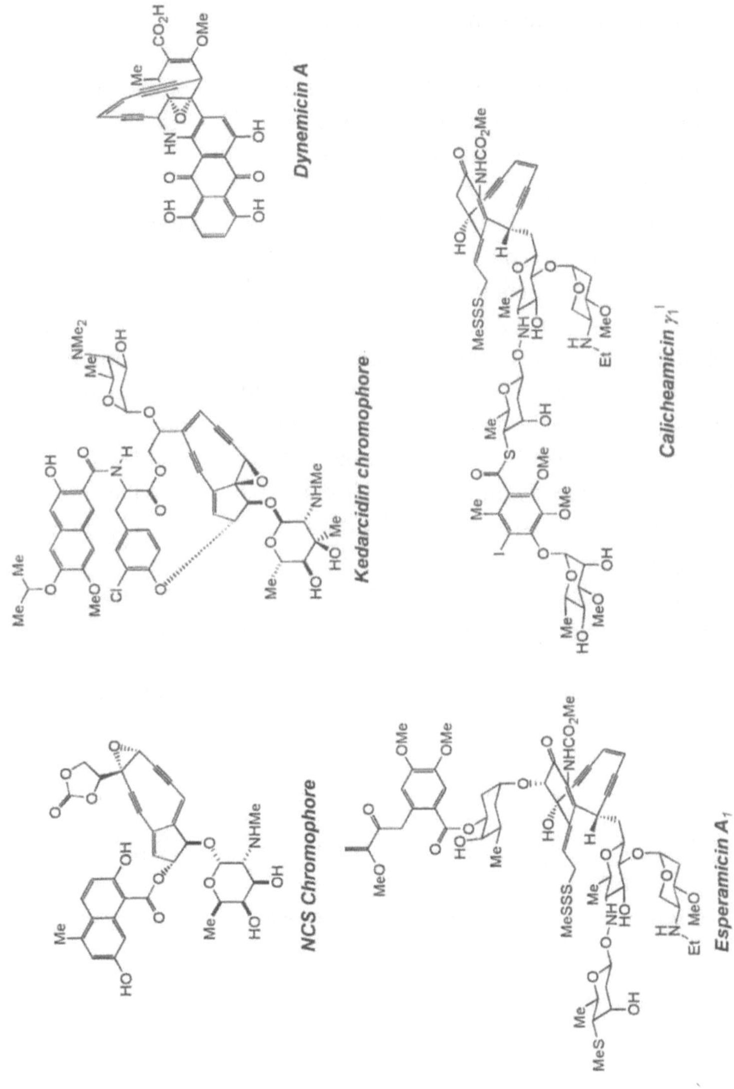

Fig. 3. Examples of enediyne antibiotics

2. Known 6-Deoxyaminohexoses

Numerous 6-deoxyaminohexoses have been described in the literature, some of them isolated from natural sources and some completely synthetic products. In Tables 1–7, the names of the reported naturally occurring compounds are shown in italics.

The amino group of 6-deoxyaminohexoses may be present in positions 2, 3, 4, and 6. Likewise, the position of the methylene carbon in tri- and tetra-deoxyhexoses can vary. This review covers all 6-deoxyaminohexoses except those that have the amino group at position 6. Aza sugars where nitrogen is in the ring, branched-chain, di- and poly-aminohexoses, and nitro-group-containing sugars are excluded from this coverage.

The common stereochemistries of the natural 6-deoxyaminohexoses are the same as for ordinary sugars: *gluco*, *galacto*, and *manno* configurations (Fig. 4). Other stereochemistries exist in addition (*allo*, *altro*, *talo*, *gulo*, and *ido*). The D-series of compounds are shown in Tables 1–7, although in nature it is normally the L-forms that appear. D-forms are conveniently obtained through synthesis. Tables 1–3 show the structures of dideoxyamino sugars, Tables 4–6 contain the structures of trideoxyamino sugars, and Table 7 shows the structures of tetradeoxyamino sugars.

Of the 2,6-dideoxy-2-amino sugars (Table 1), fucosamine has the same stereochemistry as fucose, a constituent of sialyl Lewis X expressed for example on leukocytes. Rhamnosamine is derived from the deoxy sugar rhamnose (a component of some bacterial cell walls) in

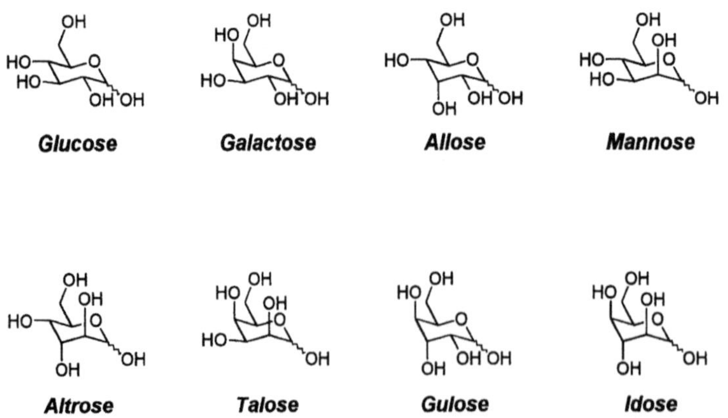

Fig. 4. Configuration of D-pyranose sugars

Table 1. *Structures of 2,6-Dideoxy-2-amino Sugars*

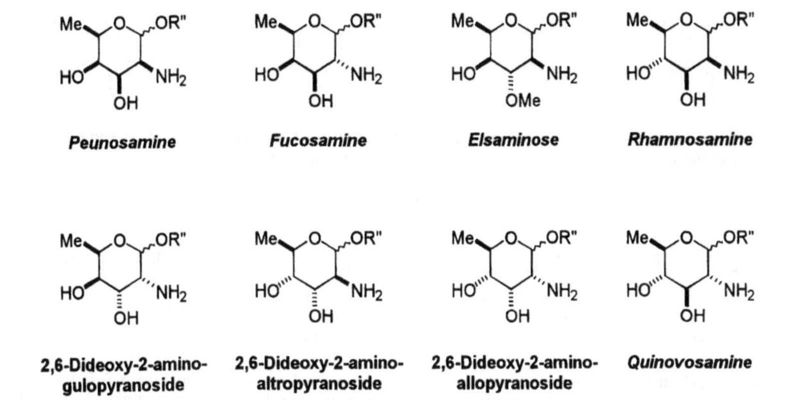

the same manner as fucosamine from fucose. Elsaminose is a constituent amino sugar of elsamicin A, an antitumor antibiotic structurally related to chartreusin. Quinovosamine has a stereo- and regio-chemistry similar to those of glucosamine and is the 2-epimer of rhamnosamine.

Of the 3,6-dideoxy-3-amino sugars, only two have been reported to occur naturally (Table 2). Mycaminose has a dimethylated amino group, and the corresponding structure with the *talo*-stereochemistry was recently described as being a part of the antifungal compound fluvirucin B1.

Table 2. *Structures of 3,6-Dideoxy-3-amino Sugars*

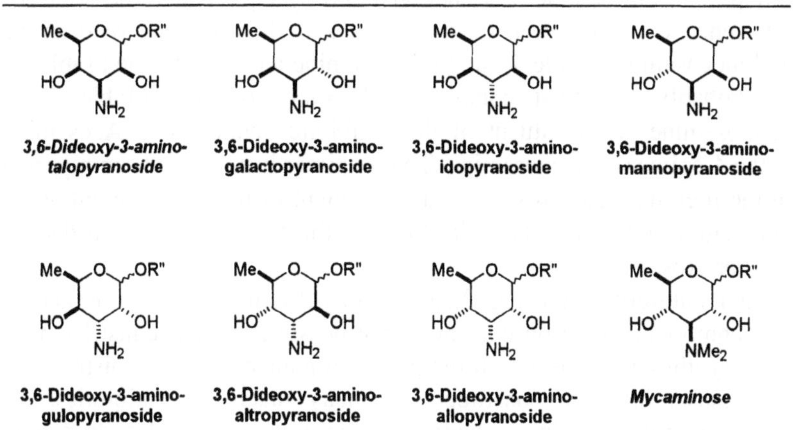

Table 3. *Structures of 4,6-Dideoxy-4-amino Sugars*

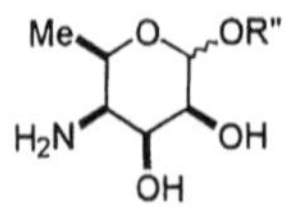	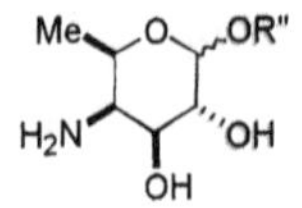	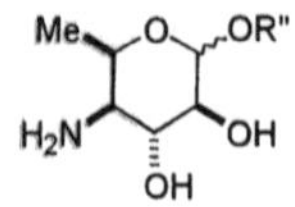	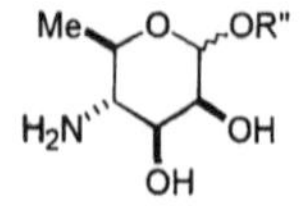
4,6-Dideoxy-4-amino-talopyranoside	***Thomosamine***	**4,6-Dideoxy-4-amino-idopyranoside**	***Perosamine***
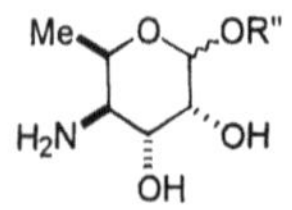	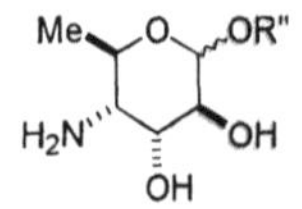	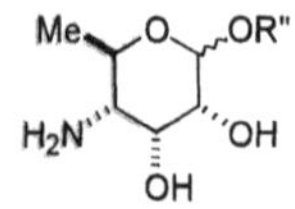	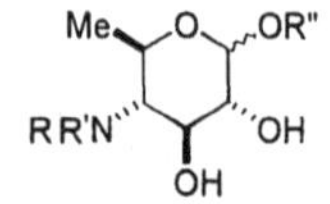
4,6-Dideoxy-4-amino-gulopyranoside	**4,6-Dideoxy-4-amino-altropyranoside**	**4,6-Dideoxy-4-amino-allopyranoside**	***Viosamine***, R,R'=H ***Bamosamine***, R,R'=H,Me ***Amosamine***, R,R'=Me

Few natural congeners of 4,6-dideoxy-4-amino sugars are known (Table 3). Perosamine, a sugar component of perimycin, is a regioisomer of rhamnosamine. Viosamine is a regioisomer of quinovosamine. Bamosamine and amosamine are partly or fully *N*-methylated analogues of viosamine. Thomosamine has the *galacto*-stereochemistry.

Many natural amino sugars have the same stereo- and regiochemistry and differ only in the substitution of the amino group (with one or two alkyls or with some other groups like acyl). The amino group may also be present in its free form. A good example of the variation is the difference between vios-, bamos-, and amos-amine: viosamine has a free amino group at position 4, while bamosamine has a methylamino group and amosamine a dimethylamino group. The same kind of difference exists between daunosamine and rhodosamine, ristosamine and megosamine (Table 4), and tolyposamine and forosamine (Table 7).

Probably the most common trideoxyaminohexose (Table 4) is daunosamine, a constituent of the antibiotic daunomycin. Acosamine is the 4-epimer of daunosamine. The 3-epimer has not yet been reported in the literature. Kedarosamine, a constituent of the enediyne antibiotic kedarcidin and first isolated in 1992, is the youngest of the trideoxyaminohexoses.

No naturally occurring 2,3,6-trideoxy-2-aminohexoses are known and compounds with only a few of the possible stereochemistries have been synthesized (those with the gulo and manno stereochemistry).

Desosamine is a constituent of many macrolide antibiotics, for example erythromycin, narbomycin, picromycin, and oleandromycin.

Table 4. *Structures of Trideoxyamino Sugars having a Methylene Group at Position 2*

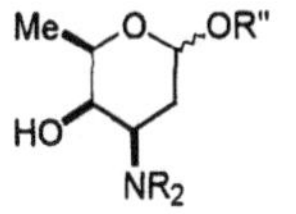	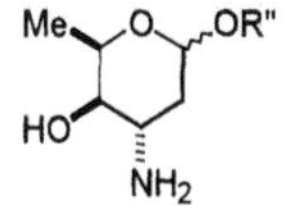	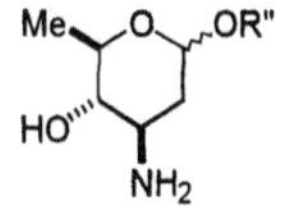	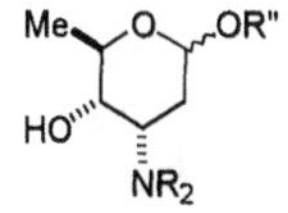
Daunosamine, R=H ***Rhodosamine***, R=Me	**2,3,6-Trideoxy-3-amino-*xylo*-hexopyranoside**	***Acosamine***	***Ristosamine***, R=H ***Megosamine***, R=Me
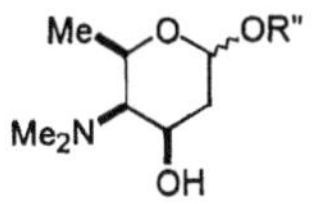	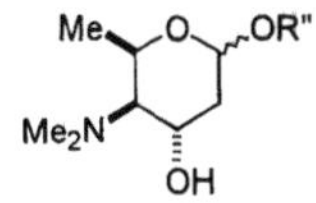	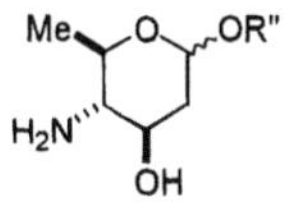	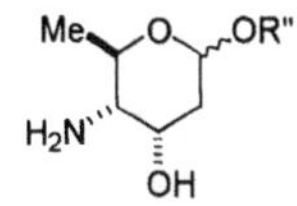
Kedarosamine	**epi-*Kedarosamine***	***Hollantosamine***	**2,4,6-Trideoxy-3-amino-*ribo*-hexopyranoside**

The 2-epimer of desosamine is known, but not the 3-epimer. As with the 2,3,6-trideoxy-2-aminohexoses, few 3,4,6-trideoxy-3-aminohexoses are known in the literature. It seems that, among the trideoxyaminohexoses, nature favors the methylene group at position 2.

Relatively few tetradeoxyaminohexoses have been isolated from natural sources and few of the possible structures are described. Evidently the 4-position for the amino group is favored in natural examples of this rare kind of sugar. Tolyposamine, forosamine and ossamine occur naturally.

Table 5. *Structures of Trideoxyamino Sugars having a Methylene Group at Position 3*

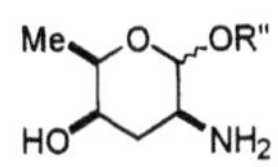	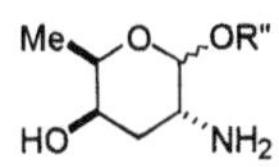	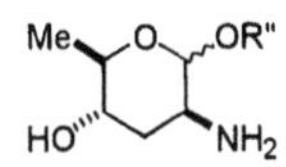	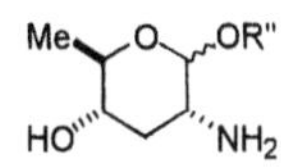
2,3,6-Trideoxy-2-amino-talopyranoside	**2,3,6-Trideoxy-2-amino-gulopyranoside**	**2,3,6-Trideoxy-2-amino-mannopyranoside**	**2,3,6-Trideoxy-2-amino-glucopyranoside**
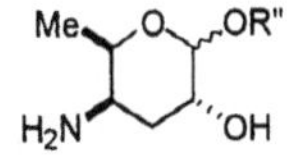	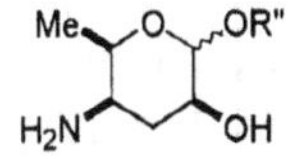	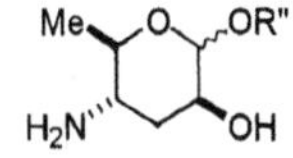	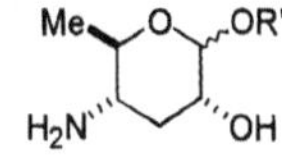
3,4,6-Trideoxy-4-amino-talopyranoside	**3,4,6-Trideoxy-4-amino-gulopyranoside**	**3,4,6-Trideoxy-4-amino-mannopyranoside**	**3,4,6-Trideoxy-4-amino-glucopyranoside**

Table 6. *Structures of Trideoxyamino Sugars having a Methylene Group at Position 4*

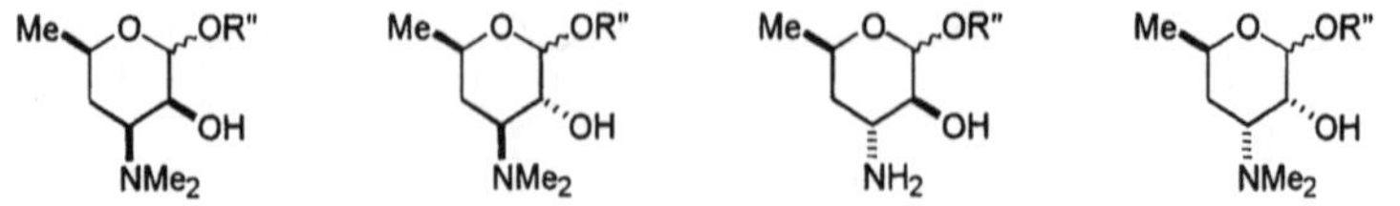

2-epi-*Desosamine* — ***Desosamine*** — **3,4,6-Trideoxy-3-amino-allopyranoside** — **3-*epi*-Desosamine**

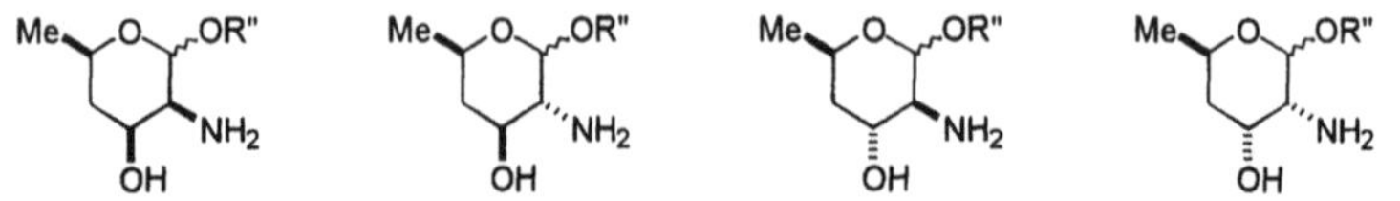

2,4,6-Trideoxy-2-amino-talopyranoside — **2,4,6-Trideoxy-2-amino-allopyranoside** — **2,4,6-Trideoxy-2-amino-idopyranoside** — **2,4,6-Trideoxy-2-amino-gulopyranoside**

Table 7. *Possible Stuctures of Tetradeoxyamino Sugars*

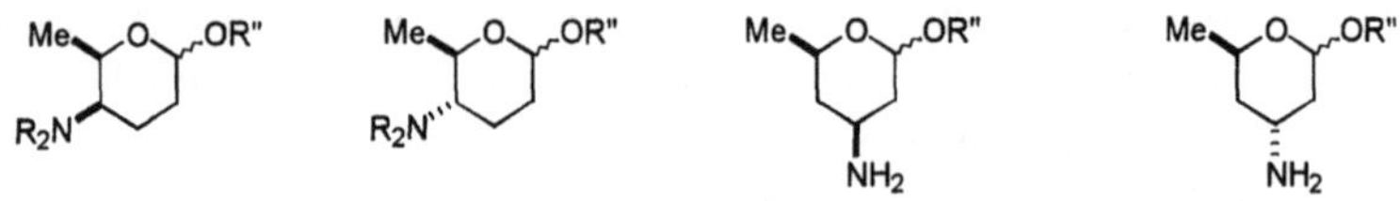

epi-*Tolyposamine*, R=H
Ossamine, R=Me

Tolyposamine, R=H
Forosamine, R=Me

2,3,4,6-Tetradeoxy-3-amino-glucopyranoside

2,3,4,6-Tetradeoxy-3-amino-allopyranoside

2,3,4,6-Tetradeoxy-2-amino-mannopyranoside — **2,3,4,6-Tetradeoxy-2-amino-glucopyranoside**

3. Synthetic Aspects

3.1. Carbohydrates as Starting Materials

The relatively inexpensive and readily available D-glucose has been used as a starting material for numerous syntheses of amino sugars. D-Glucose is easily converted to methyl α-D-glucopyranoside (also commercially available) by the Koenigs-Knorr method, by heating the carbohydrate in 2% HCl in methanol. The thermodynamically more stable α-anomer is crystallized as the monohydrate (*28*). D-Mannose and

D-galactose have also been used as starting materials for the synthesis of D-aminohexoses, while the more expensive L-rhamnose and L-fucose have commonly been used for the synthesis of L-amino sugars. When carbohydrates are used as starting materials, protections and deprotections are usually needed, which decrease the overall efficiency of the preparation. Most of these methods were developed at the beginning of this century.

3.1.1. Monoamino Dideoxyhexoses

Thomosamine, viosamine, and methyl 3,6-dideoxy-3-aminoglucopyranoside have all been synthesized from D-glucose, as the stereochemistry of the starting sugar was suitable for the target hexoses (Schemes 1, 2, 4) (*29–31*). The 6- and 4-hydroxy groups of methyl α-D-glucopyranoside were protected as the benzylidene acetal (*32*).

In the synthesis of thomosamine, Stevens *et al.* used benzyl ethers as protecting groups for 2- and 3-hydroxyls, and after cleavage of the benzylidene protection methyl 2,3-di-O-benzyl-α-D-glucopyranoside was formed in 60% overall yield (*29, 32–35*). The 4- and 6-hydroxy

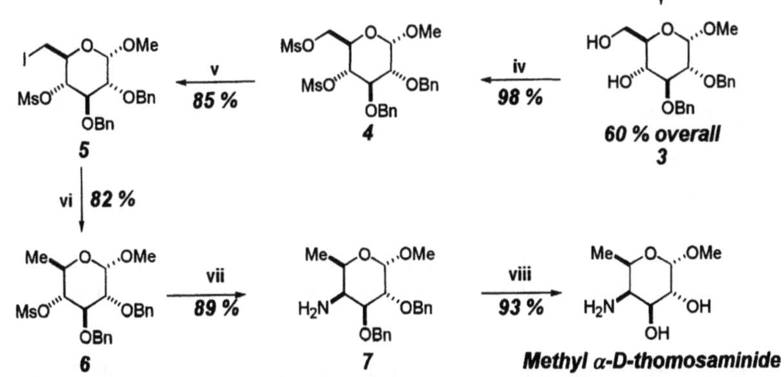

Scheme 1. Reagents: **i**, PhCHO, $ZnCl_2$; **ii**, NaH, BnBr, $Bu_4N^+I^-$, THF; **iii**, Acetone, H_2O, HCl; **iv**, MsCl, TEA; **v**, I_2, 2-butanone, heat; **vi**, $LiAlH_4$, THF; **vii**, 1) $Li^+N_3^-$ 2) $LiAlH_4$; **viii**, H_2, Pd/C, HCl

groups were removed by creating two mesyloxy groups as good leaving groups and taking advantage of the different reactivities of secondary and primary groups. The primary mesyloxy group was selectively displaced with iodide, and the intermediate was reduced to the 6-deoxy-4-mesylate. Instalment of the nitrogen function with azide ion and reduction of the azide with lithium aluminium hydride gave, after hydrogenolysis of the benzyl ethers, thomosamine hydrochloride as the α-methyl glycoside in 34% overall yield (Scheme 1).

Recently, Ward and Kaller have reported a similar synthesis of thomosamine as an intermediate of actinobolin (*36*, *37*). The only major difference between the sequence just described and their method is the use of zinc in acetic acid in the dehalogenation step.

Jary *et al.* used methyl 4,6-*O*-benzylidene-α-D-glucopyranoside as starting material in the synthesis of 3,6-dideoxy-3-aminoglucopyranoside (Scheme 2) (*38*). The C-3-stereochemistry was retained by double inversion. The first inversion was achieved by the conversion of the 3-hydroxy group to chloride (*39*). From compound **8** they prepared the 3-chloro-6-bromoallopyranoside **9** by the method of Hanessian and Plessas (*40*). The bromide was reduced by catalytic hydrogenation under basic conditions. The second inversion involved treatment of the chloride with azide. After deprotection and hydrogenation, methyl 4,6-dideoxy-4-amino-α-D-glucopyranoside was produced.

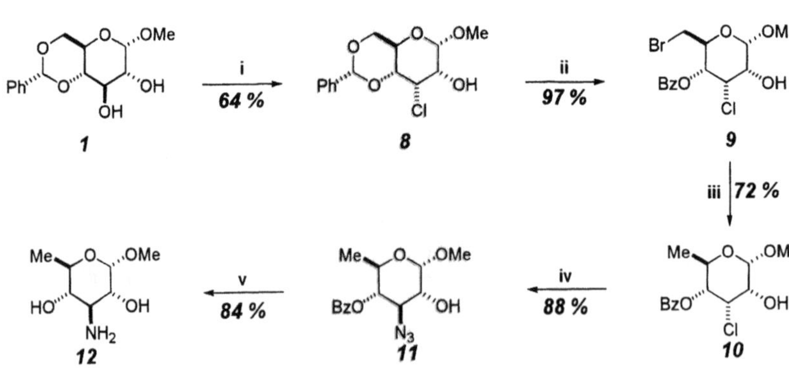

Scheme 2. Reagents: **i**, SO_2Cl_2; **ii**, NBS, $BaCO_3$; **iii**, Et_2NH, H_2, Raney Ni, MeOH; **iv**, NaN_3 DMF; **v**, 1) NaOMe, MeOH, Dowex-50 (H^+) 2) H_2, PtO_2, EtOH

Stick and Patroni attempted to use a cyclic thiocarbonate as a means of deoxygenating at C-6 and obtaining the intermediate **16** of the natural product mycaminose (*41*, *42*). However, they failed to produce the cyclic thiocarbonate **15** (Scheme 3).

References, pp. 255–263

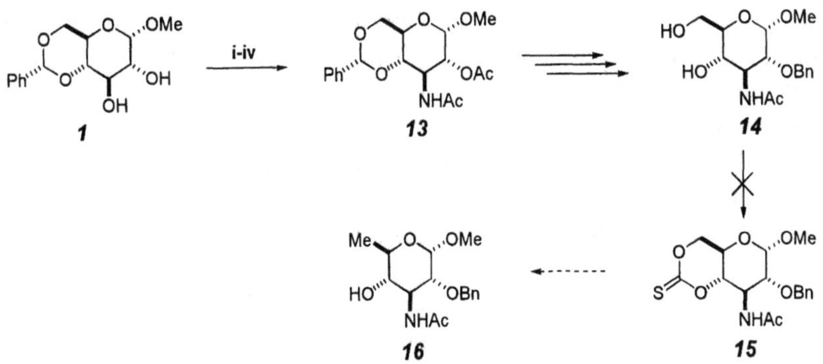

Scheme 3. Reagents: **i**, $NaIO_4$; **ii**, $PhNHNH_2$; **iii**, H_2/Raney Ni; **iv**, Ac_2O

FURSTNER *et al.* have used methyl 2,3-*O*-dimethyl-4,6-*O*-benzylidene-α-D-glucopyranoside in the synthesis of azido sugar **21**, an intermediate for the synthesis of viosamine, bamosamine, and amosamine (Scheme 4) (*43*). The starting material was synthesised by applying common carbohydrate chemistry (*44*). The synthetic sequence consists of standard protection/nucleophilic substitution steps. Azide ion perfectly discriminates between the primary tosylate and the triflate group at C-4, the latter

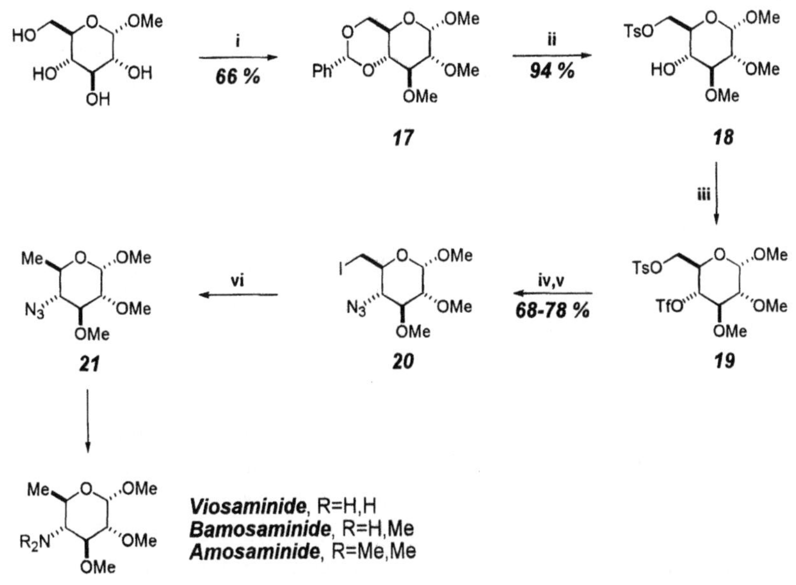

Scheme 4. Reagents: **i**, 1) PhCHO, $ZnCl_2$, 2) CH_3OSO_3Na, DMSO, MeI; **ii**, 1) *p*-TsOH · H_2O, MeOH 2) TsCl, pyr, CH_2Cl_2; **iii**, Tf_2O, pyr, CH_2Cl_2; **iv**, NaN_3, DMF; **v**, $Bu_4N^+I^-$, MeCN; **vi**, Zn/Ag-graphite, THF, rt, 25-60 min

being selectively substituted. Zinc-mediated reduction of the deoxyhalo sugar was achieved in good yield (69%). Competing dealkoxyhalogenation led to ring opening. Zinc-induced reduction was also used with 2- and 3-azido (Scheme 5) derivatives, with good results (81% and 70%) (*43*, *45*, *46*). When the formation of organozinc compounds was eliminated, ring opening was also avoided. Hydrogenation of the azido compounds would lead to the respective amino sugars.

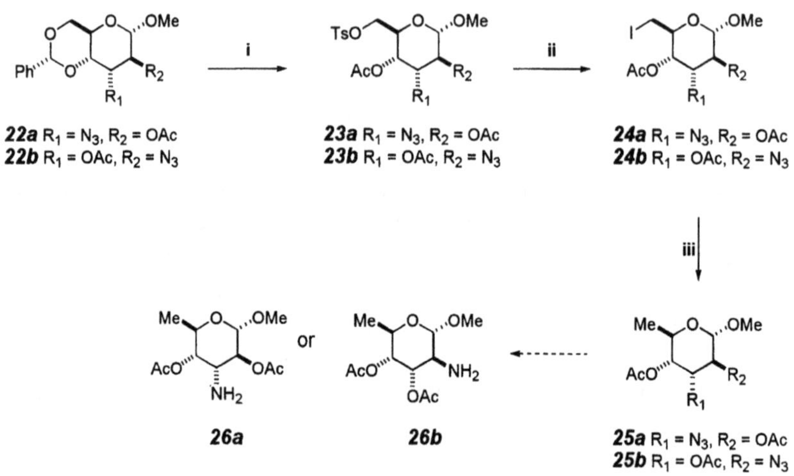

Scheme 5. Reagents: **i**, 1) p-TsOH · H_2O, MeOH 2) TsCl, pyr, CH_2Cl_2 then Ac_2O; **ii**, $Bu_4N^+I^-$, MeCN, heat; **iii**, Zn/Ag-graphite, THF, rt, 25-60 min

To synthesise mycaminose the anhydro compound **27** was treated with ethanolic dimethylamine to give a 1 : 1 mixture of 2- and 3-dimethylamino compounds (*47*, *48*). Standard tosylation and reduction of the 3-isomer produced mycaminose methyl glycoside (Scheme 6).

STEVENS *et al.* have synthesised viosamine and the related bamosamine (methylamino) and amosamine (dimethylamino) by two different routes (*29*, *30*, *49*). One route started from D-galactose with inversion of configuration at C-4 (Scheme 7) and the other one from glucose with double inversion at C-4 (Scheme 8). In both routes 6-deoxy structures were achieved with the conventional ditosyl or dimesyl method where a primary tosylate or mesylate was substituted with iodide and the alkyl iodide was dehalogenated. The tosyloxy group at C-4 of the galactose derivative was substituted with azide ion through inversion to give the correct stereochemistry. The stereochemistry at C-4 in the glucose derivative was inverted when the mesylate reacted with sodium benzoate. After hydrolysis and remesylation, treatment with azide was

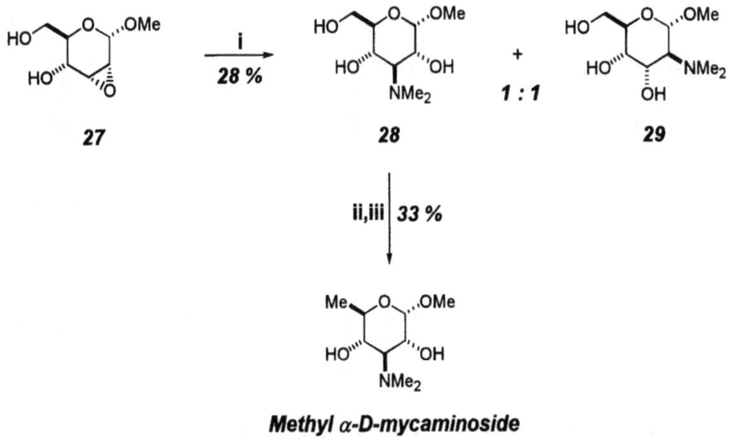

Scheme 6. Reagents: **i**, Me_2NH, EtOH; **ii**, TsCl, Pyr; **iii**, $LiAlH_4$

effected. The final steps included reduction of the azide and deprotection at C-2 and C-3 as the corresponding benzyl ethers. The dibenzylated intermediate **37** was converted to monomethyl bamosamine through an ethoxycarbamate derivative, which was reduced with lithium aluminium hydride and debenzylated. Reductive dimethylation of viosamine gave amosamine.

Another application of standard carbohydrate chemistry is the synthesis from galactose of the hydroxyamino sugar which is a structural component of calicheamicins (*31*). The synthetic sequence presented in Scheme 9 consists of protections, bromination, and dehalogenation at

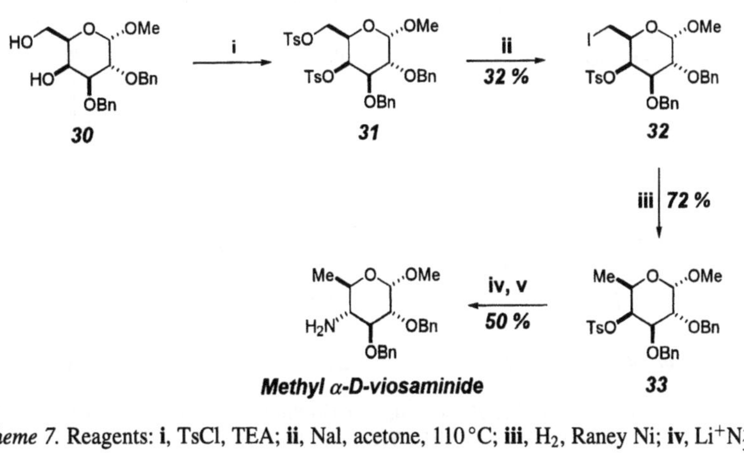

Scheme 7. Reagents: **i**, TsCl, TEA; **ii**, NaI, acetone, 110 °C; **iii**, H_2, Raney Ni; **iv**, $Li^+N_3^-$; **v**, H_2, PtO_2, MeOH

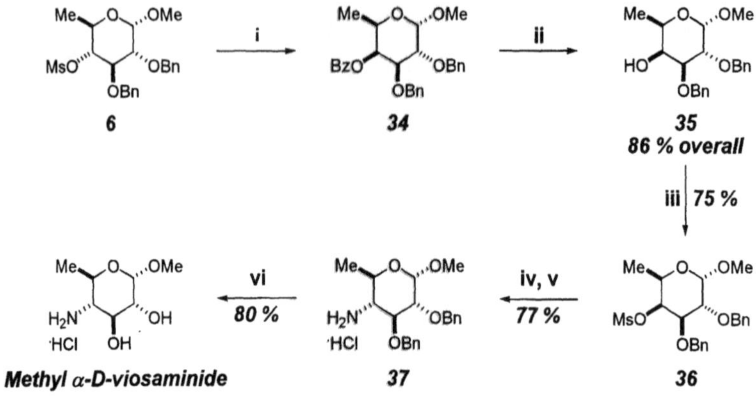

Scheme 8. Reagents: **i**, $PhCOO^-Na^+$, DMF, heat; **ii**, NaOH, EtOH, H_2O; **iii**, MsCl, Pyr.; **iv**, $Na^+N_3^-$; **v**, $LiAlH_4$; **vi**, H_2, Pd/C

C-6 as well as nucleophilic displacement with azide and reduction. If the 2,3-protection were cleaved in this step, viosamine would be produced. Instead, to build up the calicheamicin sugar unit, the amino group was converted to a hydroxyamino group *via* a nitro intermediate.

D-Perosamine, the enantiomer of naturally occurring L-perosamine (the constituent of the antibiotic perimycin), was synthesised from D-

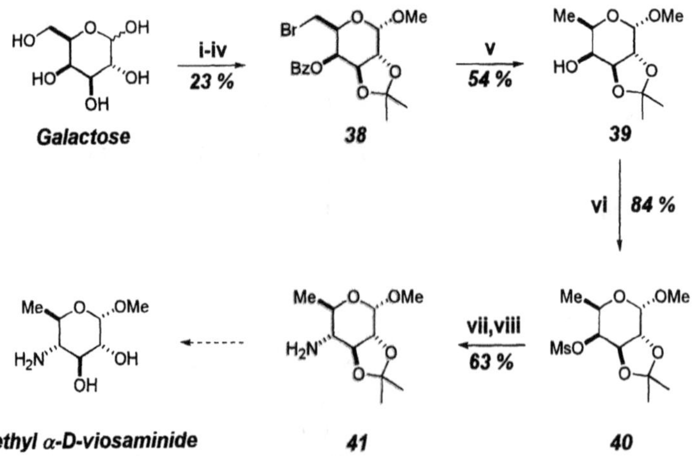

Scheme 9. Reagents: **i**, $MeOH/H^+$; **ii**, $PhCH(OMe)_2$, H^+; **iii**, $NBS/BaCO_3$, CCl_4; **iv**, $Me_2C(OMe)_2$; **v**, $LiAlH_4$, THF; **vi**, MsCl, Pyr; **vii**, NaN_3, DMF; **viii**, $NaBH_4$, DMF/MeOH

References, pp. 255–263

mannose (Scheme 10) (*50*). The stereochemistry at C-4 was inverted before nucleophilic displacement in order to retain the stereochemistry of the final product. Inversion of the C-4 hydroxy group was achieved through Swern oxidation and reduction with sodium borohydride. Equatorial attack of hydride produced the D-*talo* isomer, which was converted to the 4-mesylate. The isopropylidene group shielded the bottom face so effectively that the molecule was immune to nucleophilic displacement. It is worth noting that, in the previous case where the mesylate was of galacto stereochemistry (Scheme 9), nucleophilic substitution occurred with high yield (*31*). After removal of the isopropylidene protection, nucleophilic displacement took place. The α-methyl glycoside of D-perosamine was obtained after standard hydrogenation, in 13% overall yield. When the stereochemistry of the mesyloxy group was inverted, removal of the isopropylidene protection was not necessary.

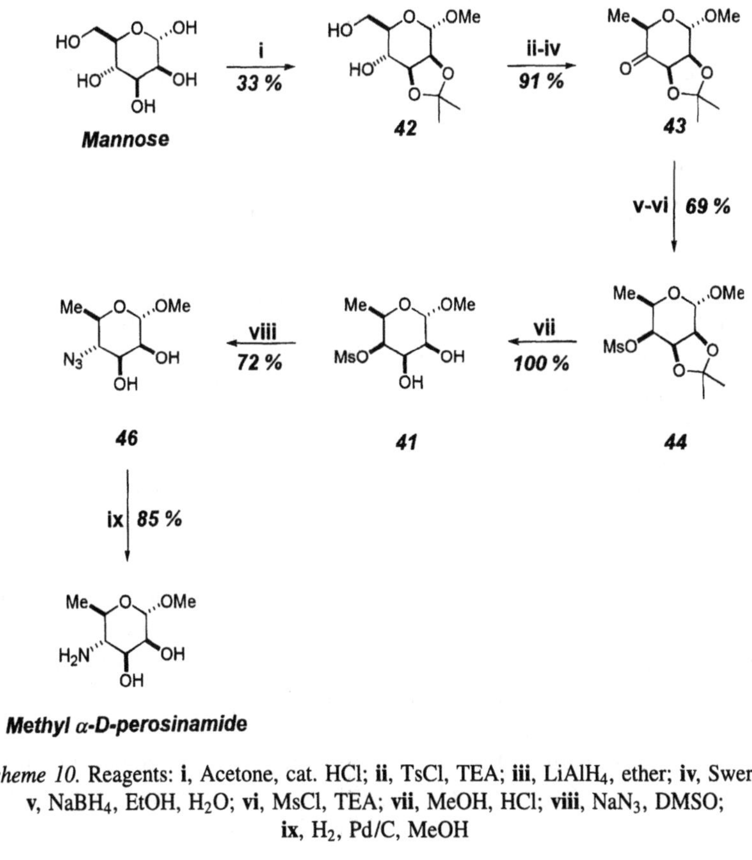

Scheme 10. Reagents: **i**, Acetone, cat. HCl; **ii**, TsCl, TEA; **iii**, $LiAlH_4$, ether; **iv**, Swern; **v**, $NaBH_4$, EtOH, H_2O; **vi**, MsCl, TEA; **vii**, MeOH, HCl; **viii**, NaN_3, DMSO; **ix**, H_2, Pd/C, MeOH

Methyl 6-deoxy-2,3-*O*-isopropylidene-α-lyxopyranosid-4-ulose **43** can also be converted into an oxime which is reduced with lithium aluminium hydride to the *talo* isomer (*51–53*). After treatment with dilute hydrochloric acid, 4-*epi*-D-perosamine **49** was obtained as the methyl glycoside. When the oxime derived from the L-enantiomer of pyranosid-4-ulose was reduced under similar conditions, the L-talo isomer was formed. Cleavage of isopropylidene gave 4-*epi*-L-perosamine (Scheme 11).

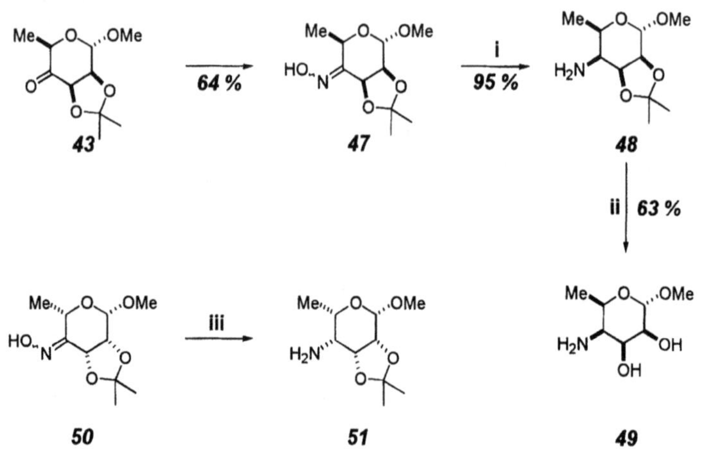

Scheme 11. Reagents: **i**, $LiAlH_4$, THF; **ii**, dil. HCl; **iii**, $LiAlH_4$

6-Deoxy-2,3-*O*-isopropylidene-α-lyxopyranosid-4-ulose **53** (*54*) derived from L-rhamnose was synthesised by a reaction sequence analogous to that of STEVENS *et al.* (*51*) reported for the D-series, with the exception that the Pfitzner-Moffatt oxidation (*55*, *56*) was employed. Reduction of the keto group with sodium borohydride followed by mesylation gave the L-talo isomer **54** (*55*). Nucleophilic displacement with azide was complicated for the L-isomer, as it was for the D-isomer. In this case only 15% of the desired azide **56** was obtained. The major constituent of the reaction mixture was the elimination product **55**. After deprotection and hydrogenation, L-perosamine was obtained in poor yield (Scheme 12). BRIMACOMBE *et al.* have also synthesised L-perosamine from methyl 6-deoxy-2,3-*O*-isopropylidene-4-*O*-methylsulfonyl-α-L-talopyranoside **54** using nucleophilic replacement. The azide **56** was reduced with lithium aluminium hydride (*52*).

A common feature of all these four syntheses is the equatorial attack of the hydride. CIEPLAK and co-workers have proposed that charge transfer stabilizes the transition state of nucleophilic addition to a

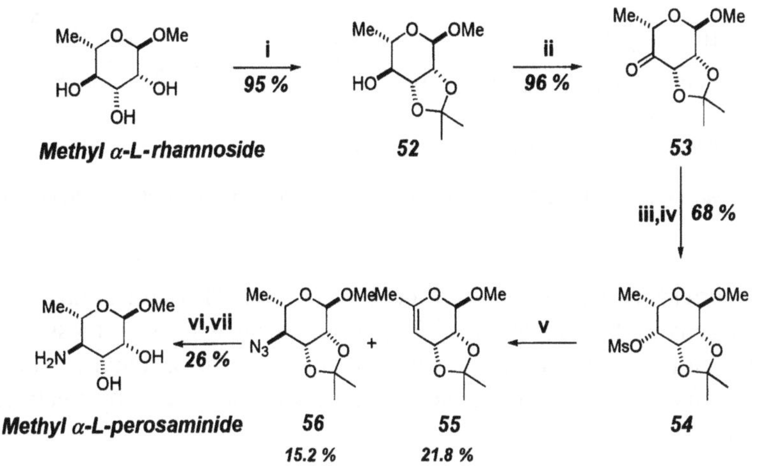

Scheme 12. Reagents: **i**, acetone, $CuSO_4$; **ii**, DMSO, benzene, pyr., TFA, $(c\text{-}C_6H_{11}N)_2C$; **iii**, $NaBH_4$, MeOH; **iv**, MsCl, pyr.; **v**, NaN_3, DMF; **vi**, 33% AcOH; **vii**, H_2, Pd/C

carbonyl group by electron donors (*57, 58*). Non-equivalence of the two faces of a carbonyl group with respect to the electron-donating power of the neighboring orbitals might create a preference for the approach that assures maximum overlap of the σ^* orbital with the most readily donating orbitals. Steric hindrance favors an equatorial approach of hydride for both ketones and oximes. For cyclic ketones without any heteroatoms, however, hydride favors the axial approach. Electron donation from the cyclohexanone σ_{CH} rather than σ_{CC} bonds into the σ^* orbital favors the axial approach since the carbon-hydrogen bonds are better donors. The electropositive α-substituent of the pyranosid-4-uloses (or oximes) derived from D-mannose or L-rhamnose favors equatorial over axial approach since the σ_{CO} bond is a better electron donor than the σ_{CH} bond.

Jary and Zobacova have reported that sometimes the replacement of tosylate or mesylate with hydrazine gives better results than replacement with azide or ammonia (*59*). As an example of hydrazinolysis, Scheme 13 presents the synthesis of 4-*epi*-perosamine. The compound can also be prepared from the corresponding oxime derivative (*53*).

Coleman *et al.* have synthesised the precursor **60** of the quinovosamine derivative **61** (Scheme 14) (*60*). Several dehalogenation methods are available to convert the iodide to the 6-deoxy derivative. The primary C-6 hydroxyl group of the carbamate **59** derived from glucosamine was selectively iodinated by heating the triol with iodine, triphenyl phosphine, and pyridine in toluene. In an alternative, longer approach,

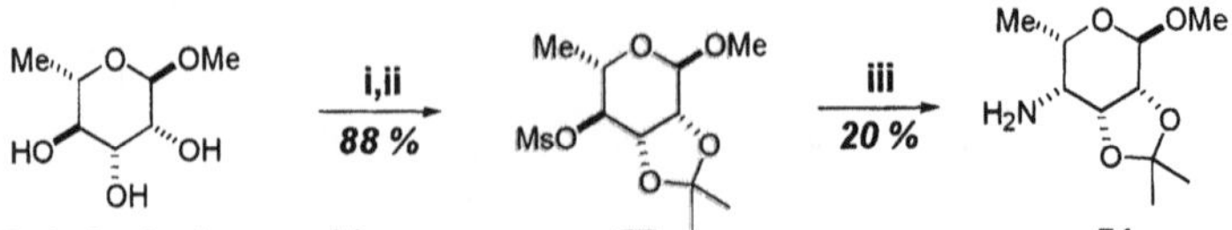

Methyl α-L-rhamnoside **57** **51**

Scheme 13. Reagents: **i**, Acetone, $CuSO_4$; **ii**, MsCl, pyr.; **iii**, NH_2NH_2, H_2, Raney Ni

the corresponding 4,6-*O*-benzylidene acetal with TBS-protection at C-3 is selectively brominated at C-6 upon treatment with *N*-bromosuccinimide-barium carbonate and a catalytic amount of azoisobutyronitrile in anhydrous carbon tetrachloride. Bromine was exchanged by treatment with sodium iodide.

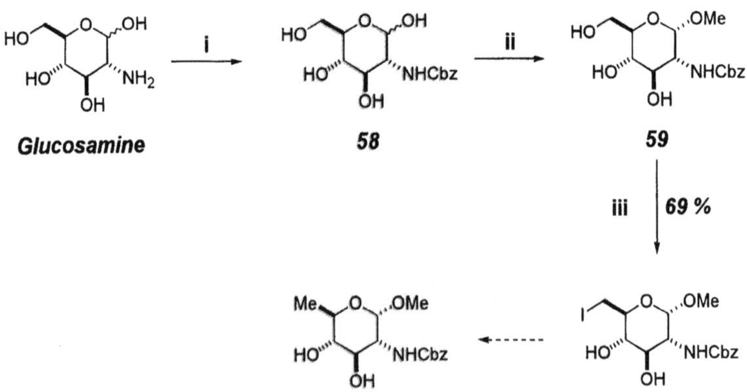

Glucosamine **58** **59** **61** **60**

Scheme 14. Reagents: **i**, $ClCO_2Bn$, $NaHCO_3$, dioxane/H_2O; **ii**, MeOH, H^+; **iii**, I_2, Ph_3P

PAULSEN *et al.* have used D-glucosamine as starting material for D-fucosamine (Scheme 15) (*61*). The synthesis requires inversion at C-4 and conversion of the primary alcohol into the deoxy form. Inversion of the C-4 hydroxy group is obtained *via* the triflate, which upon treatment with sodium nitrite in DMF gives the aminohexose **63**. A triflate is also employed in the deoxygenation of the primary alcohol. Reduction of the triflate with sodium borohydride in acetonitrile and cleavage of the acetonide protection furnishes the *O*-allyl *N*-acetylfucosaminide in 34% overall yield starting from allyl *N*-acetyl-3,6-*O*-benzoylglucosaminide **62**.

When the isopropylidene protecting group of the mesylate **67** (derived from mannose) is cleaved and the diol intermediate is treated

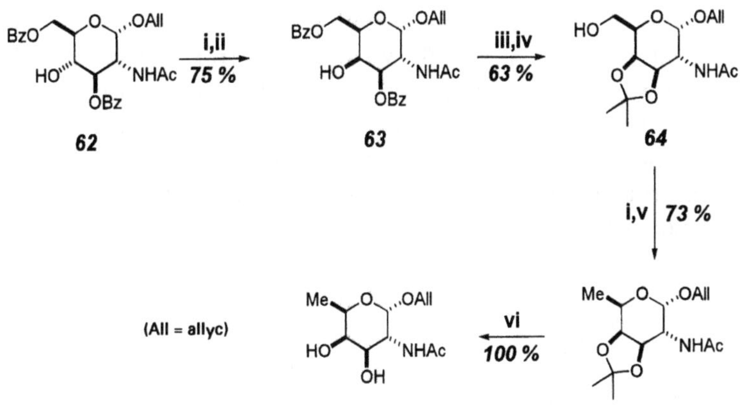

62 63 64 66 65

Scheme 15. Reagents: **i**, $(TfO)_2O$, pyr., CH_2Cl_2; **ii**, $NaNO_2$, DMF; **iii**, NaOMe, MeOH; **iv**, DMP, DMF, *p*-TsOH; **v**, $NaBH_4$, MeCN; **vi**, 80% AcOH

with sodium hydroxide, the epoxide **68** is formed in 65% yield (Scheme 16) (*62*). Introduction of nitrogen at either position 3 or 4 of the oxirane gives access to two different deoxyamino sugars, 3,6-dideoxy-3-amino-idopyranoside and perosamine, as the corresponding methyl glycosides. Usually the nucleophile attacks C-3 (*54*), but with proper modification it can be directed to C-4. When the C-2 hydroxy group is protected with bulky groups (*e.g.* benzoyl), C-4 attack of the nucleophile is favored (*62*). This is explained in Scheme 17. Because the epoxide **70** is not

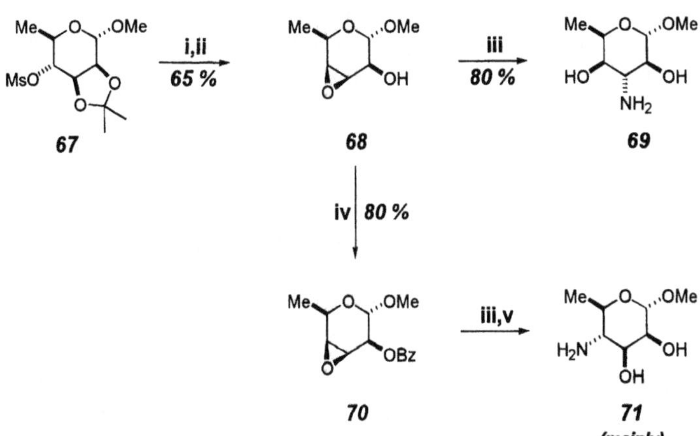

67 68 69 70 71

Scheme 16. Reagents: **i**, H^+; **ii**, NaOH; **iii**, MeOH, NH_3; **iv**, PhCOCl, pyr.; **v**, HO^-

stabilised in any way, it is considered to exist as a mixture of two half-chair conformations **A** and **B**. When the hydroxy group at C-2 is protected with a bulky group conformation **B** is destabilised by steric repulsion between the methoxy and benzoyloxy groups and thus conformation **A** is favored. For steric reasons, the nucleophile attacks C-4. In principle, larger protecting groups than benzoyl should cause even better selectivity.

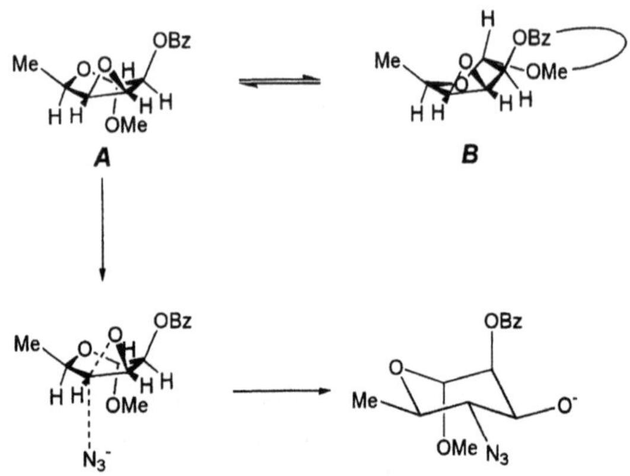

Scheme 17. Effect of C-2 *O*-benzoyl group on the nucleophilic attack

MALIK *et al.* have regioselectively opened the oxirane ring of the 2,3-anhydro sugars with silylamines to synthesise amino sugars like mycaminose (Scheme 18) (*63*). The oxirane compound was synthesised by method of HANESSIAN and PLESSAS (*40*). Trans-diaxial opening of epoxide **72** was achieved by treatment with *N,N*-dimethyltrimethylsilylamine in the presence of anhydrous aluminium chloride. No yields or detailed reaction conditions were given.

72 ***Methyl α-D-mycaminoside***

Scheme 18. Reagents: **i**, TMSDMA, $AlCl_3$

JARY *et al.* and RICHARDSON have both reported a non-stereospecific synthesis of mycaminose (Scheme 19) (*64, 65*). Periodate oxidation of methyl 6-deoxy-glucopyranoside produced the dialdehyde **74**, which underwent cyclization with nitromethane to produce the nitro pyranoside

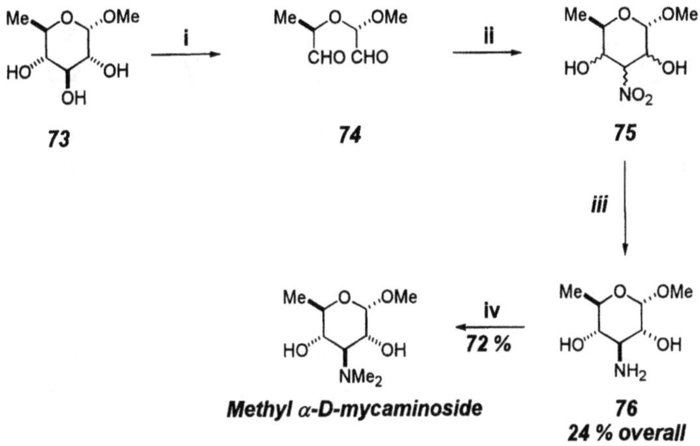

Scheme 19. Reagents: **i**, $NaIO_4$; **ii**, $MeNO_2$; **iii**, H_2/Ni; **iv**, HCHO, HCOOH

75 as a mixture of isomers. Methyl 3,6-dideoxy-3-amino-glucopyranoside was obtained after catalytic hydrogenation of the nitro group in 24% overall yield. Reductive dimethylation converted it to the mycaminoside.

In a few cases, furanosides have been used in the synthesis of aminohexoses. For example, Horton and Liav used 5-deoxyribose for the synthesis of 2-aminohexoses having either the *allo* or *altro* stereochemistry (Scheme 20) (*66*). 5-Deoxyribose was synthesised from ribose using conventional sugar chemistry in four steps in an overall yield of 27% (*67*, *68*). Treatment of 5-deoxyribose under the conditions of Kuhn and Fischer (*69*) with aniline and hydrogen cyanide produced a mixture of anilino allo- (11.5%) and altro-nitriles (25%) in poor yield. Reduction of the nitriles followed by treatment with acid produced the aminohexoses. Finally, galacto- and altro-aminohexoses were acetylated with acetic anhydride in pyridine. Starting from the appropriate anilino nitrile, 2,6-deoxy-2-amino-altrohexose **83** was produced in 47% and the allo-isomer **86** in 33% yield. In this route, seven steps are needed and overall yields starting from ribose are not very high (1.0–3.2%).

3.1.2. *Monoamino Trideoxyhexoses*

The difference between trideoxy- and dideoxy-hexoses is the presence of a methylene group in the ring. This also creates special considerations for the synthetic strategies. Deoxygenaton of a secondary alcohol by conventional S_N2 methods has invariably proved difficult. The methods used are lengthy and suffer from lack of generality. Stick and Patroni have used a free radical procedure involving the reduction of a

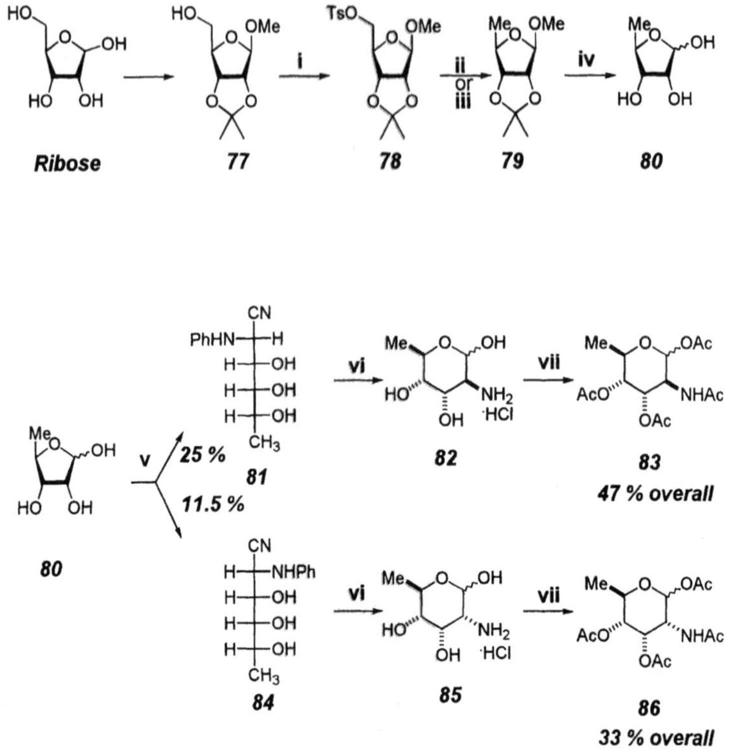

Scheme 20. Reagents: **i**, TsCl, pyr.; **ii**, $LiAlH_4$; **iii**, NaI then H_2/Pd; **iv**, H^+; **v**, $PhNH_2$, HCN; **vi**, H_2, Pd/$BaSO_4$; **vii**, Ac_2O, pyr

dithiocarbonate with tributyltin hydride (Scheme 21) (*41*). The starting methyl 3-acetylamino-4,6-*O*-benzylidene-3-deoxy-α-D-glucoside was prepared by conventional methods (*42*). Alcohol **87** was treated sequentially with sodium hydride, carbon disulfide, and methyl iodide in dimethylformamide. The dithiocarbonate **88** thus obtained was reduced with tributyltin hydride in toluene at reflux to give the desired 2-deoxysugar **89** in 65% yield. Several methods are known for deoxygenation at C–6, which would lead to the monoamino trideoxyhexoses.

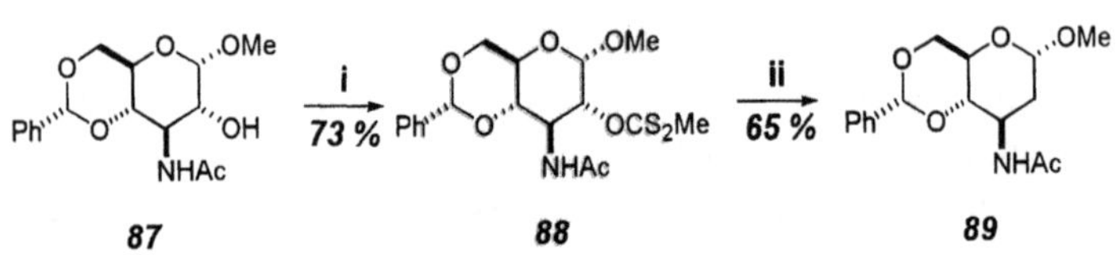

Scheme 21. Reagents: **i**, NaH, CS_2, MeI, DMF; **ii**, Bu_3SnH, toluene, heat

The methyl 3-acetamido-2,3-dideoxy-β-D-arabino-hexopyranoside **90** was converted to acosamine and daunosamine using tosylation, nucleophilic displacement, and dehalogenation (Scheme 22) (*70*). In the case of daunosamine, inversion of the C-4 hydroxy group was required. The overall yields were 22–24% for the diacetylated acosamine **93** and 18–19% for methyl *N*-acetyldaunosaminide **95**.

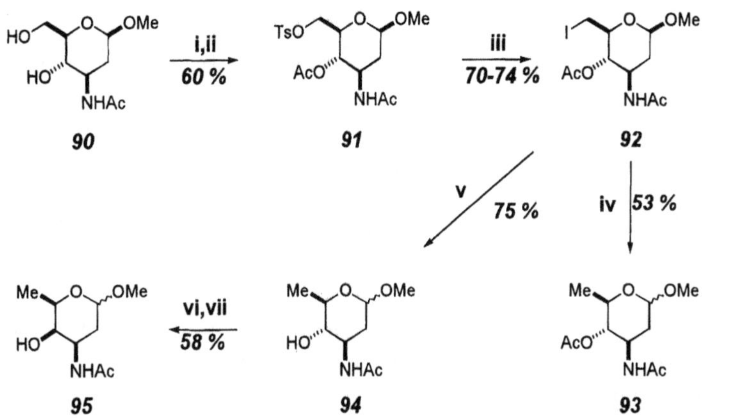

Scheme 22. Reagents: **i**, TsCl; **ii**, Ac_2O; **iii**, NaI, $NaHCO_3$, butanone; **iv**, H_2/Ni, TEA, MeOH; **v**, NaOMe, MeOH, H_2, PtO_2; **vi**, MsCl, pyr.; **vii**, NaOAc, H_2O, MeOEt, heat

Desosamine was synthesised by RICHARDSON from methyl 3-acetamido-4,6-*O*-benzylidene-3-deoxy-α-D-glucopyranoside **87** (Scheme 23) (*71*). The synthetic sequence includes protection, dimesylation, diiodination, and didehalogenation. The reaction of dimesylate **97** with sodium iodide in ethyl methyl ketone first produced the monoiodo derivative. The monoiodo derivative was slowly transformed into the diiodo compound **98** in 40% yield. Scheme 1 has presented the synthesis of thomosamine, where the same kind of dimesyl derivative **4** produced only a monohalogeno compound in high yield (85%), while the diiodo compound was formed as a minor side product. The absolute stereochemistry of the diiodo derivative **98** was not determined, but it is expected to have the configuration shown (retention), which can be explained by participation of the neighboring *N*-acetyl group *via* a bicyclic intermediate. However, the uncertainty of the configuration did not affect the synthesis owing to the subsequent dehalogenation of the diiodo compound. Finally, the acetyl protection was cleaved by basic hydrolysis.

OVEREND *et al.* have used methyl 3,4-anhydro-6-deoxy-α-L-lyxo-hexopyranosid-2-ulose **99** derived from L-rhamnose in the synthesis of

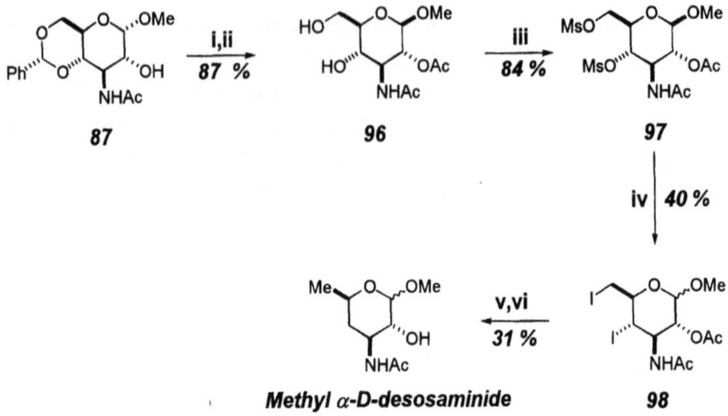

Scheme 23. Reagents: **i**, Ac_2O; **ii**, 50% AcOH; **iii**, MsCl; **iv**, NaI, 2-butanone; **v**, H_2/Ni; **vi**, dil. NaOH

methyl 2,3,6-trideoxy-2-amino-α-L-gulopyranoside **102** (Scheme 24) (72). Reaction of epoxy ketone **99** with *o*-nitrophenylhydrazine or *o,p*-dinitrophenylhydrazine in ethanol containing acetic acid produced the hydrazone derivatives **100**. On the other hand, reaction of the epoxy ketone **99** under similar conditions with phenylhydrazine or *p*-nitrophenylhydrazine yielded the azo cycloalkenes **101**, which were converted to the trideoxyaminohexose **102** by reduction of the azo moiety with sodium borohydride and hydrogenation of the double bond. When the reactions with phenyl- and *p*-nitrophenyl-hydrazine were run under weakly alkaline conditions, the formation of similar hydrazone derivatives was observed. This indicates that azo cycloalkenes are formed *via* hydrazone derivatives which undergo acid catalysed opening

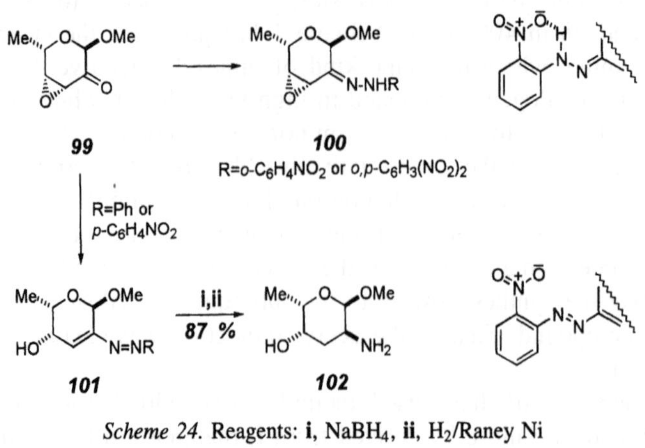

Scheme 24. Reagents: **i**, $NaBH_4$, **ii**, H_2/Raney Ni

of the epoxide. An *ortho*-nitro substituent in the hydrazone inhibits the epoxide opening for at least two reasons: an intramolecular hydrogen bond stabilises the hydrazone form and destabilises the azo form due to the electronic interaction between the oxygen and nitrogen lone pairs. In this way a C-3 methylene group is formed without recourse to the typical S_N2 mechanism.

Glycals are common precursors for the synthesis of trideoxyamino-hexoses having a methylene carbon at position 2 with *ribo*, *arabino*, *xylo*, and *lyxo* stereochemistries. L-Rhamnose, widely used for the synthesis of the L-series of the above-mentioned aminohexoses, can be converted to its glycal *via* tetraacetyl or tetrabenzoyl rhamnose (*73*, *74*). With slightly different procedures, tetra-protected rhamnoses are converted to the corresponding pyranosyl bromides, which are treated *in situ* with zinc dust to produce diprotected glycals. Finally, alkaline hydrolysis of protecting groups gives 1,5-anhydro-2,6-dideoxy-L-arabino-hex-1-enitol **105**. Both protected and free glycals have been used as starting materials (Scheme 25).

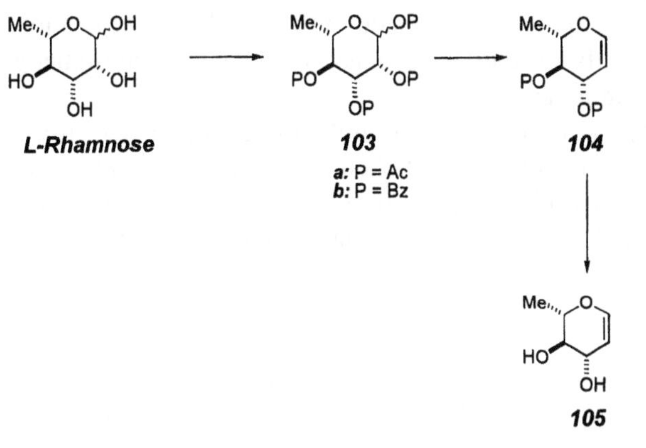

Scheme 25. Conversion of L-rhamnose to L-arabino-hex-1-enitol **105**

Methyl 2,6-dideoxy-α-L-arabino-hexopyranoside (the α-methyl glycoside of L-olivose) is another common intermediate in these syntheses. It can be obtained from either 1,5-anhydro-3,4-*O*-dibenzoyl-2,6-dideoxy-L-arabino-hex-1-enitol **104b** or L-rhamnal **105** (Scheme 26) (*74*–*77*). The 2-deoxy structure can be obtained from the glycal **104b** by treatment with methanol and cation exchange resin followed by hydrolysis with sodium methoxide. Methoxymercuration of L-rhamnal **105** followed by reduction also furnished the 2-deoxy product.

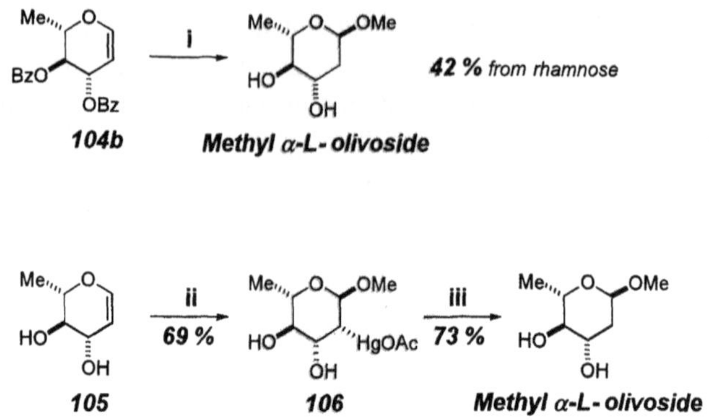

Scheme 26. Reagents: **i**, MeOH, AG 50W-X8 (H^+) then Na; **ii**, $Hg(OAc)_2$, MeOH; **iii**, KBH_4, NaOH, H_2O, MeOH

FLORENT *et al.* have heated diacetyl L-rhamnal in water to produce a mixture of pseudo-rhamnals (Scheme 27) (*78*). The addition of sodium azide in wet acetic acid to the crude mixture of **107**, **108**, followed by acetylation, gave diastereomers at C-1 and C-3 (**109**), which were converted into methyl glycosides **110** to make the separation of isomers easier. Finally standard hydrogenation of the arabino isomer gave methyl α-L-acosaminide.

GRETHE *et al.* have used L-arabinose as a precursor for a large-scale synthesis of daunosamine (Scheme 28, see also Scheme 26) *via* methyl glycoside of L-olivose (*79*). In this way a mixture of α- and β-anomers was formed. The one-carbon enlargement of L-arabinose was achieved

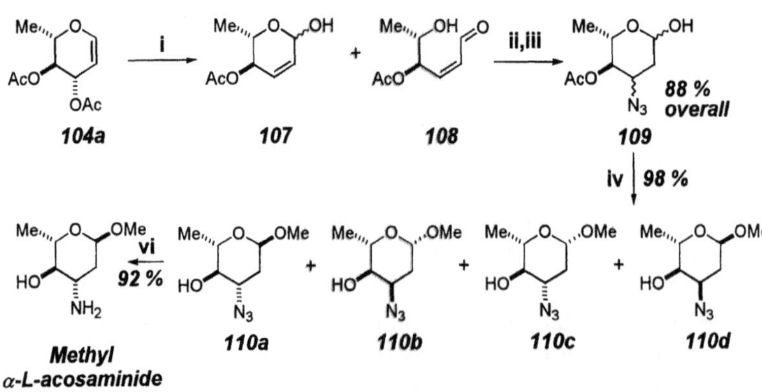

Scheme 27. Reagents: **i**, H_2O, heat; **ii**, NaN_3, H_2O, AcOH; **iii**, Ac_2O, pyr. CH_2Cl_2; **iv**, MeOH, K10 montmorillonite; **v**, NaOMe, MeOH; **vi**, H_2, Pd/C, EtOH, TEA

by condensation of the pentose with nitromethane in alkaline methanol (*80*, *81*). The intermediate formed was acetylated and converted into the nitro olefin **113** upon treatment with sodium bicarbonate in refluxing toluene. Standard hydrogenation of the olefin gave the nitro derivative **114**, which was converted into the 2-deoxy-L-arabino-hexose **115** by a modified Nef reaction using barium hydroxide and sulfuric acid (*82*, *83*). After glycosidation under standard conditions the 6-hydroxy group was removed by tosylation and reduction with lithium aluminium hydride.

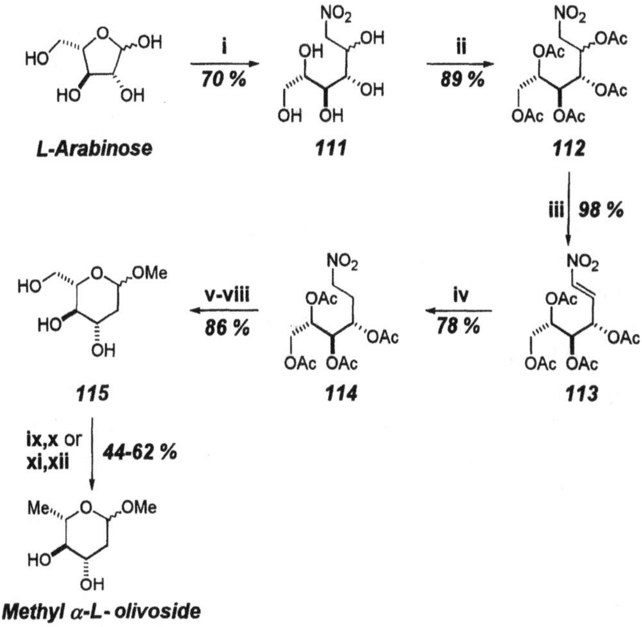

Scheme 28. Reagents: **i**, $MeNO_2$, MeONa, MeOH; **ii**, BF_3 Et_2O, Ac_2O; **iii**, $NaHCO_3$, toluene, heat; **iv**, H_2, Pd/C, EtOAc; **v**, $Ba(OH)_2 \cdot 8H_2O$; **vi**, H_2SO_4; **vii**, $BaCO_3$; **viii**, AG 50-X4(H^+), MeOH; **ix**, TsCl, pyr.; **x**, $LiAlH_4$; **xi**, NIS, PPh_3, DMF; **xii**, Raney Ni, MeOH

The 3-*O*-tosyl olivoside **116**, the 3,4-anhydro-digitoxoside **117**, and methyl α-L-oleandoside, all of which can be derived from the α-methyl glycoside of L-olivose (Scheme 29), are widely used intermediates for the synthesis of *ribo*, *arabino*, *xylo*, and *lyxo* trideoxyaminohexoses. Tosylation of methyl α-L-olivoside with tosyl chloride in pyridine produced the 3-*O*-tosyl olivoside **116** (*75*, *76*, *84*–*86*). The 3,4-anhydro-digitoxoside **117** is furnished by treatment of the 3-*O*-tosyl olivoside **116** with an alkaline solution of ethanol or methanol (*76*, *85*). Alternatively, anion exchange resins can be used for formation of the

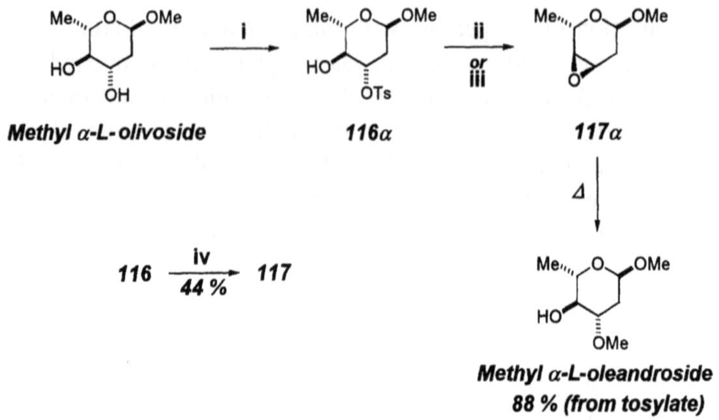

Scheme 29. Reagents: **i**, TsCl, pyr.; **ii**, NaOMe, MeOH; **iii**, NaOH, EtOH; **iv**, AG 1-X4 (OH^-)

epoxide (*79*). Upon heating compound **117** in alkaline methanol, methyl α-L-oleandroside was formed in high yield (88%) (*75*).

All these intermediates can be prepared from non-carbohydrate materials by taking advantage of acylketene [4+2] cycloadditions. COLEMAN and FRASER have recently reported on the construction of those deoxy sugars from 2,3-dihydro-4H-pyran-4-one rings which are obtained by [4+2] cycloaddition of acylketenes with electron-rich olefins (Scheme 30) (*88*). Reduction of the keto group with DIBAL-H gives the unsaturated *syn*-alcohol **120**. To synthesise the oleandroside, the 3-hydroxy group is methylated before hydroboration of the olefin.

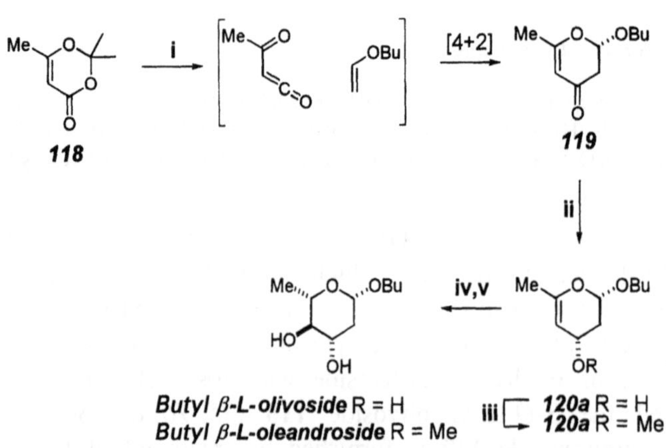

Scheme 30. Reagents: **i**, toluene, reflux; **ii**, DIBAL-H; **iii**, NaH, MeI; **iv**, BH_3SMe_2; **v**, $NaBO_3$

Hydroboration and oxidation were controlled by conformational and stereoelectronic factors giving excellent stereo- and regio-selectivity.

HADFIELD *et al.* have used methyl 2,6-dideoxy-α-L-arabino-hexopyranoside (methyl olivoside) in the synthesis of methyl 4-acetamido-2,4,6-trideoxy-α-L-lyxo-hexopyranoside **124**, which after reductive dimethylation would give L-kedarosamine (*74*). The methyl 2,6-dideoxy-α-L-arabino-hexopyranoside was monobenzoylated and the free hydroxy group was mesylated to give **122**. Finally, nucleophilic displacement with azide followed by reduction, acetylation, and debenzoylation furnished the methyl 4-acetamido-2,4,6-trideoxy-α-L-lyxo-hexopyranoside **124** (Scheme 31). Probably hexamethylphosphoric triamide is responsible for the inversion of configuration in the displacement reaction of the mesyloxy group with azide despite the presence of the neighboring benzoyloxy group.

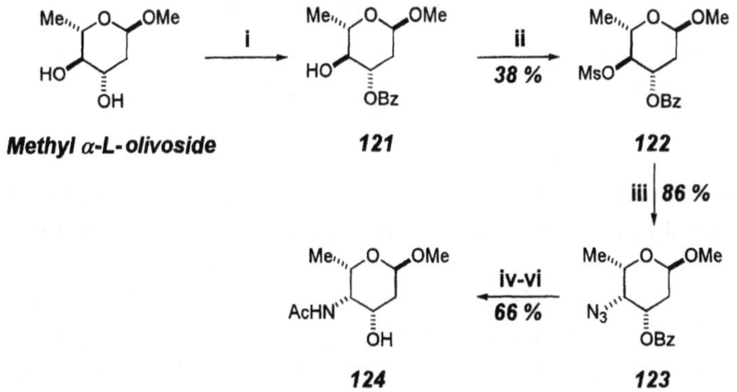

Scheme 31. Reagents: **i**, BzCl, pyr., 0 °C; **ii**, MsCl, pyr.; **iii**, HMPA, NaN_3; **iv**, H_2, Pd/C, EtOH; **v**, Ac_2O; **vi**, Na, MeOH

As a modification of the method reported by HADFIELD *et al.* (*74*) SZTARICSKAI *et al.* have reported the first synthesis of L-kedarosamine from methyl α-L-olivoside (Scheme 32) (*89*). The methyl glycoside of L-kedarosamine was obtained in 15–37% overall yield. Instead of a benzoate ester, a benzyl ether was used for the *O*-3 protection and a tosylate was used instead of mesylate as a good leaving group. The basic idea of the route is the same except for final conversion of the amino to dimethylamino group to furnish the kedarosamine. The authors do not explain the better yields when benzyl protection is cleaved immediately after tosylation, but it is clear that, without steric hindrance from the neighboring group, nucleophilic displacement is easier.

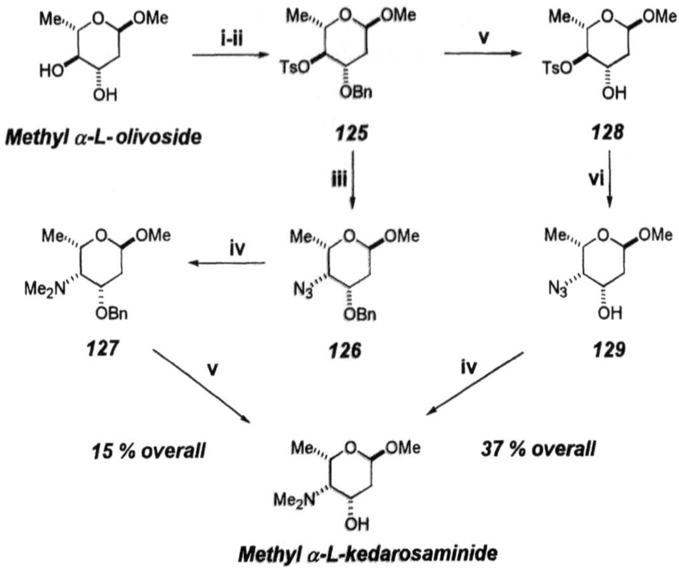

Scheme 32. Reagents: **i**, Bu_2SnO, toluene, rt then BnBr, Bu_4NI, toluene, rt; **ii**, TsCl, pyr.; **iii**, NaN_3, HMPA, 100 °C; **iv**, H_2, Pd/C, MeOH-HCHO; **v**, H_2, Pd/C, MeOH-AcOH; **vi**, NaN_3, DMSO, 130 °C

Methyl α-L-oleandroside has been used in the synthesis of two isomeric 4,6-dideoxy-4-amino-L-hexopyranosides (Scheme 33). MONNERET *et al.* oxidised methyl α-L-oleandroside with PCC and converted the ulose derivative **130** into its oxime **131** (*75*, *90*). Reduction of the oxime **131** with lithium aluminium hydride gave a 1 : 1 mixture of two aminohexoses, one the methyl glycoside of L-hollantosamine **133** (H replacing Ac) and the other its L-lyxo isomer **132** (H replacing Ac). The

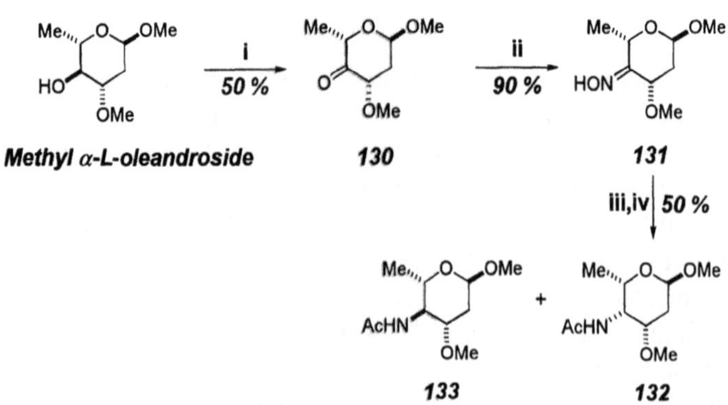

Scheme 33. Reagents: **i**, PCC; **ii**, NH_2OH, EtOH; **iii**, $LiAlH_4$; **iv**, Ac_2O, pyr

latter would give L-kedarosamine after reductive dimethylation. Finally both aminohexoses were acetylated.

In comparing the reduction of the oxime derivative **131** with the earlier mentioned reductions of oximes and ketones that had 2,3-isopropylidene protection, it is worth mentioning that, without steric bias, hydride attacks from both equatorial and axial directions. Because of the unselective hydride approach the Cieplak effect does not alone explain the selectivity in the earlier examples (Schemes 10–12); the hydride attacked only from the equatorial direction owing to steric reasons.

The methyl α-L-oleandroside can also be converted stereospecifically into the L-*lyxo* isomer of L-hollantosamine **136**, a synthetic precursor of 3-*O*-methyl L-kedarosamine (Scheme 34) (*75*). The sequence consists of the already familiar synthetic steps seen in Scheme 32.

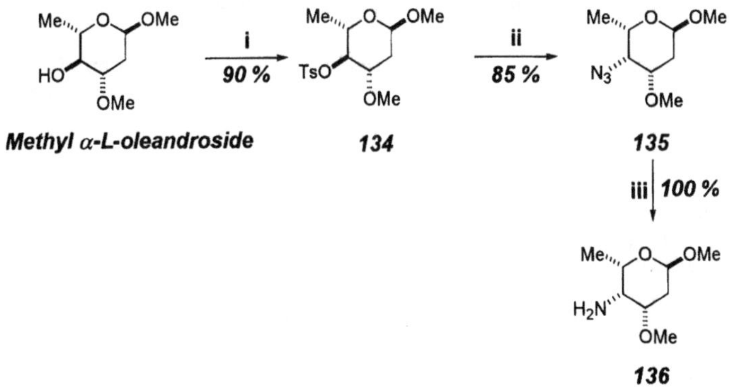

Scheme 34. Reagents: **i**, TsCl, pyr.; **ii**, HMPA, NaN_3; **iii**, H_2, Pd/C, MeOH

The oleandroside is suitable for preparation of the 3-*O*-methyl derivatives of 4-aminohexoses, but inconvenient for 3-aminohexoses. Tosylate **116** gives easy access to ristosamine (Scheme 35) *via* nucleophilic replacement with azide [also used for the synthesis of daunosamine (*76*)] followed by catalytic hydrogenation. The conditions

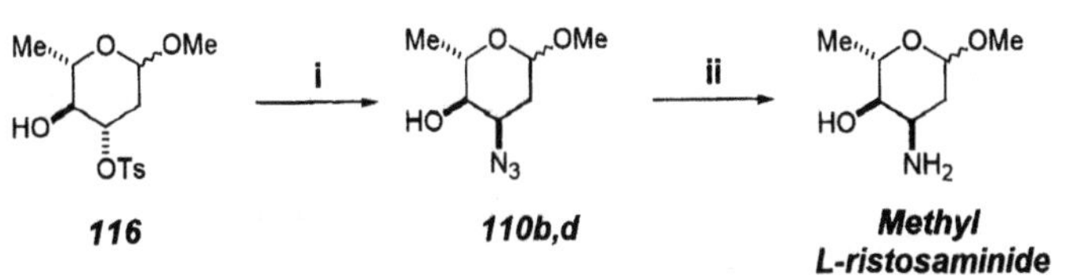

Scheme 35. Reagents: **i**, NaN_3, DMF, 110-120 °C; **ii**, H_2, Pd/C, MeOH, 20 °C

for nucleophilic substitution are drastic and the yield is low (36%). Syntheses of both the mixture of α- and β-anomers and the pure α-anomer have been reported (*76*, *84*, *86*, *87*).

Scheme 36 presents the syntheses of acosamine and daunosamine. Nucleophilic opening of the epoxy derivative **117** gives azido sugar **110a** (*85*, *86*). Catalytic hydrogenation of the azide function produces acosamine methyl glycoside. However, if the stereochemistry of the 4-hydroxy group is inverted by mesylation and substitution with sodium benzoate and the azide function is hydrogenated as earlier, the methyl glycoside of daunosamine is obtained (*47*, *76*). Similar conditions are also suitable for the large-scale synthesis of both acosamine and daunosamine as a mixture of α- and β-glycosides (*79*).

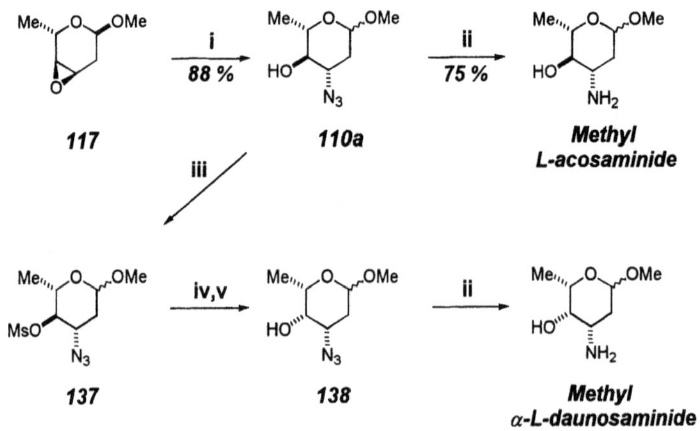

Scheme 36. Reagents: **i**, NaN_3, NH_4Cl, EtOH, H_2O; **ii**, H_2, Pd/C, MeOH; **iii**, MsCl; **iv**, $PhCO_2Na$, DMF; **v**, dil. NaOH, MeOH

BANASZEK *et al.* have reported the ring opening of oxirane **140** with dimethyl amine (Scheme 37) (*91*). Usually the 3-dimethylamino derivative predominates in the reaction mixtures, evidently due to steric reasons. The oxirane was obtained by epoxidation of olefin **139** (*92*, *93*).

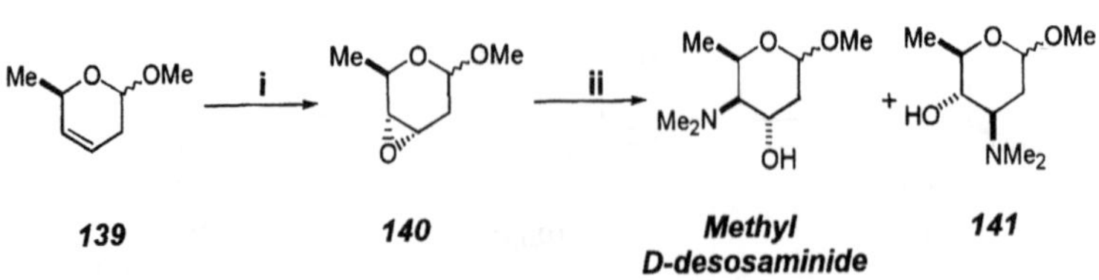

Scheme 37. Reagents: **i**, (O); **ii**, Me_2NH

References, pp. 255–263

Scheme 38 shows the conversion of lactone **142** to L-hexopyranoside **145**. Reduction of lactone **142** with Red-Al gives lactol **143**, which can be converted to the methyl glycoside **144** (*94*). After treatment with sodium hydroxide, methyl 3-amino-3,4,6-trideoxy-L-hexopyranoside **145** is formed.

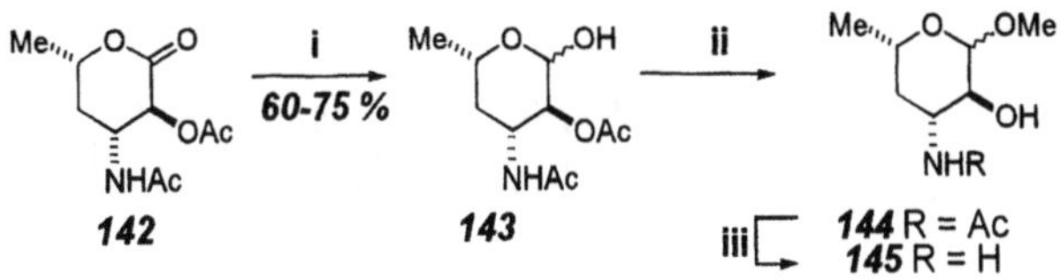

Scheme 38. Reagents: **i**, Red-Al; **ii**, MeOH, resin (H^+), heat; **iii**, 2M NaOH

3.1.3. Monoamino Tetradeoxyhexoses

Monoamino tetradeoxyhexoses have two methylene carbons. Those reported that are derived from carbohydrates have nitrogen at position 4. Amicetose and rhodinose are common starting materials; both of them are synthesised from pentaacetyl-D-glucose (Scheme 39) *via* the glycal, which upon heating in aqueous solution gives 4,6-diacetyl glucal (*97–97*). After glycosidation with triethyl formate and deacetylation the double bond is hydrogenated under standard conditions. The 6-deoxy group of amicetose is obtained *via* the already familiar mesylation, iodination, and dehalogenation sequence. Inversion of the 4-hydroxy group gives the rhodinose configuration.

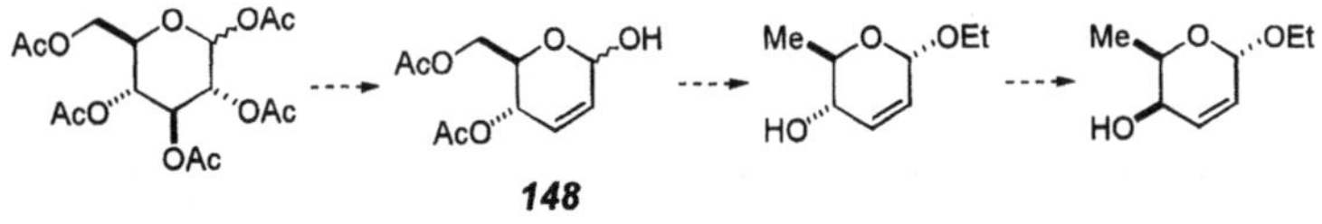

Scheme 39. Conversion of pentaacetyl D-glucose to amicetose and rhodinose

The synthetic sequence from mesylate **147** of ethyl amicetoside to forosamine (Scheme 40) involves double inversion with introduction of nitrogen on the second displacement (*98*). First the amicetose derivative is converted to the tosylate **148** of rhodinose, which undergoes nucleophilic displacement with azide to give **149**. Standard hydrogenation of azide and conversion of the amino group into the dimethylamino group furnishes ethyl α-D-forosaminide.

Panzica *et al.* have used a similar strategy in the synthesis of the L-tolyposamine derivative **152** and its epimer **153** (Scheme 41) (*99*).

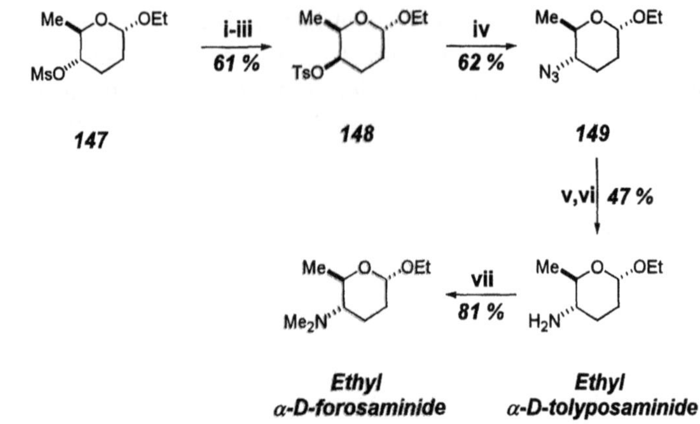

Scheme 40. Reagents: **i**, KOAc, DMF, heat; **ii**, NaOMe, MeOH, CO_2; **iii**, TsCl, pyr.; **iv**, NaN_3, DMSO; **v**, H_2, PtO_2, MeOH; **vi**, *p*-TsOH; **vii**, H_2CO, Pd/C, H_2, EtOH

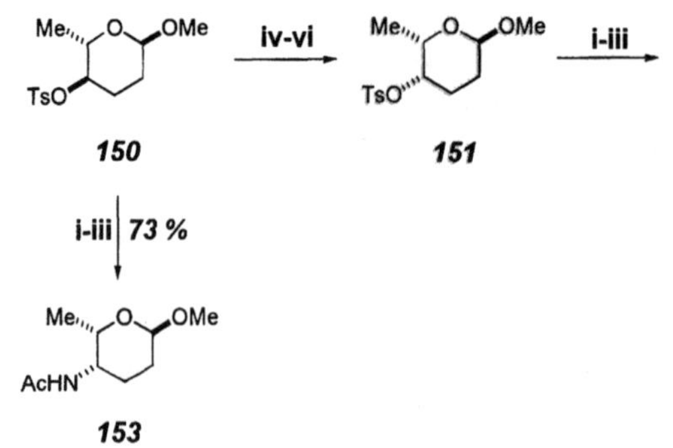

Scheme 41. Reagents: **i**, N_3^-; **ii**, H_2, Pd/C; **iii**, Ac_2O, pyr.; **iv**, PhCOONa; **v**, OH^-; **vi**, TsCl, pyr

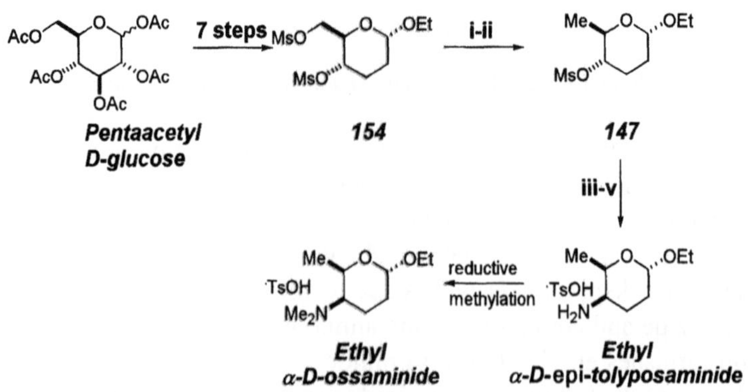

Scheme 42. Reagents: **i**, KI; **ii**, H_2, Raney Ni; **iii**, NaN_3, DMF; **iv**, $NaBH_4$; **v**, *p*-TsOH

Stevens *et al.* have synthesised ossamine from pentaacetyl D-glucose (Scheme 42) *via* the dimesylate **154**, which is obtained by a synthetic sequence similar to one employed previously (*100*). Treatment with potassium iodide followed by hydrogenolysis in the presence of Raney nickel catalyst gives the 4-*O*-mesyl derivative of amicetose. Nucleophilic displacement and reduction furnishes ethyl α-D-*epi*-tolyposaminide, which after reductive dimethylation gives the corresponding ossaminide.

3.2. Non-Carbohydrates as Starting Materials

Traditionally deoxyaminohexoses have been synthesised through transformation of other readily available carbohydrates, but recent interest has increasingly focused on non-carbohydrate starting materials. Both cyclic and acyclic precursors have been used. Amino acids which are widely used in the synthesis of natural products have not yet become common starting materials although they would appear to be ideal chiral starting materials. Introduction of hydroxy groups with the desired configuration into amino acid derivatives would be more efficient than manipulation of pre-existing hydroxy-bearing centers in common sugars. The synthesis of trideoxy monoaminohexoses from non-carbohydrates has been studied quite widely, the synthesis of other deoxyaminohexoses much less so.

3.2.1. Monoamino Dideoxyhexoses

Traditionally deoxyaminohexoses have been synthesised through transformation of other readily available carbohydrates. Panek *et al.* have reported the suitability of silyl-functionalized γ-lactones for the synthesis of 2-azidohexoses, the precursors of 2-aminohexoses (Schemes 43 and 44) (*101–103*). Earlier they had shown that the π-facial selectivity in catalytic osmylation reactions of oxygen-substituted allylsilanes is dramatically influenced by the character of the allylic substituent (*101*). γ-Lactones were synthesised from chiral (*E*)-crotylsilanes by dihydroxylation with osmium tetraoxide. Two different routes were employed to obtain enantiopure (*E*)-crotylsilanes: Claisen rearrangement of chiral [(*E*)-3-acyloxyvinyl]silanes and diastereoselective electrophilic addition to chiral β-trialkylsilyl ester enolates (*102*, *103*).

The Claisen rearrangement is one of the most predictable and widely used methods for the diastereoselective construction of vicinal stereocenters. In the rearrangement, the original asymmetric center is destroyed, while simultaneously two new ones in a vicinal relationship

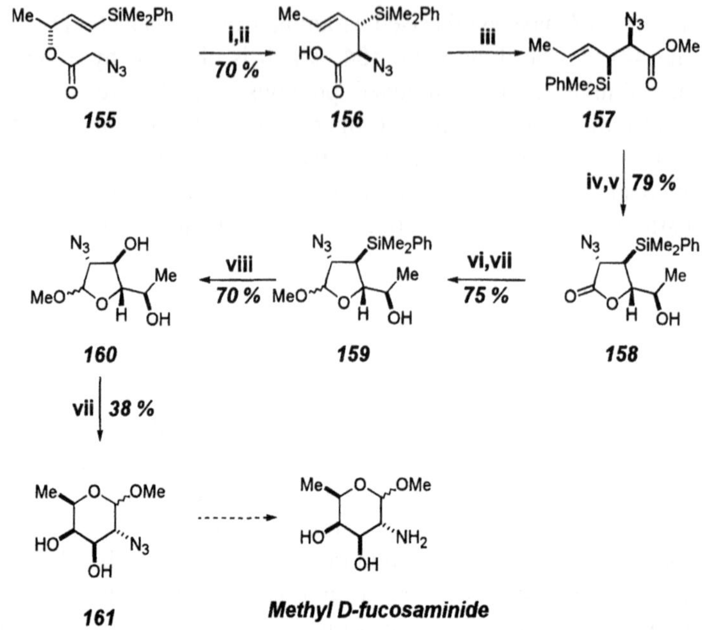

Scheme 43. Reagents: **i**, TBSTfO, TEA, CH_2Cl_2, $-78\,°C \rightarrow$ rt heat; **ii**, 10% HCl, THF, rt; **iii**, MeOH, H^+; **iv**, cat. OsO_4, TMNO, acetone, H_2O; **v**, cat. AcCl, MeOH, rt; **vi**, DIBAL-H, CH_2Cl_2, $-78\,°C$; **vii**, cat. AcCl, MeOH, heat; **viii**, $Hg(OAc)_2$, CH_3CO_3H/CH_3CO_2H, cat. H_2SO_4

are generated. Generation of the kinetic *E*-enolate of a vinylsilane in a weakly chelating solvent gives the *anti* diastereomer (*104*, *105*), whereas generation of the thermodynamic *Z*-enolate in the presence of strongly chelating HMPA leads to the *syn* diastereomer. In the case of heteroatom-substituted esters chelation controls selective enolisation by trapping and giving the thermodynamic *Z*-enolate. This leads to excellent diastereoselection in favor of the 2,3-*syn* isomer of an α-chiral-β-silyl-substituted hexenoic acid. In order to obtain the *anti*-isomer, the configuration of the enolate had to be reversed from *Z* to *E*. In the case of the strong chelating ability of the substituent in the vinylsilane, it is difficult to achieve useful levels of selectivity. Diastereoselective electrophilic addition gives access to the 2,3-*anti* isomer of an α-chiral-β-silyl-substituted hexenoic acid **164**. The trialkylsilyl group can function as an effective stereocontrolling element in electrophilic addition reactions to the derived chiral enolate.

The Claisen rearrangement of an α-azidoacetate is the first example of a Claisen rearrangement involving an α-azido group and represents a new approach to the amino sugars. Dihydroxylation of the chiral α-

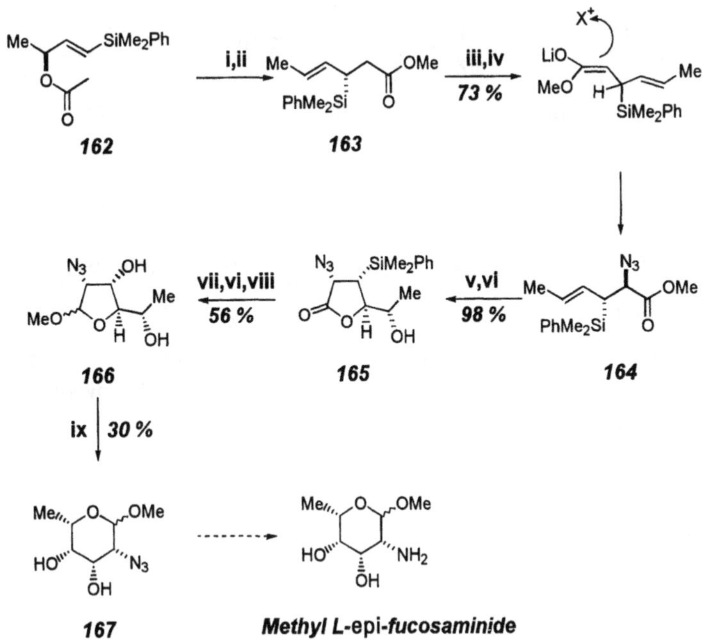

Scheme 44. Reagents: **i**, TBSTfO, TEA, CH_2Cl_2; **ii**, $SOCl_2$, MeOH; **iii**, LDA, THF, −78 °C; **iv**, Trisyl-N_3; **v**, cat. OsO_4, TMNO, acetone, H_2O; **vi**, cat. AcCl, MeOH, rt; **vii**, DIBAL-H, CH_2Cl_2, −78 °C; **viii**, $Hg(OAc)_2$, CH_3CO_3H/CH_3CO_2H, cat. H_2SO_4; **ix**, cat. AcCl, MeOH, heat

azido-β-silyl-substituted hexenoic acid methyl esters **157** and **164** followed by cyclization gives the γ-lactones **159** and **165**, which are converted to the corresponding azidohexoses having fucosamine and 2-*epi*-D-fucosamine stereo- and regio-chemistries. Osmium tetraoxide attacks opposite to the trialkylsilyl group, because the silyl group favors anti orientation to maximise the donation from high lying σ-orbitals to the transition state LUMO.

YAMADA and KOGA have reported the synthesis of methyl α-L-mycaminoside from L-alanine (Scheme 45) (*106*). Nitrous acid deamination in acetic acid gives 2-acetoxypropionic acid with retention of configuration. After formation with thionyl chloride, the acid chloride **169** is treated with an acetylenic Grignard reagent, and the resulting alkyne **170** is hydrogenated to the *cis* alkene **171**. After deacetylation the alkene undergoes cyclization in refluxing carbon tetrachloride in the presence of phosphoric acid, affording an anomeric mixture of L-hex-2-enopyranosid-4-uloses **172** and **173**. The α-anomer is reduced with lithium aluminum hydride. Axial attack of the hydride leads to the *anti*-

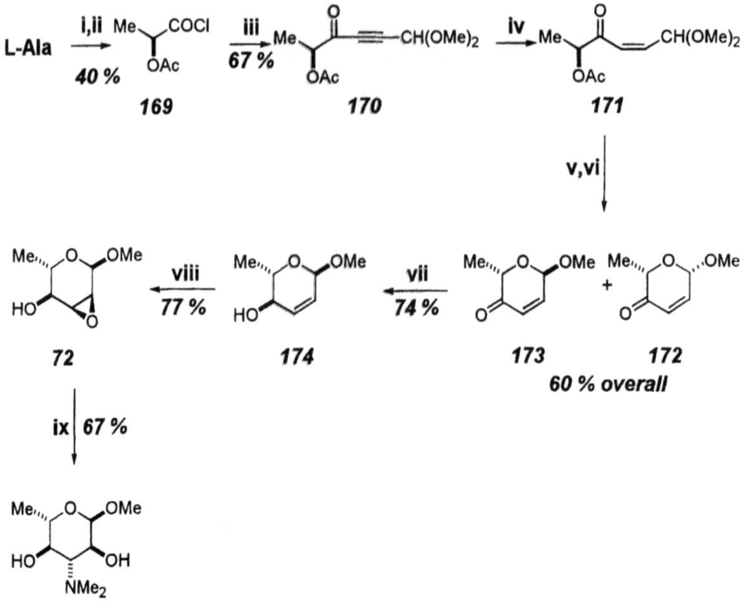

Scheme 45. Reagents: **i**, HNO_2, AcOH; **ii**, $SOCl_2$; **iii**, $BrMgCCCH(OMe)_2$; **iv**, H_2, Pd/ $BaSO_4$, EtOAc, quinoline; **v**, NaOH, dioxane; **vi**, CCl_4, H_3PO_4, heat; **vii**, $LiAlH_4$, Et_2O; **viii**, mCPBA, benzene; **ix**, aq. Me2NH

isomer **174**. Epoxidation with *m*-chloroperbenzoic acid gives the *anti-syn-syn*-isomer, which after treatment with aqueous dimethylamine gives methyl α-L-mycaminoside in 3% overall yield.

Recently, POLT and SAMES have reported an enantioselective synthesis of *N*-methylfucosamine from fully protected L-serine (Scheme 46) (*107*). Addition of propenyllithium to the L-serine-derived aldehyde gives the *syn*-amino alcohol **176** in high stereoselectivity. Catalytic osmylation in the absence of a chiral auxiliary gives a 6:1 mixture of *syn-anti-syn* and all-*syn*-aminotriols, which are acetylated *in situ* to the triacetate **179**. After reductive methylation of the *syn-anti-syn*-isomer, the silyl protection of the primary alcohol is removed and the hydroxy group is oxidized by the Swern method. Deacetylation of the aldehyde derivative with cyanide gives a mixture of pyranoside and furanoside in 20% overall yield. The aldehyde intermediate proved to be very labile towards basic conditions and thus a weak Brønsted base (potassium cyanide) is needed. Conversion of **183** to *N*-methylfucosamine is accomplished by hydrogenolysis of the benzhydryl group.

References, pp. 255–263

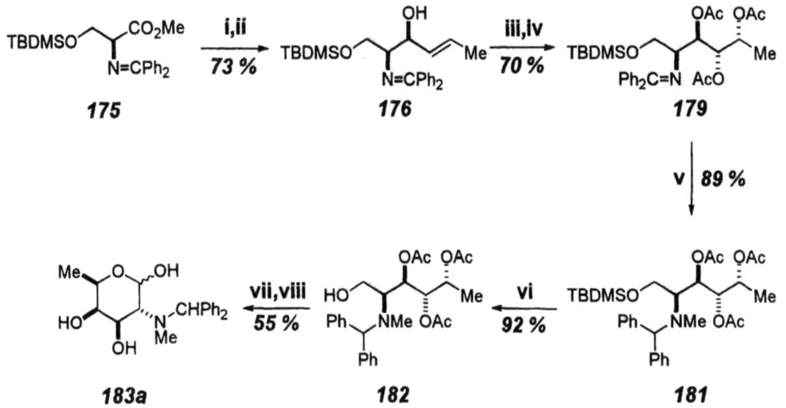

Scheme 46. Reagents: **i**, DIBAL-H, TRIBAL, CH_2Cl_2 $-78\,°C$; **ii**, $LiCHCHCH_3$, toluene, $-78\,°C \rightarrow$ rt; **iii**, $K_2OsO_2(OH)_4$, $K_2CO_3/K_3Fe(CN)_6$, t-BuOH/H_2O 1 : 1; **iv**, Ac_2O, pyr.; **v**, $NaBH_3CN$, MeCN, CH_2O, pH7; **vi**, 4% aq. HF, MeCN; **vii**, Swern oxidation; **viii**, cat. KCN, MeOH

Koskinen and Otsomaa have recently reported a synthesis of methyl 4-amino-4,6-dideoxygulopyranosides **189** from the L-threonine-derived aldehyde **184** (Scheme 47) (*108*). A modified Horner-Wadsworth-Emmons olefination led to the Z-enoate **185** in >17 : 1 selectivity. Acidic hydrolysis of the aminal protection simultaneously effected lactone formation to give **186**. Standard osmylation, reduction, and concomitant glycoside formation of the intermediate lactol gave the target **189** in 42% overall yield from aldehyde **184**.

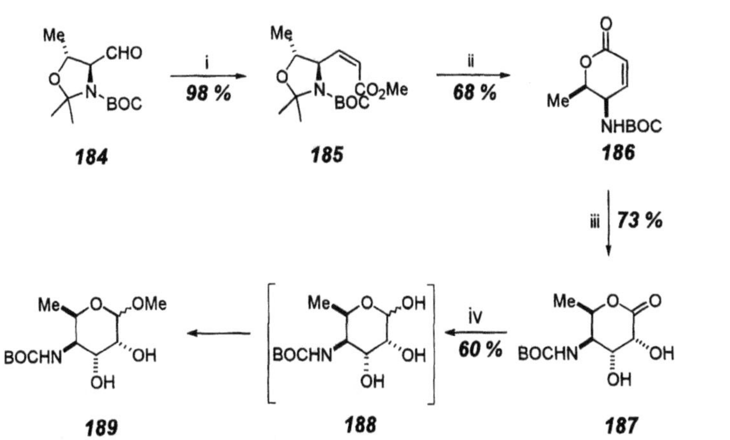

Scheme 47. Reagents: **i**, $MeO_2CCH_2P(O)(OCH_2CF_3)_2$, K_2CO_3, 18-crown-6, PhMe, $-20\,°C$ to rt; **ii**, AcOH heat; **iii**, OsO_4, t-BuOH/H_2O; **iv**, DIBAL-H, PhMe, followed by H^+, MeOH

The synthesis of elsaminose (Scheme 48) relied on the utilization of the Schöllkopf bislactim ether strategy (*109*). Thus, the valine-derived lactim ether reacted with a threonine-derived aldehyde to give **190**. Standard cleavage of the lactim ether ring system and protection gave the amino acid derivative **191**. Acidic cleavage of the acetonide led to simultaneous lactone formation, and the free secondary hydroxy group was protected as the dimethylisopropyl ether **192**. Finally reduction to the lactol **193**, followed by protecting group cleavages, gave the desired elsaminose.

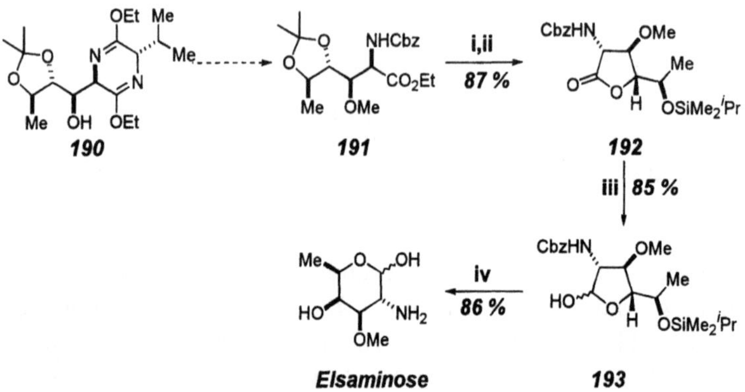

Scheme 48. Reagents: **i**, TFA:THF:H_2O 6:6:1, rt, 3h; **ii**, iPrMe_2SiCl imidazole, THF, rt, 1h; **iii**, DIBAL-H, −78 °C, PhMe : THF 2 : 1; **iv**, H_2, Pd/C, then Dowex 50 × 8–200, then HCl

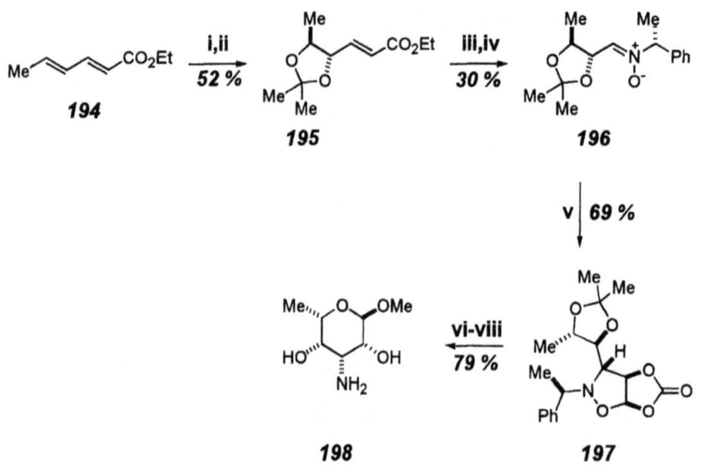

Scheme 49. Reagents: **i**, Ad-mix α, $MeSO_2NH_2$, *t*-BuOH, H_2O; **ii**, 2,2-dimethoxypropane, TsOH; **iii**, O_3, 8:1 CH_2Cl_2:MeOH −78°C; Me_2S; **iv**, (*R*)-*N*-hydroxy-α-methylbenzylamine; **v**, vinylene carbonate, PhH, 85 °C; **vi**, THF, HCl; **vii**, H_2, Pd $(OH)_2$, AcCl, MeOH; **viii**, HCl in MeOH

The synthesis of the amino sugar component of the antifungal fluvirucin (Scheme 49) relies on the introduction of the chirality through Sharpless asymmetric dihydroxylation of commercial ethyl sorbate **194** (*110*). Protection as the acetonide **195** was followed by ozonolysis of the remaining alkene and formation of the nitrone **196**. A highly diastereoselective [3 + 2] cycloaddition gave the isoxazolidine **197** with the correct stereochemistry (diastereoselectivity 20 : 1). Straightforward operations led cleanly anomer **198**.

3.2.2. Monoamino Trideoxyhexoses

The first synthesis of 6-deoxyaminohexoses from non-carbohydrate precursors was reported in 1964 by Newman (*111*). The racemic synthesis of desosamine derivatives (Scheme 50) began with nucleophilic addition of the lithium salt of propargyl aldehyde diethylacetal to propylene oxide **199**. The 1,1-diethoxy-5-hydroxyhex-2-yne **200** that forms was converted to 2-ethoxy-6-methyl-5,6-dihydro-2H-pyran **201** by reduction with one equivalent of hydrogen over palladium on charcoal, followed by addition of a small amount of hydrochloric acid. Treatment with peracid gave an epoxide, which was opened by addition of aqueous dimethylamine to give the racemic aminohexose **202**.

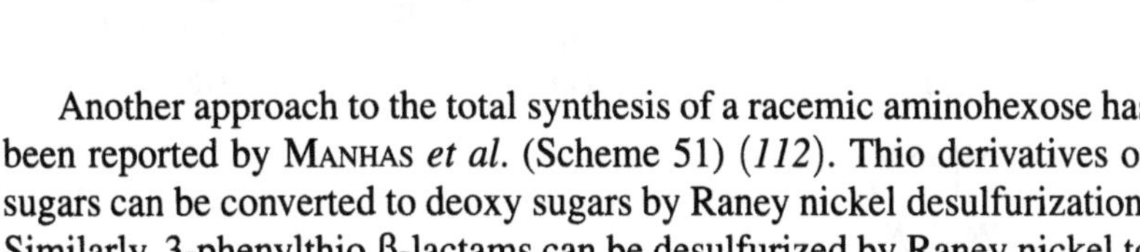

Scheme 50. Reagents: **i**, BuLi, **ii**, H_2, Pd/C; **iii**, cat. HCl; **iv**, mCPBA; **v**, sat. aq. $MeNH_2$

Another approach to the total synthesis of a racemic aminohexose has been reported by Manhas *et al.* (Scheme 51) (*112*). Thio derivatives of sugars can be converted to deoxy sugars by Raney nickel desulfurization. Similarly, 3-phenylthio β-lactams can be desulfurized by Raney nickel to give 3-unsubstituted β-lactams, which can serve as intermediates to trideoxyaminohexoses. Annelation of the Schiffs base **204** (prepared from *p*-anisidine and a suitable aldehyde) with phenylthioacetyl chloride **203** gave only the trans β-lactam **205**. After desulfurization the *N*-substituent was removed by oxidation with cerium(IV) ammonium nitrate. Hauser *et al.* have converted the β-lactam **207** to D,L-daunosamine (*113, 114*).

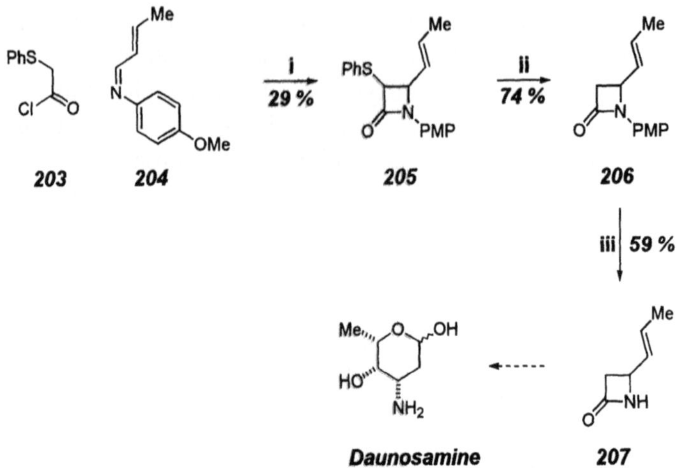

Scheme 51. Reagents: **i**, TEA; **ii**, W_2 Raney Ni, acetone; **iii**, $(NH_4)_2Ce(NO_3)_6$

WIEMANN and DYONG have utilized the Claisen rearrangement in the synthesis of daunosamine (Scheme 52) (*115*). The final product is a mixture of daunosamine and acosamine derivatives. Transetherification of 3-buten-2-ol **208** with 1,4-divinyloxybutane **209** in the presence of

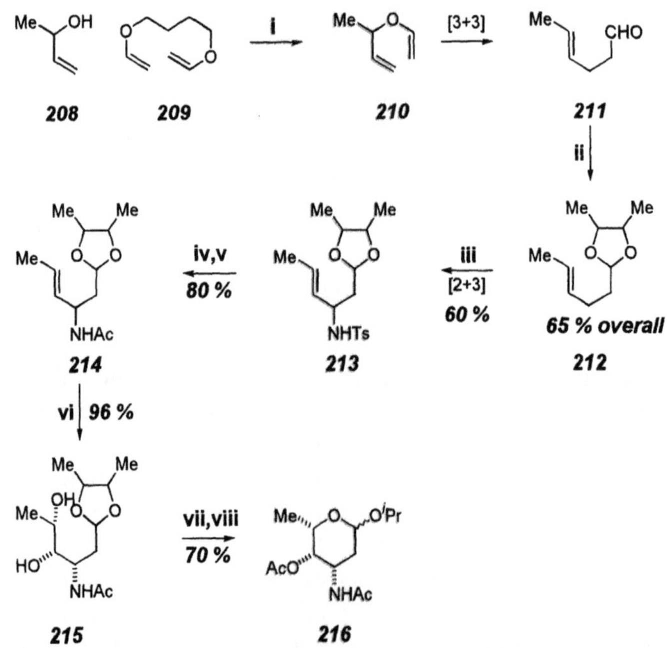

Scheme 52. Reagents: **i**, $Hg(OAc)_2$; **ii**, MeCH(OH)CH(OH)Me; **iii**, $Se(NTs)_2$; **iv**, Na, NH_3; **v**, Ac_2O, MeOH; **vi**, OsO_4, NMO; **vii**, 2N HCl, IPA; **viii**, Ac_2O, pyr

mercuric acetate gave compound **210**, which underwent Claisen rearrangement to give *trans*-4-hexenal **211**. The aldehyde was protected as the acetal **212**. Addition of the nitrogen substituent at the allylic position with selenium and chloramine-T gave the tosylamino derivative **213**. After detosylation and acetylation, the *E*-olefin was dihydroxylated with osmium tetroxide to give a mixture of diols. The all-*syn* diol **215** has the right configuration for the daunosamine synthesis. The racemate was treated with acidic, wet propan-2-ol followed by acetylation to give a racemic mixture of aminohexoses. The mixture of α- and β-anomers of daunosamine **216** was isolated from the reaction mixture by crystallization.

The first description of an asymmetric synthesis of acosamine and daunosamine utilizing non-carbohydrate precursors (Scheme 53) was that of FUGANTI *et al.* (*116*) Enzymatic, pinacol-type reaction between cinnamaldehyde and acetaldehyde gave the (2*S*, 3*R*)-diol which was protected as the acetonide **218**. Ozonolysis, and treatment with (methoxycarbonylmethylidene) triphenylphosphorane gave the *trans*-enoate **219** as the major product. The enoate was treated with dry ammonia in methanol followed by refluxing in acidic solution to give the γ-lactone **221**, and finally *N*-trifluoroacetylacosamine **223** was obtained by trifluoroacetylation and reduction with DIBAL-H.

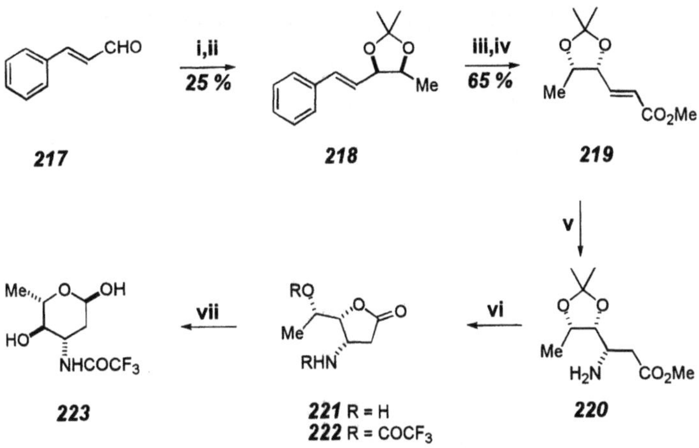

Scheme 53. Reagents: **i**, MeCHO, bakers' yeast; **ii**, DMP, PTSA; **iii**, O_3, CH_2Cl_2, Ph_3P; **iv**, $Ph_3P{=}CHCO_2Et$; **v**, NH_3, MeOH; **vi**, aq. HCl; vii, DIBAL-H

For the synthesis of daunosamine the γ-lactone **221** was converted to the δ-lactone by benzoylation. The stereochemistry of the 4-hydroxy group was inverted by a mesylation-nucleophilic substitution procedure.

The rest of the synthesis was performed as described for the acosamine (Scheme 53).

Intermolecular hetero-Diels-Alder reactions of substituted α,β-unsaturated carbonyl compounds and vinyl ethers are a useful approach in natural product synthesis. Introduction of an electron-withdrawing group at position 2 or 3 of the heterodiene greatly enlarges the scope of the reaction (*117*). However, the majority of natural 3-amino sugars contain a methyl group at C-5, which calls for a methyl group at position 2 of the oxabutadiene moiety in the employed enamino ketones (*118*). While these compounds are not reactive enough in a hetero-Diels-Alder reaction, the corresponding enamino ketones with a phenylthio group at C-3 easily undergo a cycloaddition with electron-rich dienophiles.

TIETZE *et al.* have utilized the hetero-Diels-Alder reaction in the synthesis of 3,4,6- and 2,3,6-trideoxyaminohexoses (*117, 118*). Cycloaddition of the phthalimido-protected, phenylthio-activated, enamino ketone **225** with *trans*-1-acetoxy-2-ethoxyethene yields dihydropyrans **226** and **227** with good selectivity, with the 3,4-*trans* configuration preferred (Scheme 54). Hydrogenation of the cycloadduct over Raney nickel in methanol simultaneously cleaves the thiophenyl group and reduces the olefinic bond, by hydrogen attack from the least hindered bottom face, affording the desosamine derivative **228**. Final deprotection (sodium borohydride in wet propan-2-ol) and treatment with acetic acid affords the acetate salt of desosamine ethyl glycoside **229**.

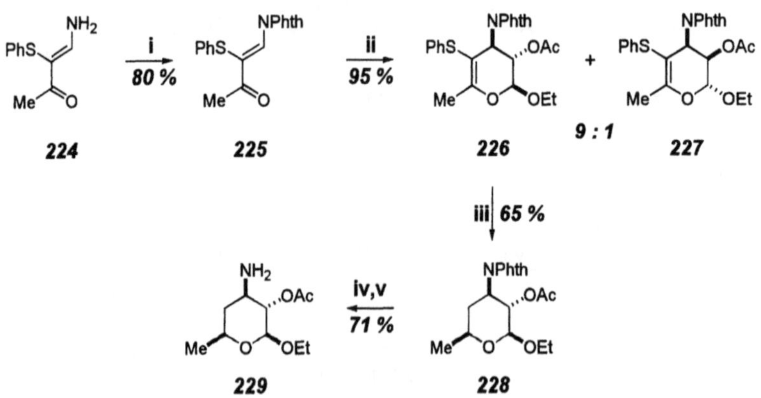

Scheme 54. Reagents: **i**, PhthCl, DMAP/TEA, CH_3Cl; **ii**, AcOCH=CHOEt, toluene/CH_2Cl_2, 120 °C; **iii**, MeOH, Raney Ni; **iv**, $NaBH_4$; v, AcOH

Cycloaddition of the same enamino ketone **225** with methyl vinyl ether yields the dihydropyrans **230** and **231**, now with the *cis* product **230** preferred (Scheme 55). This (*Z*)-enaminone forms the *cis*-

substituted dihydropyran as the main product *via* an *endo* transition state. Thus an *exo* addition with the (*E*)-enaminone would give the same product. When the cycloadduct **230** is treated with Raney nickel in anhydrous THF, desulfurization is achieved without affecting the double bond. Hydroboration of **232** leads, after oxidative work-up, to the *anti* hydroxy compound. The acetate salt of methyl acosaminide is furnished as described above for the desosaminide (Scheme 54).

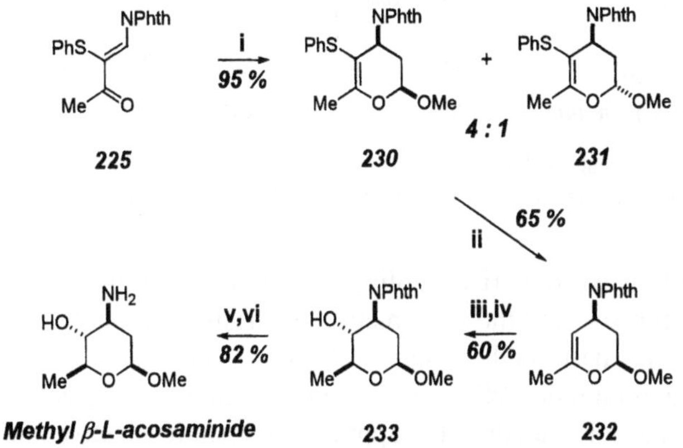

Scheme 55. Reagents: **i**, CH_2CHOMe, toluene/CH_2Cl_2, 120 °C; **ii**, THF, Raney Ni; **iii**, $H_3B\ SMe_2$; **iv**, KOH, H_2O_2; **v**, $NaBH_4$; **vi**, AcOH

GRETHE *et al.* have synthesized daunosamine (Scheme 56) from methylcyclopentadiene utilizing asymmetric hydroboration and stereoselective epoxidation to introduce the required three chiral centers (*119*). Hydroboration of the starting material **235** with (−)-di-3-pinanylborane produced the (*S*)-alcohol **236** with an enantiopurity higher than 95% ee in fair yield. The directing effect of the homoallylic hydroxy group was then utilized in the introduction of the remaining two chiral centers. Epoxidation with *m*-chloroperbenzoic acid occurred *cis* to the hydroxy group as expected. Construction of the carbohydrate skeleton was completed by a Baeyer-Villiger ring enlargement after Jones oxidation of the alcohol. Finally the δ-lactone **239** was reduced with DIBAL-H before glycosidation.

The crucial step in the synthesis was to achieve glycosidation without opening the epoxide. This was done by carrying out the reaction in anhydrous methanol using carefully purified boron trifuoride as the Lewis acid catalyst. A 2 : 1 mixture of methyl glycosides was obtained. The epoxide **240** is envisaged as a key intermediate in the synthesis of L-daunosamine.

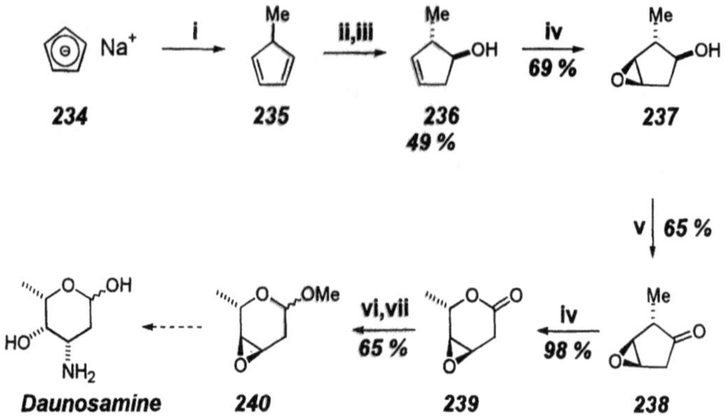

Scheme 56. Reagents: **i**, MeI; **ii**, (−)-di-3-pinanylborane; **iii**, H_2O_2, NaOH; **iv**, mCPBA, $NaHCO_3$, CH_2Cl_2; **v**, CrO_3, pyr. **vi**, DIBAL-H, toluene; **vii**, MeOH, BF_3

Dihydroisoxazoles (isoxazolines) have also been used as non-chiral cyclic precursors for the asymmetric synthesis of aminohexoses. Wade *et al.* have described a stereoselective synthesis of methyl *N,O*-diacetylacosaminide **248** from 3-nitro-4,5-dihydroisoxazole **241** and the potential for extending the approach to enantioselective amino sugar synthesis (*120*). As shown in Scheme 57 nucleophilic replacement of the nitro group of **241** with propynyllithium furnishes the alkyne **242**. The triple bond is hydrogenated using Lindlar catalyst to give an inseparable

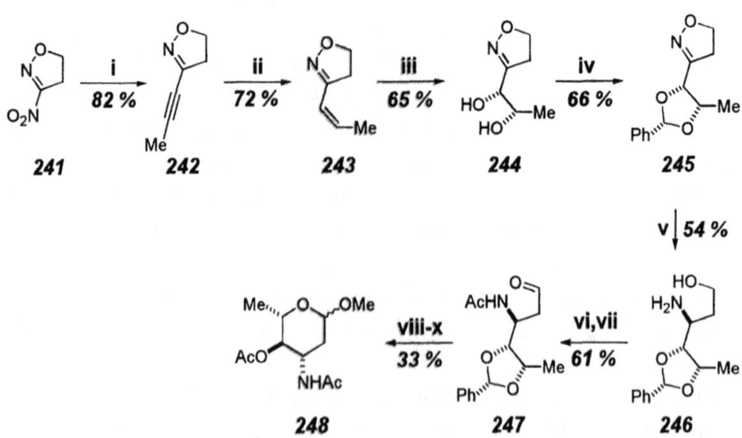

Scheme 57. Reagents: **i**, CH_3CCLi; **ii**, H_2, Lindlar catalyst, quinoline; **iii**, $Me_3N \rightarrow O$, OsO_4, wet THF; **iv**, PhCHO, $ZnCl_2$; **v**, $LiBH_4$, THF; **vi**, *p*-$AcOC_6H_4NO_2$, *N*-hydroxybenztriazole, DMSO; **vii**, $(COCl)_2$, TEA, DMSO; **viii**, AcOH; **ix**, Ac_2O, pyr.; **x**, MeOH, TsOH

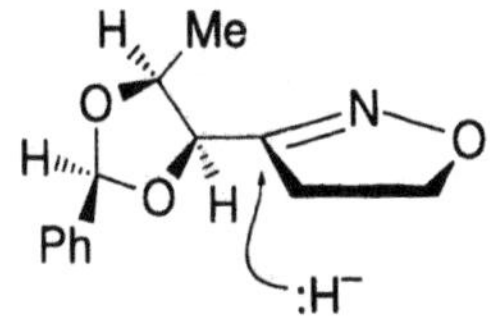

Fig. 5. Preferred approach of hydride to the isoxazoline **245**

9:1 mixture of *cis*- and *trans*-olefins. Catalytic *cis* dihydroxylation of the olefins gives the preferred diol **244** in 65% yield and the diastereomeric diol in 9% yield. Benzylidenation of the former gives the acetal **245**. Rapid reaction (30 min) provides largely the kinetic diastereomer **245** (ratio 9 : 1), whereas a longer reaction time with excess $ZnCl_2$ leads to epimerisation and formation largely of the thermodynamic, *trans*-diastereomer (ratio 3 : 7).

Reduction of the racemic isoxazoline **245** with lithium borohydride provides the γ-amino alcohol **246** with 9 : 1 stereoselectivity. In the original paper the reaction scheme was based on the wrong isoxazoline enantiomer. However, the authors' rationalization of the reduction (Fig. 5) employs the enantiomer **245** that would lead to acosamine. The terminal methyl group appears to be the stereodiscriminator protecting the upper face of the *C,N*-double bond. The synthesis is completed by *N*-acetylation, Swern oxidation to the open-chain acosamine derivative **247**, and acidic removal of benzylidene protection followed by *in situ* cyclization. Acosamine is isolated as the diacetate **248**.

In a recent report, Guanti *et al.* describe the synthesis of hollantosamine triacetate from L-*allo*-threonine *via* the cyclic aldehyde intermediate **250** (Scheme 58) (*121*). Condensation of aldehyde **250** with the lithium enolate of benzyl acetate proceeds with good stereoselection furnishing the *anti* alcohol **251**, which possesses all the carbons and the asymmetric centers of hollantosamine. Dehydration of the deprotected alcohol derivative of **251** produces the γ-lactam as expected, instead of the desired δ-lactone. The ester moiety is reduced to a primary alcohol and protected as the *p*-methoxyphenyl ether (PMP). In the next step the amino and secondary hydroxy group protections are removed, and after acetylation the triacetate intermediate **252** is obtained. A series of protecting group interchanges and oxidative removal of PMP-protection furnish the intermediate **253**. The free primary alcohol group is then oxidized with TPAP (*122*). Hydrolysis of the ketal protection gives the *N*-acetylhollantosamine **254**, which was characterized as the triacetate **255**.

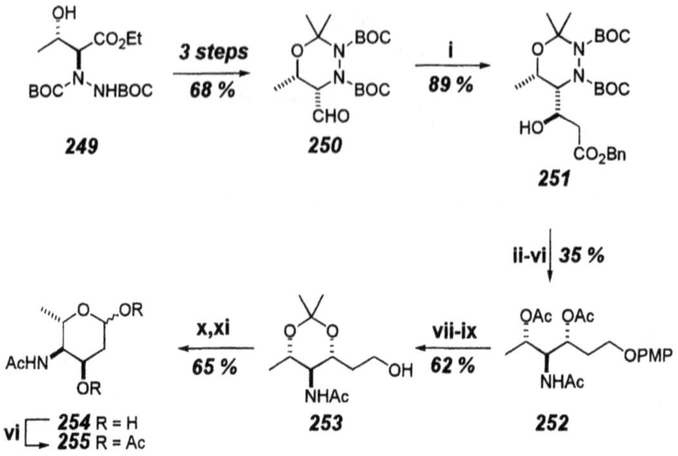

Scheme 58. Reagents: **i**, $CH_2{=}C(OLi)OBn$; **ii**, $Ca(BH_4)_2$, EtOH, THF; **iii**, $MeOC_6H_4OH$, Ph_3P, DEAD, CH_2Cl_2, **iv**, AcOH, 1N HCl; **v**, H_2, PtO_2, EtOH; **vi**, Ac_2O, pyr.; **vii**, TEA, MeOH, heat; **viii**, $MeOC(Me){=}CH_2$, PTSA, CH_2Cl_2; **ix**, $(NH_4)_2Ce(NO_3)_6$, H_2O, MeCN, pyr.; **x**, TPAP, NMO, 4Å molecular sieves, CH_2Cl_2, **xi**, AcOH, H_2O

N-Protected α-amino aldehydes are very convenient and versatile chirons. Jurczak *et al.* have utilized the α-amino aldehyde derived from L-aspartic acid (*N,O*-dibenzyl-*N*-*tert*-butoxycarbonyl-L-homoserinal) **260** in the synthesis of daunosamine (*123, 124*). In their procedure (Scheme 59), the homoserinal derivative was prepared from lactone **256**. Transesterification, followed by convenient protections and reduction of the ester, gave compound **258** after benzylation. Cleavage of the silyl functionality and subsequent oxidation lead to the key intermediate **260**.

Addition of vinylmagnesium bromide to the α-amino aldehyde afforded the *anti*-amino alcohol **262** with good diastereoselectivity (95 : 5), and epoxidation of the allylic alcohol with mCPBA lead to the *syn*-epoxide **263** with high stereoselectivity with all the desired stereocenters and carbons correctly assembled. Reductive ring-opening, protection of the resulting diol, and final Birch reduction afforded the isopropylidene derivative **264**. Oxidation of the primary alcohol followed by deketalization and *in situ* cyclization in acidic methanol furnished the anomeric mixture of methyl L-daunosaminide derivatives **265**.

The first non-carbohydrate based asymmetric synthesis of kedarosamine was recently reported by Kihlberg *et al.* (Scheme 60) (*125*). The starting *N,O*-protected D-threonine was converted into the corresponding Weinreb amide *via* the acid chloride. Coupling with the allyl Grignard reagent gave the protected α-amino ketone **267**. Non-chelation controlled reduction of the ketone intermediate with sodium borohydride

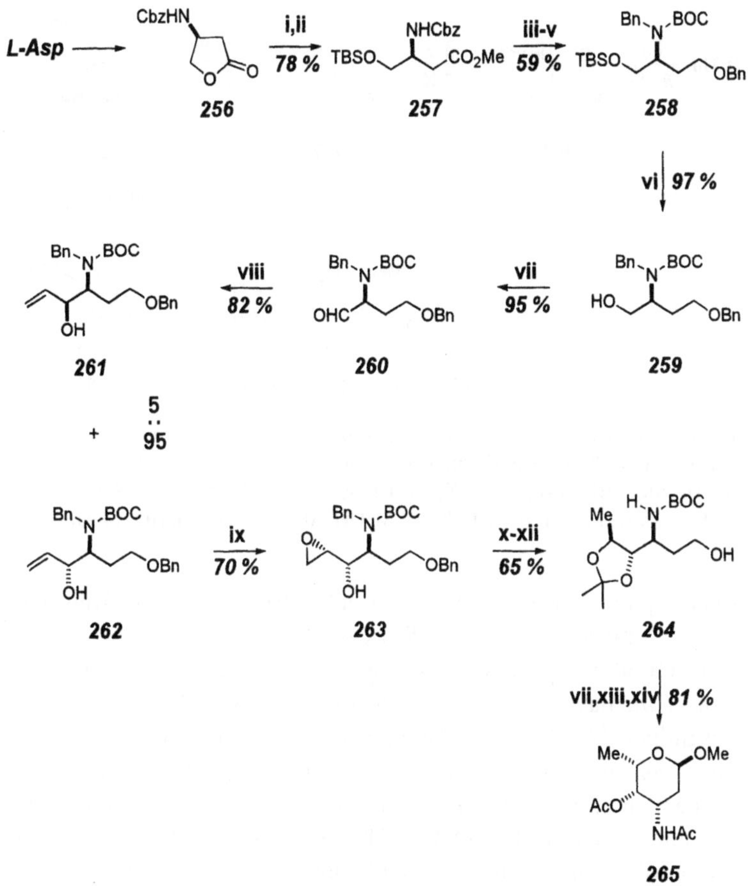

Scheme 59. Reagents: **i**, DCC, MeOH, rt; **ii**, TBSCl, imidazole, DMF; **iii**, H_2, Pd/C, BOC_2O, MeOH; **iv**, $LiAlH_4$, Et_2O; **v**, BnBr, NaH, DMF; **vi**, Bu_4NF, THF; **vii**, SO_3/pyr., DMSO; **viii**, vinyl-MgBr, Et_2O, −78 °C; **ix**, mCPBA; **x**, DIBAL-H; **xi**, DMP, H^+; **xii**, Na, NH_3; **xiii**, MeOH, H^+; **xiv**, Ac_2O, pyr

was found to be *syn*-selective, while 1,2-chelation controlled reduction with zinc borohydride was highly *anti*-selective. Unfortunately, reductive ring-opening of the isopropylidene aminal proceeded in low yield. However, after deketalization of the derivative **267**, intramolecular hydride delivery in the reduction of the resulting β-hydroxy ketone with $Me_4NBH(OAc)_3$ gave the desired *anti*-alcohol **268** as a single diastereomer (*126*). The unsaturated *anti*-alcohol **268** was then cleaved by ozonolysis, and subsequent ring closure to the corresponding hemiacetal occurred spontaneously. The final steps of the synthesis are the already familiar glycosidation, deprotection, and reductive dimethylation.

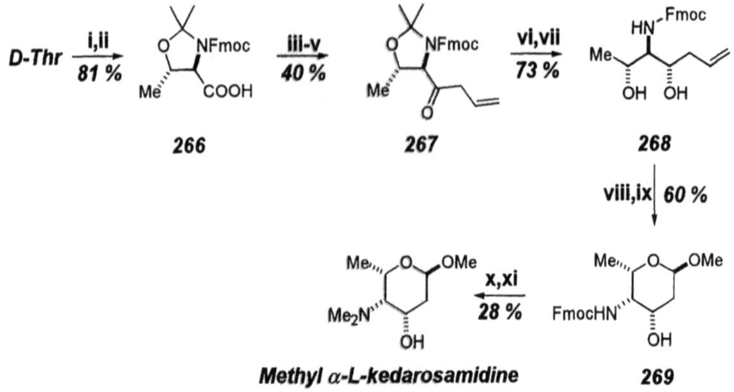

Scheme 60. Reagents: **i**, FmocCl, Na_2CO_3, dioxane; **ii**, DMP, PTSA, benzene; **iii**, cyanuric chloride, pyr., CH_2Cl_2; **iv**, Me(MeO)NH.HCl, pyr., CH_2Cl_2; **v**, allyl-MgBr, THF; **vi**, TFA, MeOH; **vii**, $Me_4NBH(OAc)_3$, MeCN, AcOH, 40 °C; **viii**, O_3, Me_2S, −78 °C; **ix**, PTSA, MeOH; **x** Pd/C, $Pd(OAc)_2$, NH_4HCO_2, MeOH; **xi**, Pd/C, H_2, HCHO, MeOH, H_2O

Enantiopure 2,3-protected 1,2,3-butanetriols derived from easily available chiral sources like $(2R,3R)$-tartaric acid or L-threonine are used as chiral precursors in the synthesis of trideoxyaminohexoses. FRONZA *et al.* have utilized them in the synthesis of L- and D-3-*epi*-daunosamines (*127*). Since the butanetriol derived from tartaric acid already has two of the three required stereocenters with correct stereochemistries for L-3-*epi*-daunosamine, only the stereocenter of the amino function had to be created. This was achieved by stereoselective amination of the *trans*-enoate **272** obtained from the Wittig reaction of the aldehyde intermediate (Scheme 61). Treatment of the *trans*-enoate with dry ammonia in methanol followed by hydrolysis and benzoylation furnished the δ-lactone **273** with 75:10 stereoselection. The lactone was finally reduced to the lactol, *N*-benzoyl-3-*epi*-daunosamine **274**. The D-enantiomer was similarly synthesized from the enantiomer of the triol **271** derived from L-threonine by deamination.

Another approach to D-*epi*-daunosamine relies on (2-thiazolylmethylene)triphenylphosphorane as a two-carbon homologating reagent with the aldehyde intermediate (Scheme 62) obtained by oxidation of the triol which is derived from L-threonine (*128*). The key steps are Wittig-type olefination, introduction of the amino function, and unmasking of the formyl group in the thiazole ring of the resulting alkylthiazole. Olefins from the unselective Wittig reaction (1:1) are enriched in the *E*-isomer **279** by isomerization with iodine ($E:Z$, 9:1). The poor electron-withdrawing character of the 2-thiazolyl group is not sufficient to make 2-alkenylthiazoles good Michael acceptors towards weak nucleophiles

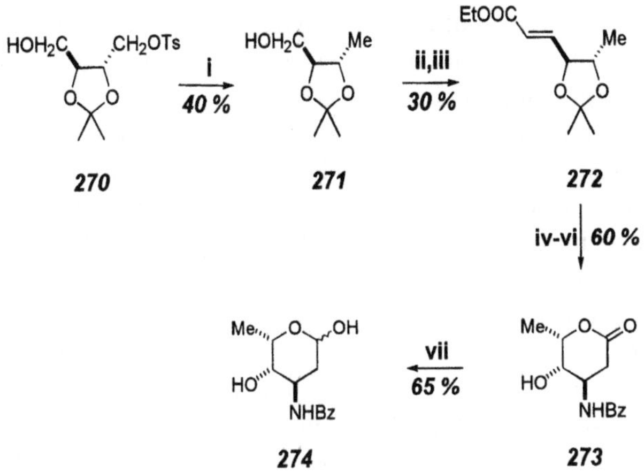

Scheme 61. Reagents: **i**, $LiAlH_4$, Et_2O; **ii**, PCC, CH_2Cl_2, AcONa; **iii**, $Ph_3P{=}CHCO_2Et$; **iv**, NH_3, MeOH, 0 °C; **v**, 2N HCl; **vi**, BzCl, pyr., CH_2Cl_2; **vii**, DIBAL-H, THF

like amines. The ability of the 2-alkenylthiazole to function as a Michael acceptor was therefore enhanced by preparing the *N*-methylthiazolium salt **279**. The salt was treated with benzylamine and then quenched with sodium borohydride to give a thiazolidine as a mixture of all-*anti*- and *anti-syn*-isomers. After acetylation and mercury-mediated hydrolysis of the thiazolidine ring, the open chain derivative of 3-*epi*-daunosamine **280**

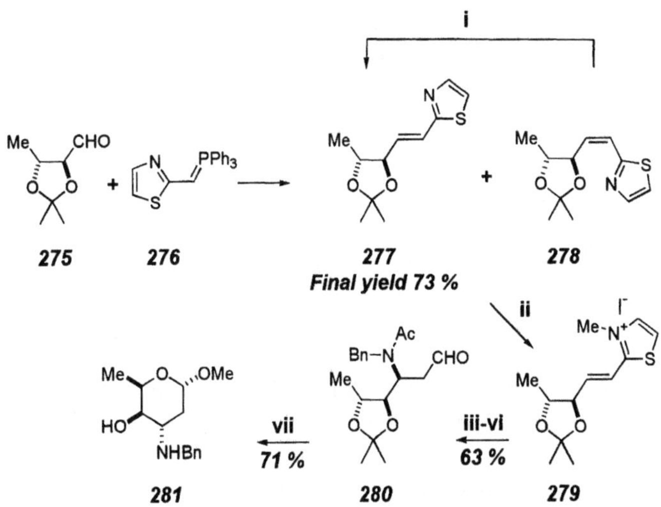

Scheme 62. Reagents: **i**, I_2; **ii**, MeI; **iii**, $BnNH_2$; **iv**, $NaBH_4$; **v**, Ac_2O; **vi**, H_2O, $Hg^{2}+$; **vii**, HCl, MeOH

was obtained. Removal of the isopropylidene and acetyl groups in acidic methanol, *in situ* cyclization, and glycosidation afforded the methyl *N*-benzyl D-3-*epi*-daunosaminide **281**.

Formation of the major *syn*-adduct upon the addition of benzylamine to **279** is consistent with a modified Felkin-Ahn transition state where the allylic alkoxy residue and the medium-sized methyleneoxy group are in the *anti* and inside positions, respectively and where the approaching nucleophile attacks the π-system from the antiperiplanar position, which is also the least hindered side.

Construction of daunosamine via amination of chiral enoates appears to be inefficient. However, the Grignard reaction of the aldehyde **282** gave the alcohol derivative **283** with fair *syn*-selectivity (Scheme 63) (*129*). Conversion of the free alcohol group into a good leaving group (with tosyl chloride) and subsequent azide displacement furnishes the azide derivative **284** with the opposite configuration. Reduction of the azide by a standard method followed by deketalization and benzoylation gave the *N*-benzoyl derivatives, with the *lyxo* configuration present in the major component **286**. Ozonolysis of the major component and treatment with dimethyl sulfide yields the *N*-benzoyl-D-daunosamine **287**.

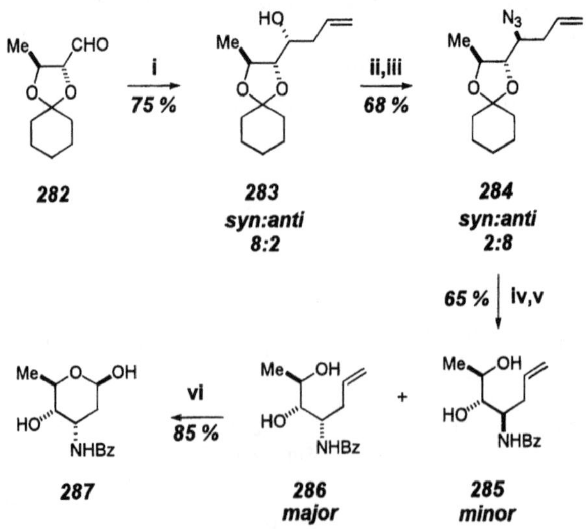

Scheme 63. Reagents: **i**, Allyl-MgBr, THF; **ii**, TsCl, pyr.; **iii**, NaN_3, NH_4Cl, DMF; **iv**, $LiAlH_4$, Et_2O; **v**, 50% AcOH then BzCl, K_2CO_3, acetone; **vi**, O_3, MeOH, Me_2S

Alkoxycarbonyloxazoles are easily obtained by direct C-acylation of isocyanoacetic esters with carboxylic acids (Scheme 64) (*130, 131*). Diphenyl phosphorazidate together with a base is a useful coupling reagent in the oxazole synthesis. The oxazole ring of compound **290** is easily cleaved under acidic conditions to give the 5-substituted 3-aminotetronic acid **291**. After *tert*-butoxycarbonyl protection of the amino group the tetronic acid was hydrogenated with outstanding stereoselectivity using rhodium on alumina as catalyst in ethyl acetate to give **293**. The α-methyl group of the tetronic acid completely blocks the attack of hydrogen from the α-face of the double bond. Reduction of the lactone with DIBAL-H gave lactol **294** (*121*), to which a C-1-unit was easily introduced by Wittig reaction with (methoxymethylene)triphenylphosphorane. The enol ether was then hydrolyzed to give L-daunosamine hydrochloride.

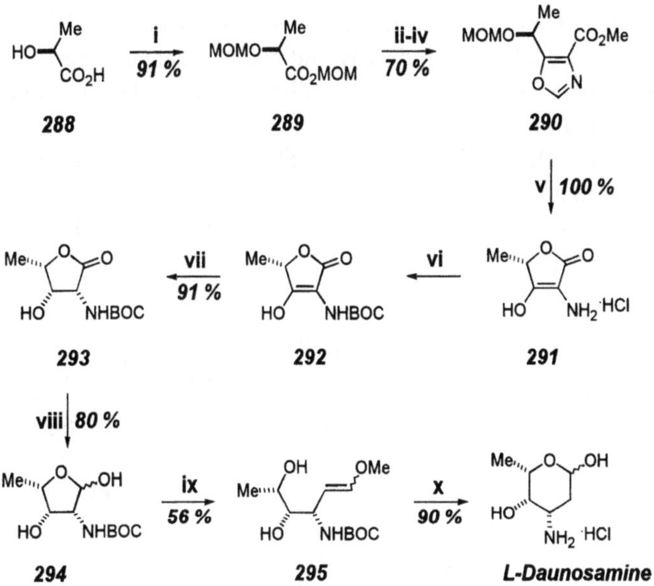

Scheme 64. Reagents: **i**, CH_3OCH_2Cl, DIPEA, CH_2Cl_2; **ii**, LiOH, H_2O-THF; **iii**, $(PhO)_2P(O)N_3$, DMF; **iv**, $CNCH_2COOMe$, NaH, DMF; **v**, 10% HCl, MeOH; **vi**, BOC_2O, $NaHCO_3$, dioxane, H_2O; **vii**, 5% Rh-Al_2O_3, H_2, EtOAc; **viii**, DIBAL-H, CH_2Cl_2; **ix**, $(Ph_3P^+CH_2OCH_3)$ Cl^-, KOtBu, glyme, toluene; **x**, 20% HCl, THF

An analogous sequence of reactions with inversion of the 4-methyl group gives access to D-ristosamine (Scheme 65) (*131*). An efficient way to invert the methyl group involves hydrolysis of the lactone with potassium superoxide in the presence of crown ether, acidification to pH 4 and estrification by the Mitsunobu method. If the 3-hydroxy group was protected as its *tert*-butyldimethylsilyl ether, two extra steps were needed and the synthetic sequence was less efficient.

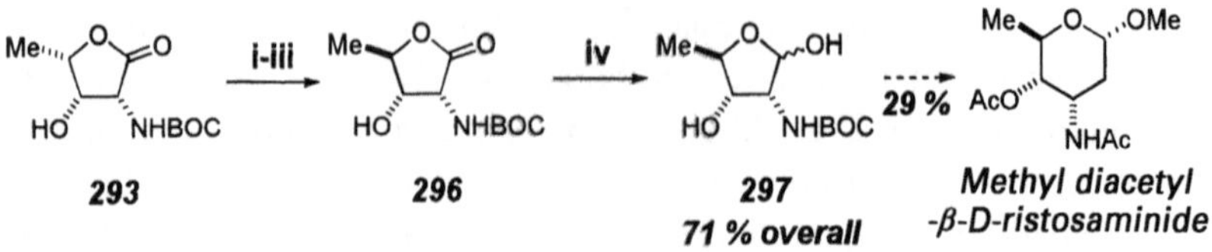

Scheme 65. Reagents: **i**, KO_2; **ii**, H^+; **iii**, Ph_3P, DEAD; **iv**, DIBAL-H

3.2.3. Monoamino Tetradeoxyhexoses

Starting in the same way as for their synthesis of hollantosamine (Scheme 58) GUANTI *et al.* reported the synthesis of *N*-acetyl-L-tolyposamine from L-*allo*-threonine *via* two different aldehyde intermediates: the acyclic aldehyde **249** and cyclic aldehyde **250** (*132*, *133*). In the first route (Scheme 66), the acyclic aldehyde **249** was extended by Wittig condensation with a stabilised phosphorane. The same Wittig conditions for the cyclic aldehyde **250** (Scheme 67) caused notable epimerization however, and the olefination was carried out instead under Roush-Masamune conditions. The choice of base had a dramatic

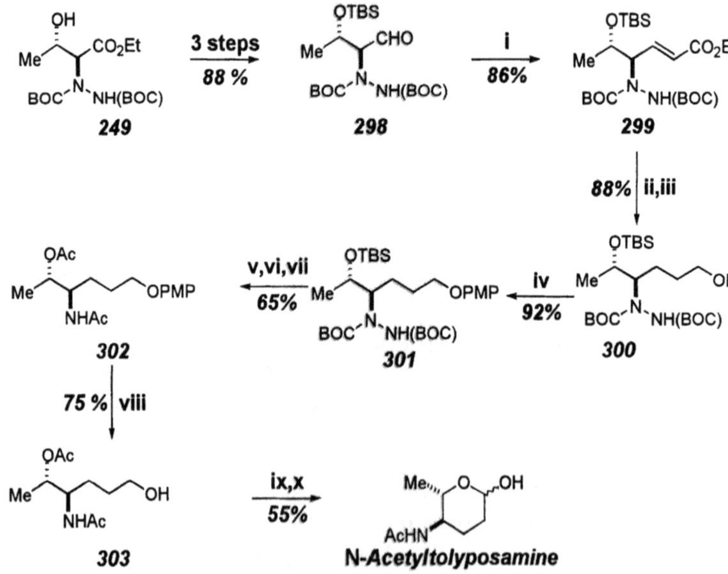

Scheme 66. Reagents: **i**, $Ph_3P{=}CHCOOEt$; **ii**, H_2, PtO_2, EtOH; **iii**, $Ca(BH_4)_2$, EtOH, THF; **iv**, $MeOC_6H_4OH$, Ph_3P, DEAD, CH_2Cl_2; **v**, AcOH, 1N HCl; **vi**, H_2, PtO_2, EtOH, H_2O; **vii**, Ac_2O, pyr. DMAP; **viii**, $(NH_4)_2Ce(NO_3)_6$, H_2O, MeCN, pyr.; **ix**, $(n\text{-}Pr)_4NRuO_4$, NMMO, CH_2Cl_2; **x**, DBU, MeOH

influence on the degree of epimerization. Both olefins were hydrogenated and the ester groups were then reduced. The primary alcohols were etherified with *p*-methoxyphenol under Mitsunobu conditions. The amino and secondary hydroxy group protections of both PMP- protected intermediates were replaced with acetyl groups, giving in both cases the intermediate **302**. Oxidative removal of the PMP-protection gave the free primary alcohol **303**, which was oxidised with TPAP (*122*). Selective removal of the *O*-acetyl protection (DBU in methanol) gave the *N*-acetyltolyposamine in an overall yield of *ca.* 15%.

Evans and Black have used an (*R*)-phenylalanine-derived oxazolidinone as a chiral auxiliary in the synthesis of *N*-Fmoc tolyposamine from glutaric anhydride (*134*). As set out in Scheme 68, acylation of the oxazolidinone with glutaric anhydride and methylation of the resulting acid gave the imide **307**. Evans' aldol reaction of the *Z* enol derivative of this imide with acetaldehyde provides a hydroxy ester with *syn*-stereochemistry. The hydroxy ester was lactonized to **308** and the imide was hydrolyzed with lithium hydroperoxide. Curtius rearrangement of the acid lactone followed by reduction and acetylation furnished the 1-*O* acetyl *N*-Fmoc tolyposamine **310** in an overall yield of 40%.

The synthesis of the *epi*-tolyposamine derivative **315** has been achieved from threonine (Scheme 69) (*108*). Olefination of aldehyde **184** followed by catalytic hydrogenation gave the saturated ester **312**. Cleavage of the aminal protection as discussed earlier (Scheme 47)

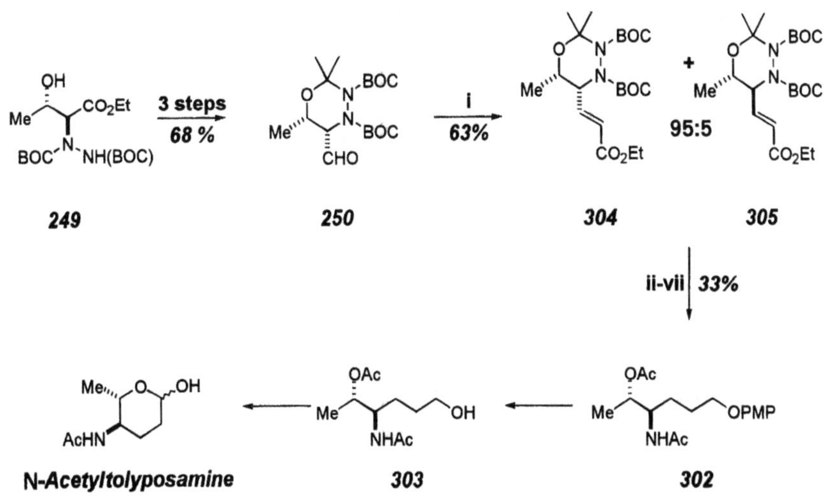

Scheme 67. Reagents: **i**, $(EtO)_2P(O)CH_2COOEt$, DIPEA, LiCl, rt, MeCN; **ii**, H_2, PtO_2, EtOH; **iii**, $Ca(BH_4)_2$, EtOH, THF; **iv**, $MeOC_6H_4OH$, Ph_3P, DEAD, CH_2Cl_2; **v**, AcOH, 1N HCl; **vi**, H_2, PtO_2, EtOH, H_2O; **vii**, Ac_2O, pyr. DMAP

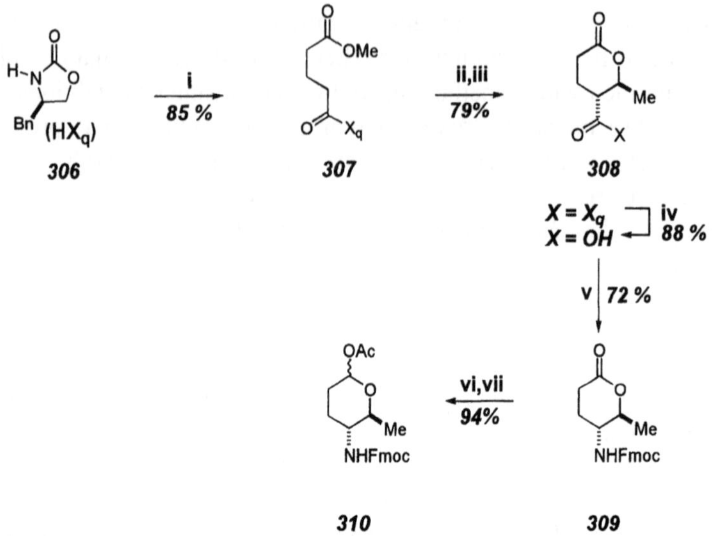

Scheme 68. Reagents: **i**, n-BuLi, THF, glutaric anhydride, CH_2N_2; **ii**, Bu_2BOTf, TEA, MeCHO, CH_2Cl_2, −78 °C; **iii**, PPTS, toluene, heat; **iv**, LiOOH, THF, 0 °C; **v**, TEA, Ph2P(O)N_3, 9-fluorenemethanol, toluene, heat; **vi**, DIBAL-H, THF; **vii**, Ac_2O, pyr

gave the lactone **313**. Further mundane manipulations led to the target **315**.

As reported above, TIETZE *et al.* employed the hetero-Diels-Alder reaction in the synthesis of two trideoxy monoaminohexoses: desosamine and acosamine (Schemes 54 and 55) (*118*). The phenylthio-substituted dihydropyran intermediates of desosamine and acosamine

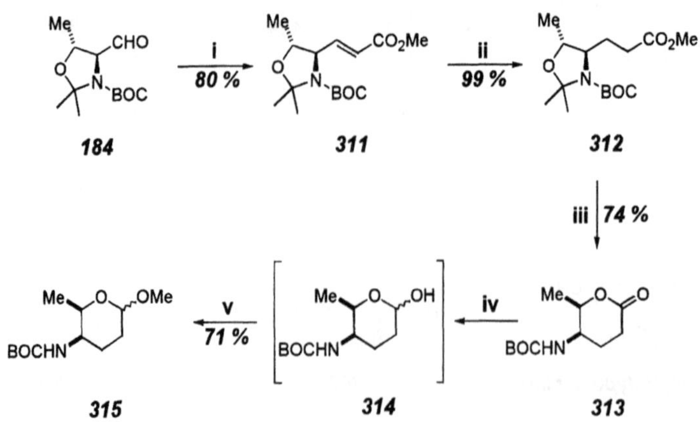

Scheme 69. Reagents: **i**, $MeO_2CCH_2P(O)(OMe)_2$, PhMe, K_2CO_3; **ii**, H_2, Pd/C, EtOAc; **iii**, AcOH, 60-80 °C; **iv**, DIBAL-H, PhMe −78 °C; **v** MeOH, H^+, $HC(OMe)_3$

References, pp. 255–263

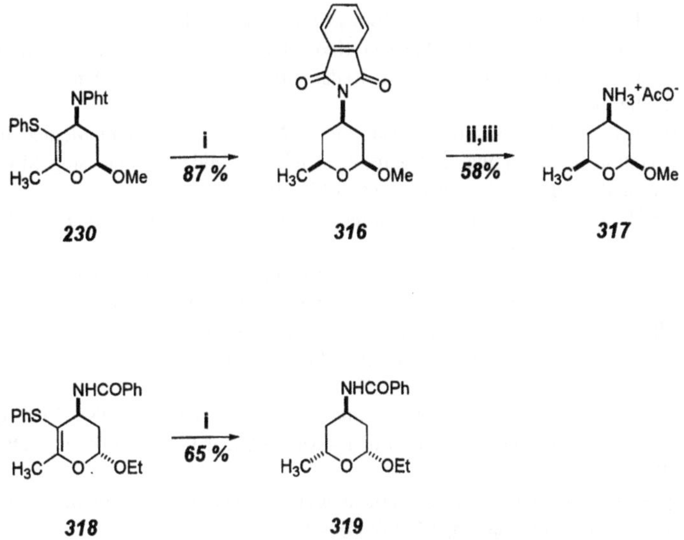

Scheme 70. Reagents: **i**, Raney Ni, MeOH; **ii**, $NaBH_4$, iPrOH; **iii**, AcOH

also give access to 4-deoxy derivatives of these known amino sugars. If the catalytic hydrogenation of the phenylthio-substituted dihydropyran derivative **230** is performed in methanol instead of THF, the all-*cis*-substituted 3-amino sugar glycoside **316** (a 4-deoxydaunosamine derivative) is obtained nearly exclusively (Scheme 70). The direction of hydrogenation is controlled by the *O*-methyl group and the bulky pseudo-equatorially oriented phthalimido group, which allow the addition of hydrogen only from the lower face. Reduction with sodium borohydride in propan-2-ol followed by addition of acetic acid gives the acetate salt of 4-deoxydaunosamine methyl glycoside **317**. Similarly, treatment of the phenylthio-substituted dihydropyran **318** with Raney nickel in methanol instead of THF gives (±)-*N*-benzoyl-4-deoxyristos-aminide **319** with a 5:1 selectivity. Hydrogenation of the olefin takes place from the upper face.

References

1. Hart, G.W.: Glycosylation. Current Options in Cell Biology, **4**, 1017 (1992).
2. Sharon, N., and H. Lis: Carbohydrates in Cell Recognition. Scientific American, **1993**, 74.
3. Höök, M., A. Woods, S. Johansson, L. Kjellen, and J.R. Couchman: Functions of Proteoglycans at Cell Surface. In: Functions of the Proteoglycans, Ciba Fundation Symposium 124, 143–157. Chichester: Wiley. 1986.

4. LEET, J.E., D.R. SCHROEDER, S.J. HOFSTEAD, J. GOLIK, K.L. COLSON, S. HUANG, S.E. KLOHR, T.W. DOYLE, and J.A. MATSON: Kedarcidin, a New Chromoprotein Antitumor Antibiotic: Structure Elucidation of Kedarcidin Chromophore. J. Am. Chem. Soc., **114**, 7946 (1992).
5. NAKAMURA, T., S. FUKATSU, S. SEKI, and T. NIIDA: A Convenient Method for the Preparation of the Acylated Macrolide Antibiotic Midecamycin using Molecular Sieves and Acylchloride. Chem. Lett., 1293 (1978).
6. LOWN, J. W.: Discovery and Development of Anthracycline Antitumour Antibiotics. Chem. Soc. Rev., 165 (1993).
7. BAUMANN, H., J.-R. TZIANABOS, D. L. KASPER, and H. J. JENNINGS: Structural Elucidation of Two Capsular Polysaccharides from One Strain of Bacteroides *fragilis* Using High-Resolution NMR Spectroscopy. Biochem., **31**, 4081 (1992).
8. LICHTENTHALER, F. W.: Emil Fischer's Proof of the Configuration of Sugars: A Centennial Tribute. Angew.Chem., **31**, 1541 (1992).
9. UESUGI, M., T. SEKIDA, S. MATSUKI, and Y. SUGIURA: Selective DNA Cleavage by Elsamicin A and Switch Function of Its Amino Sugar Group. Biochem., **30**, 6711 (1991).
10. BROCKMANN, H., and K. BAUER: Rhodomycin, ein rotes Antibioticum aus Actinomyceten. Naturwissenschaften, **37**, 492 (1950).
11. ZUNINO, F., R. GAMBETTA, A. DI MARCO, and A. ZACCARA: Interaction of Daunomycin and its Derivatives with DNA. Biochim. Biophys. Acta, **277**, 489 (1972).
12. GREIN, A., C. SPALLA, A. DI MARCO, and G. CANEVAZZI: Descrizione e classificazione di un attinomicete (*Streptomyces peucetius* ps. Nova) produttore di una sostanza ad attivita antitumorale: la daunomicina. Giorn. Microbiol., **11**, 109 (1963).
13. DUBOST, M., P. GAUTER, R. MARAL, L. NINET, S. PINNERT, J. PROUD'HOMME, and G.H. WERNER: Un nouvel antibiotique a propriétés cytostatiques: la rubidomycine. C.R. Acad. Sci. Paris, **257**, 1813 (1963).
14. BROCKMANN, H., and W. HENKEL: Pikromycin, ein neues Antibiotikum aus Actinomyceten. Naturwissenschaften, **37**, 138 (1950).
15. LARTEY, P.A., S.L. DENINNO, R. FAGHIH, D.J. HARDY, J.J. CLEMENT, and J.J. PLATTNER: Synthesis and Activity of C-21 Alkylamino Derivatives of (9R)-Erythromycylamine. J. Antibiot., **45**, 380 (1992).
16. JARET, R.S., A.K. MALLAMS, and H. REIMANN: The Megalomicins. Part IV. The Structures of Megalomicins A, B, C_1, and C_2. J. Chem. Soc. Perkin Trans., I, 1374 (1973).
17. DONIN, M.N., J. PAGANO, J.D. DUTCHER, and C.M. MCKEE: Methymycin, a New Crystalline Antibiotic. Antibiot. Annu., **1**, 179 (1953–54).
18. a) DJERASSI, C., and J.A. ZDERIC: The Structure of the Antibiotic Methymycin. J. Am. Chem. Soc., **78**, 2907 (1956); b) DJERASSI, C., and J.A. ZDERIC: The Structure of the Antibiotic Methymycin. J. Am. Chem. Soc., **78**, 6390 (1956).
19. KINOSHITA, K., S. SATOI, M. HAYASHI, and K. NAKATSU: Mycinamicins, new Macrolide Antibiotics: X. X-Ray Crystallography and the Absolute Configuration of Mycinamicin IV. J. Antibiot., **XLII 42**, 1003 (1989).
20. KIRST, H.A., and G.D. SIDES: New Directions for Macrolide Antibiotics: Pharmacokinetics and Clinical Efficacy. Antimicrob. Agents Chemother., **33**, 1419 (1989).
21. LAZAREVSKI, G., M. VINKOVIC, G. KOBREHEL, and S. DOKIC: Conformational Analysis of Azithromycin by Nuclear Magnetic Resonance Spectroscopy and Molecular Modelling. Tetrahedron, **49**, 721 (1993).
22. SHEPHARD, R.M., G.S. DUTHU, R.A. FERRAINA, and M.A. MULLINS: High-performance Liquid Chromatographic Assay with Electrochemical Detection for Azithromycin in Serum and Tissues. J. Chromatogr., **565**, 321 (1991).

23. Edo, K., M. Mizugaki, Y. Koide, H. Seto, K. Furihata, N. Otake, and N. Ishada: The Structure of Neocarzinostatin Chromophore Possessing a Novel Bicyclo-[7,3,0]dodekadiyne System. Tetrahedron Lett., **26**, 331 (1985).
24. For a review see Nicolaou, K.C., and W.-M. Dai: Chemistry and Biology of the Enediyne Anticancer Antibiotics. Angew. Chem. Int. Ed. Engl., **30**, 1387 (1991).
25. Leet, J.E., J. Golik, S.J. Hofstead, J.A. Matson, A.Y. Lee, and J. Clardy: Kedarcidin Chromophore: Structure Elucidation of the Amino Sugar Kedarosamine. Tetrahedron Lett., **33**, 6107 (1992).
26. Walker, S., K.G. Valentine, and D. Kahne: Sugars as DNA Binders: A Comment to the Calicheamicin Oligosaccharide. J. Am. Chem. Soc., **112**, 6428 (1990).
27. Smith, A.L. and K.C. Nicolaou: The Edndiyne Antibiotics. J. Med. Chem., **39**, 2103 (1996).
28. Furniss, B.S., A.J. Hannaford, P.W.G. Smith, and A.R. Tatchell: Vogels Textbook of Practical Organic Chemistry, 5th edn., 644. Essex: Longman Scientific and Technical. 1989.
29. Stevens, C.L., P. Blumbergs, and D.H. Otterbach: Synthesis and Chemistry of 4-Amino-4,6-Amino-4,6-dideoxy Sugars. I. Galactose. J. Org. Chem., **31**, 2817 (1966).
30. Stevens, C.L., P. Blumbergs, F.A. Daniher, D.H. Otterbach, and K.G. Taylor: Synthesis and Chemistry of 4-Amino-4,6-Amino-4,6-dideoxy Sugars. II. Glucose. J. Org. Chem., **31**, 2822 (1966).
31. Rainer, H., and H.-D. Scharf: Synthesis of the Hydroxyamino Sugar of Calicheamicins. Liebigs Ann. Chem., 117 (1993).
32. Furniss, B.S., A.J. Hannaford, P.W.G. Smith, and A.R. Tatchell: Vogels Textbook of Practical Organic Chemistry, 5th edn., 660. Essex: Longman Scientific and Technical. 1989.
33. Bell, D.J., and J. Lorber: Use of the Benzyl Radical in Syntheses of Methylated Sugars. Part I. 4:6-Dimethyl Glucose. J. Chem. Soc., 453 (1940).
34. Mathers, D.J., and G.J. Robertson: Optical Superposition and the 4:6-Benzylidenemethyl-1-glycosides. J. Chem. Soc., 696 (1933).
35. Freudenberg, K., and E. Plankenhorn: Zur Kenntnis der teilweise methylierten Glucose. Ber., **73B**, 621 (1940).
36. Ward, D.E., and B.F. Kaller: Diastereoselective Synthesis of Actinobolin from D-Glucose by Application of a Novel [3 + 3] Annulation. J. Org. Chem., **59**, 4230 (1994).
37. Ward, D.E., and B.F. Kaller: The Diastereoselective Synthesis of (+)-Actinobolin from D-Glucose. Tetrahedron Lett., **34**, 407 (1993).
38. Stanek, J., K. Capek, and J. Jary: Preparation of Partially Acylated Derivatives of Methyl 3-Acetamido-3,6-dideoxy-ß-D-gluco- and (β-D-Mannopyranoside. Coll. Czech. Chem. Comm., **39**, 1479 (1973).
39. Jennings, H.J., and J.K.N. Jones: Reaction of Sugar Chlorosulfates: Part V. The Synthesis of Chlorodeoxy Sugars. Can. J. Chem., **43**, 2372 (1965).
40. Hanessian, S., and N.R. Plessas: The Reaction of *O*-Benzylidene Sugars with *N*-Bromosuccinimide. III. Applications of the Synthesis of Aminodeoxy and Deoxy Sugars of Biological Importance. J. Org. Chem., **34**, 1045 (1969).
41. Stick, R.V., and J.J. Patroni: The Deoxygenation of Some Derivatives of Methyl 3-Amino-3-deoxy-α-D-glucopyranoside. Aust. J. Chem., **38**, 947 (1985).
42. Guthrie, R.D., and L.F. Johnson: Nitrogen-Containing Carbohydrate Derivatives. Part I. Methyl 4,6-Benzylidene-3-deoxy-3-phenylazo-α-D-glucoside. J. Chem. Soc., 4166 (1961).

43. FURSTNER, A., J. BAUMGARTNER, and D.N. JUMBAM: Unprecedented Influence of Azides and the Effect of Bulky Groups on Zinc-induced Reductions of Deoxy Halogeno Sugars. J. Chem. Soc. Perkin Trans., I, 131 (1993).
44. WEILER, L., and D.A. NICOLL-GRIFFITH: Introduction of a Chiral Centre on C-6 of a Carbohydrate Unit: Application to the Synthesis of the C-2 to C-15 Fragment of Ionomycin. Tetrahedron, **47**, 2733 (1991).
45. IZAWA, T., Y. NISHUMURA, and S. KONDO: 3-Amino-5-*C*-phenyl-D-altrofuranose and 3-Amino-5-*C*-[3-carboxy-4-(carboxymethyl)-2-oxo-3-cyclohexen-1-yl]-D-altrofuranose, Possible Intermediates for Synthesis of the Anthracycline Antibiotic Decilorubicin. Carbohydr. Res., **211**, 137 (1991).
46. ALI, Y., and A.C. RICHARDSON: The Reduction of Azides with Sodium Borohydride: A Convenient Synthesis of Methyl 2-Acetamido-4,6-*O*-benzylidene-2-deoxy-α-D-allopyranoside. Carbohydr. Res., **5**, 441 (1967).
47. For a review see BRIMACOMBE, J.S.: Synthesen von Antibiotica-Zuckern. Angew. Chem., **83**, 261 (1971).
48. FOSTER, A.B., T.D. INCH, J. LEHMANN, M. STACEY, and J.M. WEBBER: Carbohydrate Components of Antibiotics. Part III. Synthesis of 3,6-Dideoxy-3-dimethylamino-ß-D-glucose Hydrochloride Monohydrate: the Absolute Configuration of Mycaminose. J. Chem. Soc., 2116 (1962).
49. REIST, E.J., R.R. SPENCER, and B.R. BAKER: Potential Anticancer Agents. XXIX. Inversion of a Ring Carbon of a Glycoside. J. Org. Chem., **24**, 1618 (1959).
50. EIS, M.J., and B. GANEM: An Improved Synthesis of D-Perosamine and Some Derivatives. Carbohydr. Res., **176**, 316 (1988).
51. STEVENS, C.L., R.P. GLINSKI, and K.G. TAYLOR: 4,6-Dideoxy-4-(*N*,*N*-dimethylamino)-D-talopyranose Hydrochloride. J. Org. Chem., **33**, 1586 (1968).
52. BRIMACOMBE, J.S., A.K. AL-RADHI, and L.C.N. TUCKER: The Deamination of Methyl 4-Amino-4,6-dideoxy-2,3-*O*-isopropylidene-α-L-talo- and -manno-pyranosides with Nitrous Acid. J. Chem. Soc. Perkin Trans., **I**, 315 (1992).
53. OVEREND, W.G., S.W. GUNNER, and N.R. WILLIAMS: The Preparation of Amino Sugars from Methyl Glycopyranosiduloses: Methyl 4-Acetamido-4,6-dideoxy-α-L-talopyranoside. Carbohydr. Res., **4**, 498 (1967).
54. JARY, J., K. CAPEK, and J. KOVAR: Synthese von Derivaten der 3,6-Didesoxy-3-amino-L-idose. Coll. Czech. Chem. Comm. **28**, 2171 (1963).
55. BRIMACOMBE, J.S., O.A. CHING, and M. STACEY: Nucleophilic Displacement Reactions in Carbohydrates. Part XI. Reaction of Methyl 6-Deoxy-2,3-isopropylidene-4-*O*-methylsulphonyl-α-L-talopyranoside with Sodium Azide: a Synthesis of L-Perosamine (4-Amino-4,6-didexy-L-mannose) Derivatives. J. Chem. Soc. (C), 1270 (1969).
56. PFITZNER, K.E., and J.G. MOFFATT: A New and Selective Oxidation of Alcohols. J. Am. Chem. Soc., **85**, 3027 (1963).
57. CIEPLAK, A.S.: Stereochemistry of Nucleophilic Addition to Cyclohexanone. The Importance of Two-Electron Stabilizing Interactions. J. Am. Chem. Soc., **103**, 4540 (1981).
58. CIEPLAK, A.S., B.D. TAIT, and C.R. JOHNSON: Reversal of π-facial Diastereoselection upon Electronegative Substitution of the Substrate and the Reagent. J. Am. Chem. Soc., **111**, 8447 (1989).
59. JARY, J., and A. ZOBACOVA: Amino Sugars via Displacement of Sulfonyloxy Groups with Hydrazine. Methods in Carbohydrate Chemistry, **6**, 229-235 (1972).
60. COLEMAN, R.S., Y. DONG, and A.J. CARPENTER: A Convenient Preparation of Terminally Differentiated, Selectively Protected Six-Carbon Synthons from D-glucosamine. J. Org. Chem., **57**, 3732 (1992).

61. PAULSEN, H., V. RUTZ, and I. BROCKHAUSEN: Synthese von modifizierten Derivaten der 2-Acetamido-2-desoxy-D-galactose zur Untersuchung der Substratspezifität der Core-1-β3-Gal-Transferase und der Core-1-β3-GlcNAc-Transferase der Biosynthese von *O*-Glycoproteinen. Liebigs Ann. Chem., 735 (1992).
62. STEVENS, C.L., R.P. GLINSKI, K.G. TAYLOR, P. BLUMBERGS, and S. K. GUPTA: The Synthesis and Proof of Structure of Perosamine (4-Amino-4,6-dideoxy-D-mannose) Derivatives. J. Am. Chem. Soc., **92**, 3160 (1970).
63. MALIK, A., S.N.-H. KASMI, and Z. AHMAD: A Regioselective One-pot Synthesis of *N,N*-Dialkylamino Sugars *via* Aminosilanes. J. Chem. Res. (S), 124 (1992).
64. JARY, J., K. CAPEK, and J. STANEK jr.: Preparation of Methyl 3-amino-3,6-dideoxy-β-D-hexopyranosides by Condensation of (*2R, 4R*)-2-Methoxy-4-methyl-3-oxapentane-1,5-dial with Nitrometane. Coll. Czech. Chem. Comm., **39**, 1462 (1974).
65. RICHARDSON, A.C.: The Synthesis of D- and L-Mycaminose Hydrochlorides. J. Chem. Soc., 2758 (1962).
66. HORTON, D., and A. LIAV: A Synthesis of 2-Amino-2,6-dideoxy-D-allose and -D-altrose Hydrochlorides and their Tetraacetyl Derivatives. Carbohydr. Res., **47**, 326 (1976).
67. LEVENE, P.A., and E.T. STILLER: The Synthesis of Ribose-5-phosphoric Acid. J. Biol. Chem., **104**, 299 (1934).
68. SHUNK, C.H., J.B. LAVIGNE, and K. FOLKERS: Studies on Carcinolytic Compounds. V. 6,7-Dimethyl-9-[1-(5desoxy-D-ribityl)]-isoalloxazine. J. Am. Chem. Soc., **77**, 2210 (1955).
69. KUHN, R., and H. FISCHER: Aminozucker-Synthesen XI: D-Altrosamin und D-Allosamin. Liebigs Ann. Chem., **615**, 88 (1958).
70. BAER, H.H., K. CAPEK, and M.C. COOK: Synthesis of 3-Acetamido-2,3,6-trideoxy-D-*lyxo*-hexose (*N*-Acetyl-D-daunosamine) and its D-*Arabino* Isomer. Can. J. Chem., **47**, 89 (1969).
71. RICHARDSON, A.C.: The Synthesis of Desosamine Hydrochloride. J. Chem. Soc., 5364 (1964).
72. OVEREND, W.G., G.S. HAJIVARNAVA, and N.R. WILLIAMS: Arylazo-Glycenosides. Part 7. Synthesis of Amino-sugars from Methyl Arylazo-hexenopyranosides. J. Chem. Soc. Perkin Trans., **I**, 205 (1982).
73. ISELIN, B., and T. REICHSTEIN: Krystallisierte 2-Desoxy-*l*-rhamnose (2-Desoxy-*l*-chinovose). Helv. Chim. Acta, **27**, 1146 (1944).
74. HADFIELD, A.F., L. CUNNINGHAM, and A.C. SARTORELLI: The Synthesis and Cytotoxic Activity of 1,3,4,-tri-*O*-Acetyl-2,6-dideoxy-L-*arabino*-and -L-*lyxo*-hexopyranose. Carbohydr. Res., **72**, 93 (1979).
75. MONNERET, C., C. CONREUR, and Q. KHUONG-HUU: Synthesis of Methyl 4-Amino-2,4,6,trideoxy-3-*O*-methyl-L-*arabino*-hexopyranosides (Methyl α- and β-L-Hollantosaminide) and of Methyl 4-Amino-2,4,6-trideoxy-3-*O*-methyl-β-L-*lyxo*-hexopyranoside (Methyl α-L-3-epiHollantosaminide). Carbohydr. Res., **65**, 35 (1978).
76. GOODMAN, L., J. P. MARSH, C. W. MOSHER, and E. M. ACTON: 6-(3-Methylbut-3-enylamino)purine: a Highly Active Cytokinin. Chem. Comm., 973 (1967).
77. INGLIS, G.R., J.C.P. SCHWARZ, and L. MCLAREN: The Methoxymercuration of D-Glucal and Tri-*O*-acetyl-D-glucal: a New Route to 2-Deoxyglucopyranosides. J. Chem. Soc., 1014 (1962).
78. FLORENT, J.-C., B. ABBACI, and C. MONNERET: Addition of Hydrazoic Acid to Pseudoglycals Stereoselective Synthesis of L-Acosamine and L-Daunosamine. Bull. Soc. Chim. Fr., 667 (1989).

79. Grethe, G., T. Mitt, T.H. Williams, and M.R. Uskokovic: Synthesis of Daunosamine. J. Org. Chem., **48**, 5309 (1983).

80. Sowden, J.C., and H.O.L. Fischer: The Condensation of Nitromethane with D- and L-Arabinose: Preparation of L-Glucose and L-Mannose. J. Am. Chem. Soc., **69**, 1963 (1947).

81. Sowden, J.C., and H.O.L. Fischer: Carbohydrate C-Nitroalcohols: the Acetylated Nitroolefins. J. Am. Chem. Soc., **69**, 1048 (1947).

82. Satoh, C., and A. Kiyomoto: Studies on Nitrogen-Containing Sugars. I. Synthesis of *N*-Acetyl-D-mannosamine from 1-Nitro-1-deoxy-D-mannitol Pentaacetate. Chem. Pharm. Bull. Jpn., **12**, 615 (1964).

83. Perr, M.B., and A.C. Webb: Synthesis of 2-Acetamido-2-deoxy-D-gulose, 2-Acetamido-2-deoxy-D-idose, and 2-Deoxy-D-*xylo*-hexose from 3,4,5,6-Tetraacetoxy-D-*xylo*-1-nitro-1-hexene. Can. J. Chem., **47**, 1245 (1969).

84. Sztaricskai, F., I. Pelyvas, L. Szilagyi, R. Bognar, J. Tamas, and A. Neszmelyi: A Synthesis of L-Ristosamine and a Derivative of its C-4 Epimer. Carbohydr. Res., **65**, 193 (1978).

85. Gupta, S.K.: The Synthesis of Methyl 3-Amino-2,3,6-trideoxy-α-L-*arabino*-hexopyranoside, a Structural Analog of Daunosamine. Carbohydr. Res., **37**, 381 (1974).

86. Lee, W.W., H.Y. Wu, J.J. Marsh, C.W. Mosher, E.M. Acton, L. Goodman, and D.W. Henry: Confirmation by Synthesis of Ristosamine as 3-Amino-2,3,6-trideoxy-L-ribohexose. J. Med. Chem., **18**, 767 (1975).

87. Sztaricskai, F., I. Pelyvas, R. Bognar, and G. Bujtas: The Synthesis of *N*-Benzoylristosamine. Tetrahedron Lett., 1111 (1975).

88. Coleman, R.S., and J.R. Fraser: Acylketene [4 + 2] Cycloadditions: Divergent de Novo Synthesis of 2,6-Dideoxy Sugars. J. Org. Chem., **58**, 385 (1993).

89. Sztaricskai, F. J., M. Hornyak, and I. F. Pelyvas: Definitive Synthesis of Methyl α-kedarosaminide, a Sugar Component of the Antitumor Antibiotic Kedarcidin. Tetrahedron Lett., **34**, 4087 (1993).

90. Overend, W.G., P.J. Beynon, and P.M. Collins: Aspects of the Chemistry of Oximes Derived from Methyl Hexopyranosiduloses. J. Chem. Soc. (C), 272 (1969).

91. Banaszek, A., A. Zamojski, and E. Baranowska: Mass Spectra of Methyl 3,4-Dideoxy-3-dimethylamino- and 2,4-Dideoxy-2-dimethylamino-α-DL-glycopyranosides. Pol. J. Chem., **58**, 439 (1984).

92. Banaszek, A., and A. Zamojski: Derivatives of 2-Alkoxy-5,6-dihydro-2*H*-pyran as Starting Materials in the Synthesis of Monosaccharides. Part IX. Total Synthesis of Methyl 3,4-Dideoxy- and 3,4,6-Trideoxy-D,L-hex-3-enopyranosides. Carbohydr. Res., **25**, 453 (1972).

93. Banaszek, A., and A. Zamojski: Derivatives of 2-Alkoxy-5,6-dihydro-α-pyran as Substrates in the Synthesis of Monosaccharides. Part V. Directions of the Oxirane Ring Opening in Methyl 2,3-Anhydro-4,6-dideoxy-α-DL-*lyxo*- and *ribo*-hexopyranosides. Roczniki Chem., **45**, 391 (1971).

94. Jary, J., K. Kefurt, K. Capek, J. Capkova, and Z. Kefurtova: Amino Sugars. XXVII. The Synthesis of Derivatives of 3-Amino-3,4,6-trideoxy-D or L-*xylo*-Hexopyranose. Coll. Czech. Chem. Comm., **37**, 2985 (1972).

95. Overend, W. G., S. Laland, and M. Stacey: Deoxy-Sugars. Part X. Some Methanesulphonyl and Toluene-p-sulphonyl Derivatives of α-Ethyl-2:3-dideoxy-D-glucoside. J. Chem. Soc., 738 (1950).

96. Foster, A.B., R.Harrison, J. Lehmann, and J.M. Webber: Aspects of Stereochemistry. Part XVI. Intramolecular Hydrogen Bonding in Ethyl 2,3-Dideoxy-α-D-*erythro*- and -*threo*-hexopyranosides and Related Compounds. J. Chem. Soc., 4471 (1963).

97. Brimacombe, J.S., L.W. Doner, and A.J. Rollins: Syntheses of Methyl 2,3,6-Trideoxy-α-L-*erythro*-hexopyranoside (Methyl α-L-Amicetoside) and Methyl 2,3,4,6-Tetradeoxy-4-(dimethylamino)-α-L-*threo*-hexopyranoside (Methyl α-L-Ossaminide). J. Chem. Soc. Perkin Trans., **I**, 2977 (1972).
98. Stevens, C.L., and C.P. Bryant: Synthesis of Amino Sugars with Retention of Configuration. Methods in Carbohydrate Chemistry, **6**, 225-228 (1972).
99. Panzica, R.P., E. Abushanab, G. Bastian, M. Bessodes, S.-F. Chen, J.D. Stoeckler, and R.E. Parks: Adenoside Deaminase Inhibitors. Conversion of a Single Chiral Synthon into *erythro*- and *threo*-9-(2-Hydroxy-3-nonyl)adenines. J. Med. Chem., **24**, 1385 (1981).
100. Stevens, C.L., G.E. Gutowski, C.P. Bryant, and R.P. Glinski: The Isolation and Synthesis of Ossamine, the Aminosugar Fragment from the Fungal Metabolite Ossamycin. Tetrahedron Lett., 1181 (1969).
101. Panek, J.S., and J. Zhang: Diastereoselectivity in the Osmium Tetraoxide Promoted Dihydroxylation of Chiral (*E*)-Crotylsilanes: Asymmetric Synthesis of Silyl-Functionalized γ-Lactones. J. Org. Chem., **58**, 294 (1993).
102. Panek, J.S., and M.A. Sparks: Claisen Rearrangements of Enantiomerically Pure C3-(Acyloxy)-(*E*)-vinylsilanes. J. Org. Chem., **56**, 3431 (1991).
103. Panek, J.S., R. Beresis, F. Xu, and M.Yang: Diastereoselective Electrophilic Addition Reactions to chiral β-Dimethylphenylsilyl Ester Enolates. Synthesis of 2,3-Anti-α-substituted-β-silyl-(*E*)-hex-4-enoates. J. Org. Chem., **56**, 7341 (1991).
104. Ireland, R.E., R.H. Mueller, and A.K. Willard. The Ester Enolate Claisen Rearrangement. Stereochemical Control through Stereoselective Enolate Formation. J. Am. Chem. Soc., **98**, 2868 (1976).
105. Ireland, R.E.: Synthetic Methdology in the Context of Natural Product Total Synthesis. Aldrichimica Acta, **21**, 59 (1988).
106. Yamada, S.-I., K. Koga, M. Yoh, and T. Mizoguchi: Stereoselective Total Synthesis of 6-Deoxy-L-hexose Derivatives from L-Alanine without a Resolution Step. Carbohydr. Res., **36**, C9 (1974).
107. Polt, R., and D. Sames: An Enantioselective Synthesis of *N*-Methylfucosamine *via* Tandem C-C/C-O Bond Formation. J. Org. Chem., **59**, 4596 (1994).
108. Koskinen, A.M.P., and L.A. Otsomaa: A New Access to Enantiomerically Pure Deoxy Aminohexoses: Methyl 4-Amino-4,6-dideoxy Gulopyranoside and *epi*-Tolyposamine. Tetrahedron, **53**, 6473 (1997).
109. Ruiz, M., V. Ojea, and J.M. Quintela: Amino Acid Based Diastereoselective Synthesis of Elsaminose. Tetrahedron Lett., **37**, 5743 (1996).
110. Xu, Z., C.W. Johannes, S.S. Salman, and A.H. Hoveyda: Enantioselective Total Synthesis of Antifungal Agent Sch 38516. J. Am. Chem. Soc., **118**, 10926 (1996).
111. Newman, H.: Degradation and Synthesis of Desosamine. J. Org. Chem., **29**, 1461 (1964).
112. Manhas, M.S., V.R. Hegde, D.R. Wagle, and A.K. Bose: Studies on Lactams. Part 74. An Approach to the Total Synthesis of Amino Sugars *via* β-Lactams. J. Chem. Soc. Perkin Trans., **I**, 2045 (1985).
113. Hauser, F.M., and R.P. Rhee: A Brief Total Synthesis of *N*-Benzoyl-DL-daunosamine. J. Org. Chem., **46**, 227 (1981).
114. Hauser, F.M., R.P. Rhee, and S.R. Ellenberger: Total Synthesis of Optically Active *N*-Benzoyldaunosamine from an Azetidinone. J. Org. Chem., **49**, 2236 (1984).
115. Wiemann, R., and I. Dyong: Ein leistungsfähiges Syntheseprinzip für Aminozucker vom Daunosamin-Typ. Angew. Chem., **90**, 728 (1978).

116. Fuganti, C., G. Fronza, and P. Grasselli: Synthesis of *N*-Trifluoroacetyl-L-acosamine and -L-daunosamine. J. Chem. Soc. Chem. Comm., 442 (1980).

117. Tietze, L.F., and U. Hartfield: Hetero-Diels Alder Reaction of Substitued 1-Oxabutadienes And 2-Ethoxyvinylacetate: An Entry To Various Natural Occurring Carbohydrates. Tetrahedron Lett., **31**, 1697 (1990).

118. Tietze, L.F., U. Hartfield, T. Hubsch, E. Voss, K. Bogdanowicz-Szwed, and J. Wichmann: Syntheses of the 3-Amino Sugar Glycosides *rac*-4-Deoxydaunosaminide, *rac*-4-Deoxyristosaminide, and *rac*-Acosaminide. Liebigs Ann. Chem., 275 (1991).

119. Grethe, G., J. Sereno, T.H. Williams, and M.R. Uskokovic: Asymmetric Synthesis of Daunosamine. J. Org. Chem., **48**, 5315 (1983).

120. Wade, P.A., J.A. Rao, J.F. Bereznak, and C.-K. Yuan: A Dihydroisooxazole-based Route to 2,3,6-Trideoxy-3-aminohexose Derivatives. Tetrahedron Lett., **30**, 5969 (1989).

121. Guanti, G., L. Banfi, E. Narisano, and S. Thea: Stereoselective Synthesis of 4-Acetylamino-2,4,6-trideoxy-L-*ribo*-hexose from Ethyl (*S*)-β-Hydroxybutyrate. Synlett, **1992**, 311.

122. Griffith, W.P., and S.V. Ley: TPAP: Tetra-*n*-propylammonium Perruthenate, A Mild and Convenient Oxidant for Alcohols. Aldrichimica Acta, **23**, 13 (1990).

123. Jurczak, J., and A. Golebiowski: α-Amino-β-hydroxy Acids in the Total Synthesis of Amino Sugars. Synlett, 241 (1993).

124. Jurczak, J., J. Kozak, and A. Golebiowski: The Total Synthesis of L-Daunosamine. Tetrahedron, **48**, 4231 (1992).

125. Kihlberg, J., T. Vuljanic, and P. Somfai: Diastereoselective Synthesis of Methyl α-Kedarosaminide, a Carbohydrate Moiety of the Enediyne Antitumor Antibiotic Kedarcidin Chromophrore. Tetrahedron Lett., **35**, 6937 (1994).

126. Saksena, A.K., and P. Mangiaracina: Recent Studies on Veratrum Alkaloids: A New Reaction of Sodium Triacetoxyborohydride. Tetrahedron Lett., **24**, 273 (1983).

127. Fronza, G., C. Fuganti, P. Grasselli, and G. Marinoni: Synthesis of *N*-Benzoyl L- and D-2,3,6-Trideoxy-3-amino-*xylo*-hexose from Non-carbohydrate Precursors. Tetrahedron Lett., 3883 (1979).

128. Dondoni, A., G. Fantin, M. Fogagnolo, and P. Merino: Regio- and Stereoselective Conjugate Addition of Nitrogen Nucleophiles to 2-Alkenyl *N*-methylthiazolium Iodides. Synthesis of D-3-*epi*-Daunosamine and Some Lincosamine Analogues. Tetrahedron, **46**, 6167 (1990).

129. Fuganti, C., P. Grasselli, and G. Pedrocchi-Fantoni: Synthesis of *N*-Benzoyl-L-daunosamine from D-Threonine. Tetrahedron Lett., **22**, 4017 (1981).

130. Hamada, Y., A. Kawai, and T. Shioiri: New Methods and Reagents in Organic Synthesis. 49. A Highly Efficient Stereoselective Synthesis of L-Daunosamine through Direct C-Acylation Using Diphenyl Phosphorazidate (DPPA). Tetrahedron Lett., **25**, 5409 (1984).

131. Hamada, Y., A. Kawai, T. Matsui, O. Hara, and T. Shioiri: 4-Alkoxycarbonyloxazoles as Hydroxy – amino Acid Synthons: Efficient, Stereoselective Syntheses of 3-Amino-2,3,6-trideoxyhexoses and a Hydroxy Amino Acid Moiety of AI-77-B. Tetrahedron, **46**, 4823 (1990).

132. Guanti, G., L. Banfi, E. Narisano, and R. Riva: Stereoselective Synthesis of *N*-Acetyl-L-Tolyposamine from (S)-Ethyl β-Hydroxybutyrate. Tetrahedron Lett., **33**, 2221 (1992).

133. GUANTI, G., L. BANFI, and E. NARISANO: Enantiospecific and Diastereoselective Preparation of Synthetic Equivalents of 2,4-Deoxy-2-amino-L-threose and -L- erythrose from (S)-Ethyl β-Hydroxybutyrate. Stereochemical Course of their Condensations with C-Nucleophiles. Tetrahedron Lett., **30**, 5511 (1989).
134. EVANS, D.A., and W.C. BLACK: Total Synthesis of (+)-A83543A [(+)-Lepicidin A]. J. Am. Chem. Soc., **115**, 4497 (1993).

(Received April 3, 1997)

Author Index

Page numbers printed in *italics* refer to References

Subject Index

Springer-Verlag
and the Environment

WE AT SPRINGER-VERLAG FIRMLY BELIEVE THAT AN international science publisher has a special obligation to the environment, and our corporate policies consistently reflect this conviction.

WE ALSO EXPECT OUR BUSINESS PARTNERS – PRINTERS, paper mills, packaging manufacturers, etc. – to commit themselves to using environmentally friendly materials and production processes.

THE PAPER IN THIS BOOK IS MADE FROM NO-CHLORINE pulp and is acid free, in conformance with international standards for paper permanency.

MIX
Papier aus verantwortungsvollen Quellen
Paper from responsible sources
FSC® C105338

If you have any concerns about our products,
you can contact us on
ProductSafety@springernature.com

In case Publisher is established outside the EU,
the EU authorized representative is:
Springer Nature Customer Service Center GmbH
Europaplatz 3, 69115 Heidelberg, Germany

Printed by Libri Plureos GmbH
in Hamburg, Germany